# THE BEST TEST PREPARATION FOR THE ADVANCED PLACEMENT EXAMINATION IN

# CHEMISTRY

**Philip E. Dumas, Ph.D.**
Chairperson of Chemistry Department
Trenton State College, Trenton, NJ

**Ronald M. Fikar, Ph.D.**
Research Associate
Rutgers University, New Brunswick, NJ

**Jay Templin, Ph.D.**
Assistant Professor of Biology
Widener University, Wilmington, DE

*AP Chemistry Course Review by*

**William Uhland**
Research Scientist
Cytogen Corporation, Princeton, NJ

Research and Education Association
61 Ethel Road West
Piscataway, New Jersey 08854

The Best Test Preparation for the
ADVANCED PLACEMENT EXAMINATION
IN CHEMISTRY

REVISED PRINTING 1994



Printed in the United States of America

Library of Congress Catalog Card Number 88-90706

International Standard Book Number 0-87891-648-2

Research & Education Association
61 Ethel Road West
Piscataway, New Jersey 08854

REA supports the effort to conserve and protect environmental resources by printing on recycled papers.

# CONTENTS

# PREFACE

This book provides an accurate and complete representation of the Advanced Placement Examination in Chemistry. The six practice exams provided are based on the most recently administered Advanced Placement Chemistry Exams. Each exam is three hours in length and includes every type of question that can be expected on the actual exam. Following each exam is an answer key complete with detailed explanations designed to clarify the material to the student. By completing all six exams and studying the explanations which follow, students can discover their strengths and weaknesses and thereby become well prepared for the actual exam.

# ABOUT THE TEST

The Advanced Placement Chemistry Examination is offered each May at participating schools and multischool centers throughout the world.

The Advanced Placement Program is designed to allow high school students to pursue college-level studies while attending high school. The participating colleges, in turn, grant credit and/or advanced placement to students who do well on the examinations.

The Advanced Placement Chemistry course is designed to be the equivalent of a college introductory chemistry course, often taken by chemistry majors in their first year of college.

The exam is divided into two sections: 1) **Multiple-choice:** composed of 85 multiple-choice questions designed to test the student's ability to recall and understand a broad range of chemical concepts and calculations. About 90 to 105 minutes are allowed for this section of the exam. Since the test covers a broad range of topics, no student is expected to answer all of the questions correctly.

2) **Free-response section**: composed of several comprehensive problems and essay topics. This section constitutes 55% of the final grade and the student is allowed 75 to 90 minutes for this portion of the exam. The student may choose from the questions provided. These problems and

essays are designed to test the student's ability to think clearly and to present ideas in a logical and coherent fashion.

The student is allowed to bring an electronic hand-held calculator. However, no programs are allowed and any memory must be erased.

Students may find these exams considerably more difficult than most classroom exams. The Advanced Placement Exams are, in fact, designed to have average scores of approximately 50% of the maximum possible score for the multiple-choice section and the free-response section, in order to provide information about differences in students' achievements in chemistry.

The multiple-choice section of the exam is scored by crediting each correct answer with one point, and deducting only partial credit (one-fourth of a point) for each incorrect answer. Questions omitted receive neither a credit nor a deduction. The essay section is scored by a group of over one thousand college and high school educators familiar with the AP Program, who evaluate the accuracy and coherence of the essays accordingly. The grades given for the essays are combined with the results of the multiple-choice section, and the total raw score is then converted to the Program's five-point scale: 5 – extremely qualified, 4 – well qualified, 3 – qualified, 2 – possibly qualified, 1 – no recommendation. Colleges participating in the Advanced Placement Program usually recognize grades of 3 or higher.

## TABLE OF MEASUREMENTS FOR SECTION II

Universal gas constant: R = 8.31 joules/ (mole-K) = 0.0821 liter-atm/(mole-K)
= 62.4 liter-mm Hg/ (mole-K) = 1.99 calories/ (mole-K)
= 8.31 (volt) (coulombs) / (mole-K)

1 faraday ( $\mathcal{F}$ ) = 96,500 coulombs = 23,060 calories/volt

1 calorie = 4.184 joules

1 electron volt/atom = 23.1 kilocalories/mole
= 96.5 kilojoules/mole

Speed of light in vacuum - $2.998 \times 10^8$ m/sec

$\ln_e = 2.303 \log_{10}$

Planck's constant $h = 6.63 \times 10^{-34}$ joule-sec

Boltzmann's constant $k = 1.38 \times 10^{-23}$ joule/K

Avogadro's number = $6.022 \times 10^{23}$ molecules/mole

At 25° C, $\frac{RT}{n\mathcal{F}} \ln Q = \frac{0.0591}{n} \log Q$

The Table above contains information that will be useful in solving the problems in Section II of each exam.

# PERIODIC TABLE OF THE ELEMENTS

ATOMIC NUMBER ➤ 46 +2 ◄ OXIDATION STATES
SYMBOL ➤ Pd +4
ATOMIC WEIGHT ➤ 106.4
-18-18-0 ◄ ELECTRON CONFIGURATION

TRANSITION ELEMENTS

GROUP 8

| 1a | 2a | 3b | 4b | 5b | 6b | 7b | 8 | 8 | 8 | 1b | 2b | 3a | 4a | 5a | 6a | 7a | 0 | Orbit |
|---|---|---|---|---|---|---|---|---|---|---|---|---|---|---|---|---|---|---|
| 1 H +1 −1<br>1.0079<br>1 | | | | | | | | | | | | | | | | | 2 He 0<br>4.00260<br>2 | K |
| 3 Li +1<br>6.94<br>2-1 | 4 Be +2<br>9.01218<br>2-2 | | | | | | | | | | | 5 B +3<br>10.81<br>2-3 | 6 C +2 +4 −4<br>12.011<br>2-4 | 7 N +1 +2 +3 +4 +5 −1 −2 −3<br>14.0067<br>2-5 | 8 O −2<br>15.9994<br>2-6 | 9 F −1<br>18.998403<br>2-7 | 10 Ne 0<br>20.17$_9$<br>2-8 | K-L |
| 11 Na +1<br>22.98977<br>2-8-1 | 12 Mg +2<br>24.305<br>2-8-2 | | | | | | | | | | | 13 Al +3<br>26.98154<br>2-8-3 | 14 Si +2 +4 −4<br>28.0855<br>2-8-4 | 15 P +3 +5 −3<br>30.97376<br>2-8-5 | 16 S +4 +6 −2<br>32.06<br>2-8-6 | 17 Cl +1 +5 +7 −1<br>35.453<br>2-8-7 | 18 Ar 0<br>39.948<br>2-8-8 | K-L-M |
| 19 K +1<br>39.0983<br>-8-8-1 | 20 Ca +2<br>40.08<br>-8-8-2 | 21 Sc +3<br>44.9559<br>-8-9-2 | 22 Ti +2 +3 +4<br>47.90<br>-8-10-2 | 23 V +2 +3 +4 +5<br>50.9415<br>-8-11-2 | 24 Cr +2 +3 +6<br>51.996<br>-8-13-1 | 25 Mn +2 +3 +4 +7<br>54.9380<br>-8-13-2 | 26 Fe +2 +3<br>55.847<br>-8-14-2 | 27 Co +2 +3<br>58.9332<br>-8-15-2 | 28 Ni +2 +3<br>58.71<br>-8-16-2 | 29 Cu +1 +2<br>63.546<br>-8-18-1 | 30 Zn +2<br>65.38<br>-8-18-2 | 31 Ga +3<br>69.735<br>-8-18-3 | 32 Ge +2 +4<br>72.59<br>-8-18-4 | 33 As +3 +5 −3<br>74.9216<br>-8-18-5 | 34 Se +4 +6 −2<br>78.96<br>-8-18-6 | 35 Br +1 +5 −1<br>79.904<br>-8-18-7 | 36 Kr 0<br>83.80<br>-8-18-8 | -L-M-N |
| 37 Rb +1<br>85.467$_8$<br>-18-8-1 | 38 Sr +2<br>87.62<br>-18-8-2 | 39 Y +3<br>88.9059<br>-18-9-2 | 40 Zr +4<br>91.22<br>-18-10-2 | 41 Nb +3 +5<br>92.9064<br>-18-12-1 | 42 Mo +6<br>95.94<br>-18-13-1 | 43 Tc +4 +6 +7<br>98.9062<br>-18-13-2 | 44 Ru +3<br>101.07<br>-18-15-1 | 45 Rh +3<br>102.9055<br>-18-16-1 | 46 Pd +2 +4<br>106.4<br>-18-18-0 | 47 Ag +1<br>107.868<br>-18-18-1 | 48 Cd +2<br>112.41<br>-18-18-2 | 49 In +3<br>114.82<br>-18-18-3 | 50 Sn +2 +4<br>118.69<br>-18-18-4 | 51 Sb +3 +5 −3<br>121.75<br>-18-18-5 | 52 Te +4 +6 −2<br>127.60<br>-18-18-6 | 53 I +1 +5 +7 −1<br>126.9045<br>-18-18-7 | 54 Xe 0<br>131.30<br>-18-18-8 | -M-N-O |
| 55 Cs +1<br>132.9054<br>-18-8-1 | 56 Ba +2<br>137.33<br>-18-8-2 | 57 La +3<br>138.9055<br>-18-9-2 | 72 Hf +4<br>178.49<br>-32-10-2 | 73 Ta +5<br>180.947$_9$<br>-32-11-2 | 74 W +6<br>183.85<br>-32-12-2 | 75 Re +4 +6 +7<br>186.207<br>-32-13-2 | 76 Os +3 +4<br>190.2<br>-32-14-2 | 77 Ir +3 +4<br>192.22<br>-32-15-2 | 78 Pt +2 +4<br>195.09<br>-32-16-2 | 79 Au +1 +3<br>196.9665<br>-32-18-1 | 80 Hg +1 +2<br>200.59<br>-32-18-2 | 81 Tl +1 +3<br>204.37<br>-32-18-3 | 82 Pb +2 +4<br>207.2<br>-32-18-4 | 83 Bi +3 +5<br>208.9804<br>-32-18-5 | 84 Po +2 +4<br>(209)<br>-32-18-6 | 85 At<br>(210)<br>-32-18-7 | 86 Rn 0<br>(222)<br>-32-18-8 | -N-O-P |
| 87 Fr +1<br>(223)<br>-18-8-1 | 88 Ra +2<br>226.0254<br>-18-8-2 | 89 Ac +3<br>(227)<br>-18-9-2 | 104 Rf +4<br>(260)<br>-32-10-2 | 105 Ha<br>(260)<br>-32-11-2 | 106<br>(263)<br>-32-12-2 | | | | | | | | | | | | | O P Q |

NOBLE GASES

Numbers in parentheses are mass numbers of most stable isotope of that element

| | | | | | | | | | | | | | | | Orbit |
|---|---|---|---|---|---|---|---|---|---|---|---|---|---|---|---|
| LANTHANIDES | 58 Ce +3 +4<br>140.12<br>-20-8-2 | 59 Pr +3<br>140.9077<br>-21-8-2 | 60 Nd +3<br>144.24<br>-22-8-2 | 61 Pm +3<br>(145)<br>-23-8-2 | 62 Sm +2 +3<br>150.4<br>-24-8-2 | 63 Eu +2 +3<br>151.96<br>-25-8-2 | 64 Gd +3<br>157.25<br>-25-9-2 | 65 Tb +3<br>158.9254<br>-27-8-2 | 66 Dy +3<br>162.50<br>-28-8-2 | 67 Ho +3<br>164.9304<br>-29-8-2 | 68 Er +3<br>167.26<br>-30-8-2 | 69 Tm +3<br>168.9342<br>-31-8-2 | 70 Yb +2 +3<br>173.04<br>-32-8-2 | 71 Lu +3<br>174.967 ± 0.003<br>-32-9-2 | N O P |
| ACTINIDES | 90 Th +4<br>232.0381<br>-18-10-2 | 91 Pa +5 +4<br>231.0359<br>-20-9-2 | 92 U +3 +4 +5 +6<br>238.029<br>-21-9-2 | 93 Np +3 +4 +5 +6<br>237.0482<br>-22-9-2 | 94 Pu +3 +4 +5 +6<br>(244)<br>-24-8-2 | 95 Am +3 +4 +5 +6<br>(243)<br>-25-8-2 | 96 Cm +3<br>(247)<br>-25-9-2 | 97 Bk +3 +4<br>(247)<br>-27-8-2 | 98 Cf +3<br>(251)<br>-28-8-2 | 99 Es +3<br>(254)<br>-29-8-2 | 100 Fm +3<br>(257)<br>-30-8-2 | 101 Md +2 +3<br>(258)<br>-31-8-2 | 102 No +2 +3<br>(259)<br>-32-8-2 | 103 Lr +3<br>(260)<br>-32-9-2 | O P Q |

# AP CHEMISTRY

# COURSE REVIEW

# CHAPTER 1

# THE STRUCTURE OF MATTER

## A. ATOMIC PROPERTIES

### 1. THE ATOMIC THEORY AND EVIDENCE FOR THE ATOMIC THEORY

John Dalton, in the early 19th century, proposed that all matter is composed of small subunits that he called "atoms." This was to explain such phenomena as the behavior of gases, and the changes in the physical state of matter. For example, the atoms of solids are packed quite closely together, where the only possible motion is vibration. As the solid is heated, the vibrations become more intense, causing the atoms to move further apart (remember, heat makes things expand). If the solid is heated enough, the atoms will finally break loose from one another, and we say the solid has melted, or become a liquid. In liquids the atoms are free to move around, and travel fairly rapidly. This is why liquids flow, and unlike solids, they have no definite shape; they assume the shape of their container. As the liquid is heated, the atoms move even faster, and further apart, until the liquid becomes a gas. In gases, the atoms are quite far apart, and travel very fast. As a result, unlike solids and liquids, gases have no definite volume. Their atoms are so far apart they can be pushed closer together with relatively little effort. Hence, a gas will assume the volume of its container, as well as the container's shape.

### 2. CHEMICAL AND PHYSICAL APPROACHES TO ATOMIC WEIGHT DETERMINATION

The gram-atomic weight of any element is defined as the mass, in grams which contains one mole of atoms of that element.

For example, approximately 12.0 g of carbon, 16.0 g oxygen, and 32.1 g sulfur, each contain 1 mole of atoms.

The term "one mole" in the above definition is a certain number of atoms. Just as a dozen always means 12, and a gross is always 144, a mole is always $6.02 \times 10^{23}$. The value $6.02 \times 10^{23}$ is also known as Avogadro's number, which by definition is the number of atoms in one gram of pure hydrogen.

### PROBLEM:

What is the atomic weight of lead if 207.2 grams of the metal are found to have as many atoms as 1 gram of hydrogen (Avogadro's number)?

### SOLUTION:

If we have Avogadro's number of atoms present, that is $6.02 \times 10^{23}$, we then have 1 mole of atoms present. This gives us 207.2 grams/per 1 mole, which satisfies the above definition of atomic weight. The answer is 207.2 Daltons, or 207.2 a.m.u. (atomic mass units). Note the types of units used.

## 3. ATOMIC NUMBER AND MASS NUMBER, ISOTOPES, MASS SPECTROSCOPY

The atoms that make up matter can be subdivided into outer layers containing negatively charged electrons that are a great distance from a central nucleus, which is quite dense. The nucleus is made up of two types of particles — protons, which are positively charged, and neutrons which have no charge. The general term for a particle in a nucleus is "nucleon." The atomic weight tells how many nucleons are in the nucleus of that type of atom. For example, carbon has an atomic weight of 12, therefore it has 12 nucleons in its nucleus. (Remember, the atomic weight also tells us that 12 grams of carbon would contain 1 mole ($6.02 \times 10^{23}$) of atoms.)

**Atomic Number** — The number of protons found in a nucleus is the atomic number, and it is the number of protons found in a nucleus that determines what element we have. For example, the atomic number of hydrogen is 1, and it will always be 1 (1 proton); if it were 2, the atom would be helium, not hydrogen. Likewise, lead will always have atomic number 82; if it were 83 it would be bismuth; if it were 81 it would be thallium, and not lead in either case. Atomic numbers will always be whole numbers, and never involve fractions.

To find the number of neutrons in a nucleus, one simply subtracts the atomic number from the atomic weight. For example, carbon is atomic number 6, and atomic weight 12. How many protons and neutrons are found in its nucleus. The atomic number tells us right away how many protons we have ... 6. The number of neutrons is not much more difficult to determine: $12 - 6 = 6$. Therefore, we have 6 protons (which make it carbon) and 6 neutrons (which give it a total weight of 12).

**Isotopes** — Refer to the example in Section A.2. on Page 1, where we calculated lead to have an atomic weight of 207.2. Can a nucleus have two-tenths of a nucleon? Obviously, it cannot. All atoms of lead have 82 protons; however, in nature, lead atoms can exist with 122, 124, 125 and 126 neutrons, resulting in atoms with atomic weights of 204, 206, 207, and 208. These atoms, with the same number of protons, but a different number of neutrons are called "isotopes." Hence, we could say that in nature we find four isotopes of lead. To determine a value to place on a chart of atomic weights, we take a weighted average, based upon the natural abundance of each lead isotope. The abundances are as follows:

| | |
|---|---|
| Pb–204 | 1.42 % abundant |
| Pb–206 | 24.1% |
| Pb–207 | 22.1% |
| Pb–208 | 52.4% |

Using this data we can calculate the average atomic weight of lead as:

$$(0.0142)(204) + (0.241)(206) + (0.221)(207) + (0.524)(208) = 207.2$$

Some elements, such as gold, exist as one isotope. All gold has 79 protons and 118 neutrons, and hence an atomic weight of 197. Other elements, such as lead, can have two or more isotopes of different abundances. (Tin is the "winner," with 10 different isotopes, with abundances ranging from 0.38% up to 32.4%, giving an average atomic weight of 118.69.) As can be seen, atomic weights need not be whole numbers.

**Mass Spectroscopy** — To determine the number of isotopes that exist of an element, the atoms are first charged electrically (a charged atom is called an ion), and passed through a magnetic field. Heavier isotopes are deflected to a lesser extent than lighter ones by the magnetic field (provided their charges are equal), and isotopes of lesser abundance will produce a less intense ion beam than isotopes of greater abundance. From the ion beam position, the atomic weight can be determined. An instrument used for making such measurements is called a "mass spectrometer." For example, if natural thallium is placed in a mass spectrometer, one will observe two ion beams with the beam being deflected the most by the magnetic field also being the weakest. From this scant data one can tell that thallium exists as two isotopes in nature, with the heavier one more abundant than the lighter one. An actual investigation would further show that thallium exists as Tl–203 with abundance of 29.5% and Tl–205 of 70.5% abundance.

On a larger scale this same principle can be used to actually separate the isotopes that occur in nature. Such a device is called a "calatron."

## 4. ELECTRON ENERGY LEVELS

The ground state is the lowest energy state available to the atom.

The excited state is any state of energy higher than that of the ground state.

The formula for changes in energy ($\Delta E$) is

$$\Delta E_{\text{electron}} = E_{\text{final}} - E_{\text{initial}}$$

When the electron moves from the ground state to an excited state, it absorbs energy.

When it moves from an excited state to the ground state, it emits energy.

This exchange of energy is the basis for atomic spectra.

Bohr applied to the hydrogen atom the concept that the electron can exist only in certain stable energy levels and that when the electronic state of the atom changes, it must absorb or emit exactly that amount of energy equal to the difference between the final and initial states:

$$\Delta E = E_a - E_b$$

$$E_b - E_a = \frac{z^2 e^2}{2 a_0} \left[ \frac{1}{n_a^2} - \frac{1}{n_b^2} \right]$$

measures the energy difference between states $a$ and $b$, where $n$ = the (quantum) energy level, $E$ = energy, $e$ = charge on electron, $a_0$ = Bohr radius, and $z$ = atomic number.

The Rydberg-Ritz equation permits calculation of the spectral lines of hydrogen:

$$\frac{1}{\lambda} = R \left[ \frac{1}{n_a^2} - \frac{1}{n_b^2} \right]$$

where $R$ = 109678 $cm^{-1}$ (Rydberg constant), $n_a$ and $n_b$ are the quantum numbers for states $a$ and $b$, and $\lambda$ is the wavelength of light emitted or absorbed.

Light behaves as if it were composed of tiny packets, or quanta, of energy (now called "photons").

$$E_{\text{photon}} = h\upsilon$$

where $h$ is Planck's constant and $\upsilon$ is the frequency of light.

$$E = \frac{hc}{\lambda}$$

where $c$ is the speed of light and $\lambda$ is the wavelength of light.

The electron is restricted to specific energy levels in the atom. Specifically,

$$E = -\frac{A}{n^2}$$

where $A = 2.18 \times 10^{-11}$ erg, and $n$ is the quantum number.

## Basic Electron Charges

Cathode rays are made up of very small negatively-charged particles named electrons. The cathode is the negative electrode and the anode is the positive electrode.

The nucleus is made up of small positively-charged particles called protons and of neutral particles called neutrons. The proton mass is approximately equal to the mass of the neutron, and is 1,837 times the mass of the electron.

## Components of Atomic Structure

The number of protons and neutrons in the nucleus is called the mass number, which corresponds to the isotopic atomic weight. The atomic number is the number of protons found in the nucleus.

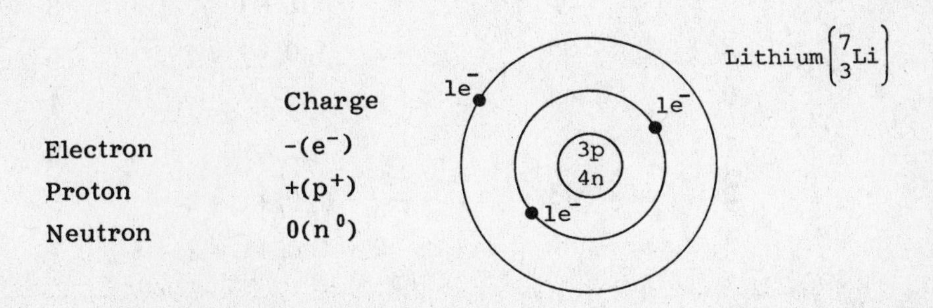


The electrons found in the outermost shell are called valence electrons. When these electrons are lost or partially lost (through sharing), the oxidation state is assigned a positive value for the element. If valence electrons are gained or partially gained by an atom, its oxidation number is taken to be negative.

**EXAMPLE:**

$_{17}Cl$ = ●)2)8)7 ← valence electrons.

↘ nucleus

$(\cdot\ddot{\underset{..}{Cl}}:)^-$ This is called the Lewis dot structure of the chloride ion. Its oxidation number is – 1.

## The Wave Mechanical Model

Each wave function corresponds to a certain electronic energy and describes a region about the nucleus (called an orbital) where an electron having that energy may be found. The square of the wave function, $|\psi|^2$, is called a probability density, and equals the probability per unit volume of finding the electron in a given region of space.

## Summary of Quantum Numbers

| Principal Quantum Number, *n* (Shell) | Azimuthal Quantum Number, 1 (Subshell) | Subshell Designation | Magnetic Quantum Number, *m* (Orbital) | Number of Orbitals in Subshell |
|---|---|---|---|---|
| 1 | 0 | 1s | 0 | 1 |
| 2 | 0 | 2s | 0 | 1 |
| | 1 | 2p | – 1, 0, +1 | 3 |
| 3 | 0 | 3s | 0 | 1 |
| | 1 | 3p | –1, 0, + 1 | 3 |
| | 2 | 3d | – 2, – 1, 0, +1, + 2 | 5 |
| 4 | 0 | 4s | 0 | 1 |
| | 1 | 4p | –1, 0, +1 | 3 |
| | 2 | 4d | – 2, – 1, 0, +1, + 2 | 5 |
| | 3 | 4f | – 3, – 2, – 1, 0, +1, + 2, + 3 | 7 |

## Subshells and Electron Configuration

The Pauli exclusion principle states that no two electrons within the same atom may have the same four quantum numbers.

## Table — Subdivision of Main Energy Levels

| main energy level | 1 | 2 | 3 | 4 |
|---|---|---|---|---|
| number of sublevels($n$) | 1 | 2 | 3 | 4 |
| number of orbitals ($n^2$) | 1 | 4 | 9 | 16 |
| kind and no. of orbitals per sublevel | s<br>1 | s p<br>1 3 | s p d<br>1 3 5 | s p d f<br>1 3 5 7 |
| maximum no. of electrons per sublevel | 2 | 2 6 | 2 6 10 | 2 6 10 14 |
| maximum no. of electrons per main level ($2n^2$) | 2 | 8 | 18 | 32 |

## Table — Electron Arrangements

| Main Levels | 1 | 2 | | | | 3 | Summary |
|---|---|---|---|---|---|---|---|
| Sublevels | s | s | p | | | s | |
| H | (↑) | | | | | | $1s^1$ |
| He | (↑↓) | | | | | | $1s^2$ |
| Li | (↑↓) | (↑) | | | | | $1s^2 2s^1$ |
| Be | (↑↓) | (↑↓) | | | | | $1s^2 2s^2$ |
| B | (↑↓) | (↑↓) | (↑) | ( ) | ( ) | | $1s^2 2s^2 2p^1$ |
| C | (↑↓) | (↑↓) | (↑) | (↑) | ( ) | | $1s^2 2s^2 2p^2$ |
| N | (↑↓) | (↑↓) | (↑) | (↑) | (↑) | | $1s^2 2s^2 2p^3$ |
| O | (↑↓) | (↑↓) | (↑↓) | (↑) | (↑) | | $1s^2 2s^2 2p^4$ |
| F | (↑↓) | (↑↓) | (↑↓) | (↑↓) | (↑) | | $1s^2 2s^2 2p^5$ |
| Ne | (↑↓) | (↑↓) | (↑↓) | (↑↓) | (↑↓) | | $1s^2 2s^2 2p^6$ |
| Na | (↑↓) | (↑↓) | (↑↓) | (↑↓) | (↑↓) | (↑) | $1s^2 2s^2 2p^6 3s^1$ |
| Mg | (↑↓) | (↑↓) | (↑↓) | (↑↓) | (↑↓) | (↑↓) | $1s^2 2s^2 2p^6 3s^2$ |

Hund's rule states that for a set of equal-energy orbitals, each orbital is occupied by one electron before any orbital has two. Therefore, the first electrons to occupy orbitals within a sublevel have parallel spins. The rule is shown in the table above.

Transition elements are elements whose electrons occupy the d sublevel.

Transition elements can exhibit various oxidation numbers. An example of this is manganese, with possible oxidation numbers of +2, +3, +4, +6 and +7.

Groups IB through VIIB and Group VIII constitute the transition elements.

## 5. THE PERIODIC TABLE AND PERIODIC RELATIONSHIPS: SYMBOLS, RADII, IONIZATION ENERGY, ELECTRON AFFINITY, OXIDATION STATES

### Periodic Table

Periodic law states that chemical and physical properties of the elements are periodic functions of their atomic numbers.

Vertical columns are called groups, each containing a family of elements possessing similar chemical properties.

The horizontal rows in the periodic table are called periods.

The elements lying in two rows just below the main part of the table are called the inner transition elements.

In the first of these rows are elements 58 through 71, called the lanthanides or rare earths.

The second row consists of elements 90 through 103, the actinides.

Group IA elements are called the alkali metals.

Group IIA elements are called the alkaline earth metals.

Group VIIA elements are called the halogens, and the Group O elements, the noble gases.

The metals in the first two groups are the light metals, and those toward the center are the heavy metals. The elements found along the dark line in the chart are called metalloids. They have characteristics of both metals and nonmetals. Some examples of metalloids are boron and silicon.

### Properties Related to the Periodic Table

The most active metals are found in the lower left corner. The most active nonmetals are found in the upper right corner.

Metallic properties include high electrical conductivity, luster, generally high melting points, ductility (ability to be drawn into wires), and malleability (ability to be hammered into thin sheets). Nonmetals are uniformly very poor conductors of electricity, do not possess the luster of metals and form brittle solids. Metalloids have properties intermediate between those of metals and nonmetals.

### Atomic Radii

The atomic radius generally decreases across a period from left to right. The atomic radius increases down a group.

### Electronegativity

The electronegativity of an element is a number that measures the relative strength with which the atoms of the element attract valence electrons in a chemical bond. This electronegativity number is based on an arbitrary scale from 0 to 4. Metals have electronegativities less than 2. Electronegativity increases from left to right in a period and decreases as you go down a group.

### Ionization Energy

Ionization energy is defined as the energy required to remove an electron from an isolated atom in its ground state. As we proceed down a group, a decrease in ionization energy occurs. Proceeding across a period from left to right, the ionization energy increases. As we proceed to the right, base-forming properties decrease and acid-forming properties increase.

# B. BONDING

## 1. TYPES OF BONDS

An ionic bond occurs when one or more electrons are transferred from the valence shell of one atom to the valence shell of another.

The atom that loses electrons becomes a positive ion (cation), while the atom that acquires electrons becomes a negatively-charged ion (anion). The ionic bond results from the coulomb attraction between the oppositely-charged ions.

The octet rule states that atoms tend to gain or lose electrons until there are eight electrons in their valence shell.

A covalent bond results from the sharing of a pair of electrons between atoms.

In a nonpolar covalent bond, the electrons are shared equally.

Nonpolar covalent bonds are characteristic of homonuclear diatomic molecules. For example, the Fluorine molecule:

$$\cdot\ddot{\underset{\cdot\cdot}{F}}: \quad \cdot\ddot{\underset{\cdot\cdot}{F}}: \quad \rightarrow \quad :\ddot{\underset{\cdot\cdot}{F}}:\ddot{\underset{\cdot\cdot}{F}}:$$

Fluorine atoms → Fluorine molecule

Where there is an unequal sharing of electrons between the atoms involved, the bond is called a polar covalent bond. An example:

$$H \ \overset{\times}{\cdot}\ddot{\underset{\cdot\cdot}{Cl}}:$$

× hydrogen electron
• chlorine electrons

$$H \ \overset{\times}{\cdot}\ddot{\underset{\cdot\cdot}{O}}\ :$$

× hydrogen electron
• chlorine electrons

Because of the unequal sharing, the bonds shown are said to be polar bonds (dipoles). The more electronegative element in the bond is the negative end of the bond dipole. In each of the molecules shown here, there is also a non-zero molecular dipole moment, given by the vector sum of the bond dipoles.

A dipole consists of a positive and negative charge separated by a distance. A dipole is described by its dipole moment, which is equal to the charge times the distance between the positive and negative charges:

$$\text{net dipole moment} = \text{charge} \times \text{distance}$$

In polar molecular substances, the positive pole of one molecule attracts the negative pole of another. The force of attraction between polar molecules is called a dipolar force.

When a hydrogen atom is bonded to a highly electronegative atom, it will become partially positively-charged, and will be attracted to neighboring electron pairs. This creates a hydrogen bond. The more polar the molecule, the more effective the hydrogen bond is in binding the molecules into a larger unit.

The relatively weak attractive forces between molecules are called Van der Waals forces. These forces become apparent only when the molecules approach one another closely (usually at low temperatures and high pressure). They are due to the way the positive charges of one molecule attract the negative charges of another molecule. Compounds of the solid state that are bound mainly by this type of attraction have soft crystals, are easily deformed, and vaporize easily. Because of the low intermolecular forces, the melting points are low and evaporation takes place so easily that it may occur at room temperature. Examples of substances with this last characteristic are iodine crystals and napthalene crystals.

**Metallic Bonds** — In a metal the atoms all share their outer electrons in a manner that might be thought of as an electron "atmosphere." The electrons hop freely from one atom to another, and it is this property that make metals good conductors of electricity.

## 2. EFFECTS OF BONDING FORCES ON STATES, STRUCTURES, AND PROPERTIES OF MATTER

Ionic substances are characterized by the following properties:

1. Ionic crystals have large lattice energies because the electrostatic forces between them are strong.
2. In the solid phase, they are poor electrical conductors.
3. In the liquid phase, they are relatively good conductors of electric current; the mobile charges are the ions (in contrast to metallic conduction, where the electrons constitute the mobile charges).
4. They have relatively high melting and boiling points.
5. They are relatively non-volatile and have low vapor pressure.
6. They are brittle.
7. Those that are soluble in water form electrolytic solutions that are good conductors of electricity.

The following are general properties of molecular crystals and/or liquids:

1. Molecular crystals tend to have small lattice energies and are easily deformed because their constituent molecules have relatively weak forces between them.
2. Both the solids and liquids are poor electrical conductors.
3. Many exist as gases at room temperature and atmospheric pressure; those that are solid or liquid at room temperature are relatively volatile.
4. Both the solids and liquids have low melting and boiling points.
5. The solids are generally soft and have a waxy consistency.
6. A large amount of energy is often required to chemically decompose the solids and liquids into simpler substances.

## 3. POLARITY AND ELECTRONEGATIVITY

Polarity is covered in Section B.1. on Page 9. In brief, if we neglect the row of inert gases on the periodic table, the further to the right and up an element is, the greater its electronegativity. (Electronegativity is the ability of an atom to pull electrons off another atom.) Looking at the periodic chart, one can easily see that fluorine is the most electronegative element, while cesium is the least. Elements with very great differences in electronegativity (i.e., from opposite sides of the periodic table) are more likely to combine by an ionic bond, e.g., $C_sF$. Elements with *no* difference in electronegativity will combine by a non-polar covalent bond, e.g., $Cl_2$. Elements with slight differences in electronegativity (usually from the same side of the periodic table) will combine by a polar covalent bond, e.g., HCL, BrCl, $SO_2$.

## 4. GEOMETRY OF IONS, MOLECULES, AND COORDINATION COMPLEXES

The process of mixing different orbitals of the same atom to form a new set of equivalent orbitals is termed hybridization. The orbitals formed are called hybrid orbitals.

Valence Shell Electron Pair Repulsion (VSEPR) theory permits the geometric arrangement of atoms, or groups of atoms, about some central atom to be determined solely by considering the repulsions between the electron pairs present in the valence shell of the central atom.

Based on VSEPR, the general shape of any molecule can be predicted from the number of bonding and non-bonding electron pairs in the valence shell of the central atom, recalling that nonbonded pairs of electrons (lone pairs) are more repellent than bonded pairs.

### Table — Summary of Hybridization

| Number of Bonds | Number of Unused e pairs | Type of Hybrid Orbital | Angle between Bonded Atoms | Geometry | Example |
|---|---|---|---|---|---|
| 2 | 0 | sp | 180° | Linear | $BeF_2$ |
| 3 | 0 | $sp^2$ | 120° | Trigonal planar | $BF_3$ |
| 4 | 0 | $sp^3$ | 109.5° | Tetrahedral | $CH_4$ |
| 3 | 1 | $sp^3$ | 90° to 109.5° | Pyramidal | $NH_3$ |
| 2 | 2 | $sp^3$ | 90° to 109.5° | Angular | $H_2O$ |
| 6 | 0 | $sp^3d^2$ | 90° | Octahedral | $SF_6$ |

From the geometry of these hybridized bonds we can predict the shape of the resultant molecule. For example, methane hybridizes the s orbitals from hydrogen with 3 p orbitals from carbon. The resulting bond is referred to as "$sp^3$," its shape is tetrahedral, as is the shape of the resulting methane molecule. Note also that the slight differences in electronegativity cause a polar covalent bond to be formed between the carbon and each hydrogen. Each such bond has its associated dipole moment, but as shown below, each dipole cancels:

```
                                H
                                |↓
methane: no net dipole    H  —  C  —  H
                            →   |↑  ←
                                H
```

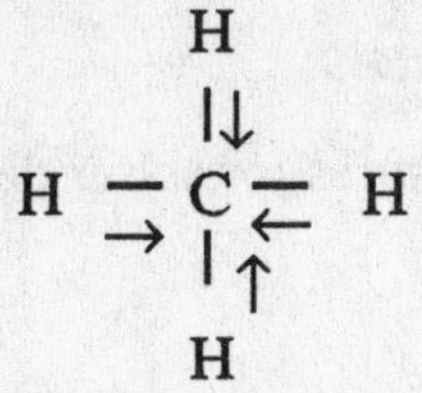

Hence, a molecule that contains polar bonds may not be polar. One must check to see whether or not the dipoles cancel.

## 5. MOLECULAR MODELS

### Double and Triple Bonds

Sharing two pairs of electrons produces a double bond. An example:

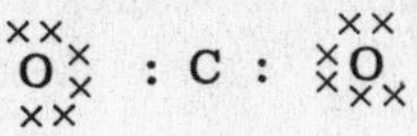

```
 ××          ××
 O×  : C :  ×O        or,    O = C = O
 ××          ××
```

The sharing of three electron pairs results in a triple bond. An example:

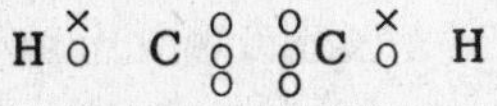

```
   ×   o  o  ×
 H o C o  o C o  H     or,   H — C ≡ C — H
       o  o
```

Greater energy is required to break double bonds than single bonds, and triple bonds are harder to break than double bonds. Molecules which contain double and triple bonds have smaller interatomic distances and greater bond strength than molecules with only single bonds. Thus, in the series,

$$H_3C-CH_3,\ H_2C=CH_2,\ HC\equiv CH,$$

the carbon-carbon distance decreases, and the C-C bond energy increases because of increased bonding.

### Sigma and Pi Bonds

A molecular orbital that is symmetrical around the line passing through two nuclei is called a sigma ($\sigma$) orbital. When the electron density in this orbital is concentrated in the bonding region between two nuclei, the bond is called a sigma bond.

The bond that is formed by the sideways overlap of two p orbitals, and that provides electron density above and below the line connecting the bound nuclei, is called a $\pi$ bond (Pi bond).

Pi bonds are present in molecules containing double or triple bonds.

Of the sigma and Pi bonds, the former has greater orbital overlap and is generally the stronger bond.

## Resonance Structures

The resonance structures for sulfur dioxide are as follows:

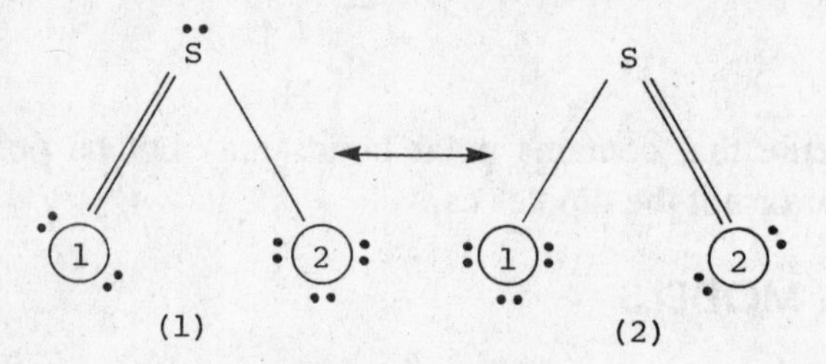


The actual electronic structure of $SO_2$ does not correspond to either 1 or 2, but, instead, to an "average" structure somewhere in between. This true structure is known as a resonance hybrid of the contributing structures 1 and 2.

## Structural Isomerism

There are 2 types of butanes — normal butane and iso-butane. They have the same molecular formula, $C_4H_{10}$, but have different structures. N-Butane is a straight chain hydrocarbon whereas isobutane is a branched-chain hydrocarbon. N-Butane and isobutane are structural isomers and differ in their physical and chemical properties.

```
                                    H  H  H  H
                                    |  |  |  |
C4H10 ≡ CH3CH2CH2CH3 ≡           H-C--C--C--C-H      n-butane
                                    |  |  |  |        (straight chained)
                                    H  H  H  H

                                    H  H  H
                                    |  |  |
C4H10 ≡ CH3CH(CH3)CH3 ≡          H-C--C--C-H          isobutane
                                    |  |  |           (branched)
                                    H  |  H
                                       |
                                    H-C-H
                                       |
                                       H
```

# C. NUCLEAR CHEMISTRY, NUCLEAR EQUATIONS, HALF-LIVES, RADIOACTIVITY

There are three main types of particles that can escape from an atom's nucleus, named after the first three letters of the Greek alphabet, $\alpha$, $\beta$ and $\gamma$ (alpha, beta, and gamma). $\alpha$ particles consist of 2 neutrons and 2 protons bound together. This arrangement might remind one of a helium nucleus, and that is exactly what it is. After an $\alpha$ particle slows down, it can capture 2 electrons, and can become a helium atom. Decay by $\alpha$ emission results in a decrease of atomic mass by 4 and atomic number by 2.

## EXAMPLE:

The radioactive gas radon-222 decays by alpha emissions. What does it decay to?

We can write the equation as

$$^{222}_{86}\text{Rn} \rightarrow \alpha + ?$$

It is easier to write the equation using the symbol for helium, rather than $\alpha$, this gives us:

$$^{222}_{86}\text{Rn} \rightarrow {}^{4}_{2}\text{He} + ?$$

We can now balance our equation, as the mass numbers (superscript) must have the same total on each side of the equation. If we subtract 4 from 222 we get 218, this is the atomic mass (weight) of the new atom. Likewise, the atomic number (subscripts) must have the same total on each side of the arrow. If we subtract 2 from 86 we get 84. This is our new atomic number, and by looking at the periodic table we can see it belongs to polonium. We can now properly write the equation as

$$^{222}_{86}\text{Rn} \rightarrow {}^{4}_{2}\text{He} + {}^{218}_{84}\text{Po} \quad \text{or} \quad {}^{222}_{86}\text{Rn} \rightarrow \alpha + {}^{218}_{84}\text{Po}.$$

Beta particles are electrons that are emitted from a nucleus when a neutron is converted to a proton and an electron. The electron is shot out of the nucleus sometimes at a speed very close to that of light. Once it slows down, outside of the nucleus, it becomes an ordinary electron, and behaves accordingly. Beta ($B^-$) decay does not cause any change in atomic weight, but due to the conversion of a neutron into a proton, it increases the atomic number by 1.

## EXAMPLE:

The Polonium-218 from the previous example decays by $B^-$ emissions. What does it decay to?

We would start our equation with what we know.

$$^{218}_{84}\text{Po} \rightarrow \text{B}^- + ?$$

It is useful to substitute the electron symbol for $\text{B}^-$, resulting in the following:

$$^{218}_{84}\text{Po} \rightarrow {}^{0}_{-1}\text{e} + ?$$

Notice that the atomic weight (mass) of the electron is 0, (it isn't exactly, it's really $^1/_{1832}$, but for this purpose 0 is close enough) and the "atomic number" is –1. The atomic number is the measure of positive charges (protons); an electron has a negative charge; hence –1. Again, as in a decay, the atomic numbers and atomic weights (mass) must total the same on both sides of the arrow. If we subtract 0 from 218, we still have 218, our mass remains unchanged. If we subtract –1 from 84 we get 84 –(–1) = 84 +1 = 85. If we look at the periodic table, element no. 85 is astatine. We can then write the correct equation as:

$$^{218}_{84}\text{Po} \rightarrow {}^{0}_{-1}\text{e} + {}^{218}_{85}\text{At, or } {}^{218}_{84}\text{Po} \rightarrow {}^{-}\text{B} + {}^{218}_{85}\text{At}.$$

Gamma ($\gamma$) rays are made up of massless particles (this time the mass really is 0), and resemble X-rays very closely. $\gamma$ rays are usually emitted along with $\text{B}^-$ and $\alpha$ radiation (but not always), and they are just not shown in the equations. Pure $\gamma$ decay can occur from a nucleus in a "metastable" state, when it undergoes a transformation from a higher energy state to a lower one. The resultant atom may or may not be radioactive. The metastable state is designated by the lowercase letter "m" after the atomic mass (e.g., Tc–99m, as opposed to Tc–99). Since the $\gamma$s emitted from the nucleus possess neither charge nor mass, they change neither the atomic number nor mass.

**EXAMPLE:**

How would one write the decay of indium-113m to stable (non-radioactive) indium?

$$^{113m}_{49}\text{In} \rightarrow \gamma + {}^{113}_{49}\text{In}$$

**Penetrating Power** — $\alpha$ particles penetrate the least, since they are relatively slow moving and large. A sheet of paper can stop $\alpha$s, as does the outer layer of one's skin. $\text{B}^-$ particles are smaller and more energetic, usually a sheet of lead or several inches of wood will stop $\text{B}^-$s. Normal clothing will stop most $\text{B}^-$ rays. $\gamma$ rays are much harder to stop. They require several inches of lead or several feet of concrete. Since they can easily penetrate plant and animal tissues, they are useful in "tracer" studies.

**Half-Life** — A useful term in nuclear chemistry is "half-life," which is the amount of time it takes for half of the material to decay away. Half-lives can range from fractions of second to trillions of years. They are constant for each isotope, and cannot be altered by changes in temperature, oxidation state, or any other factors.

**EXAMPLE:**

Strontium-85 has a half-life of 65.2 days. If we start with 10 grams of Sr-85, how much will be left after 130.4 days?

130.4 ÷ 65.2 = 2, or two half-lives. After one half-life 5 grams would be left; after two half-lives, half of that would be left, or 2.5 grams. If we waited another half-life, or a total of 195.6 days, there would be 1.25 grams left.

# CHAPTER 2

# STATES OF MATTER

## A. GASES

### 1. IDEAL GAS LAWS

The following laws apply to ideal gases, which do not exist. The molecules of an ideal gas would have zero volume, and their collisions would be totally elastic. Real gases behave very much like ideal gases under low pressures and high temperatures.

A gas has no shape of its own; rather, it takes the shape of the container. It has no fixed volume, but is compressed or expanded as its container changes in size. The volume of a gas is the volume of the container in which it is held.

Pressure is defined as force per unit area. Atmospheric pressure is measured using a barometer.

Atmospheric pressure is directly related to the length (h) of the column of mercury in a barometer and is expressed in mm or cm of mercury (hg).

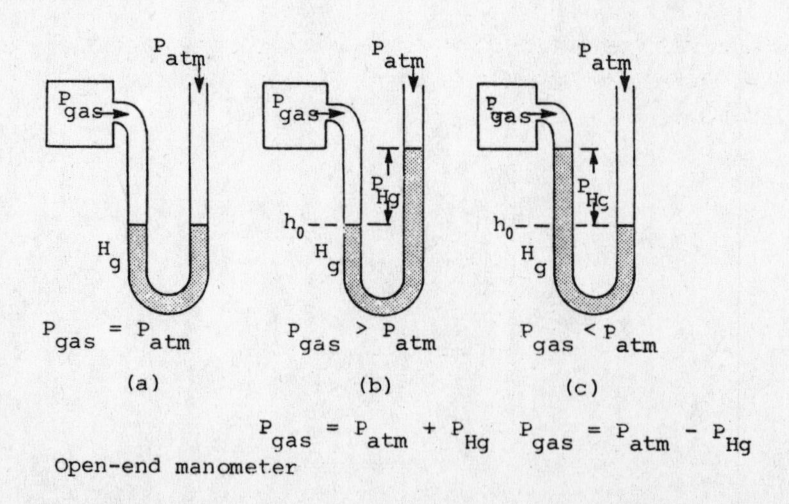


Open-end manometer

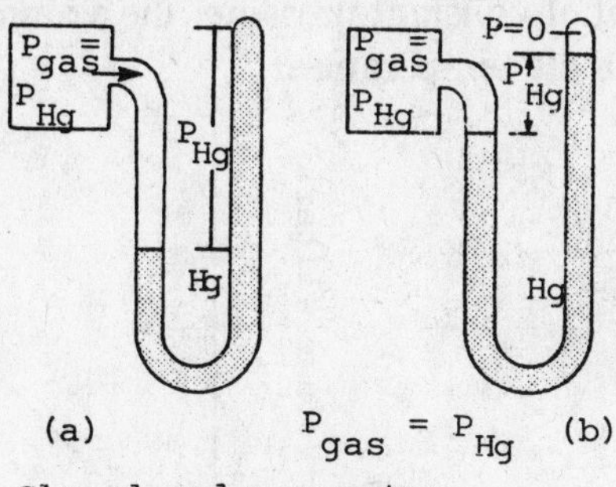

Closed-end manometer

Standard atmospheric pressure is expressed in several ways: 14.7 pounds per square inch (psi), 760 mm of mercury, 760 torr or simply 1 "atmosphere" (1 atm).

## Boyle's Law

Boyle's law states that, at a constant temperature, the volume of a gas is inversely proportional to the pressure:

$$V \propto \frac{1}{P} \text{ or } V = \text{constant} \cdot \frac{1}{P} \text{ or } PV = \text{constant.}$$

$$P_i V_i = P_f V_f$$

$$V_f = V_i \left( \frac{P_i}{P_f} \right)$$

A hypothetical gas that would follow Boyle's law under all conditions is called an ideal gas. Deviations from Boyle's law that occur with real gases represent non-ideal behavior.

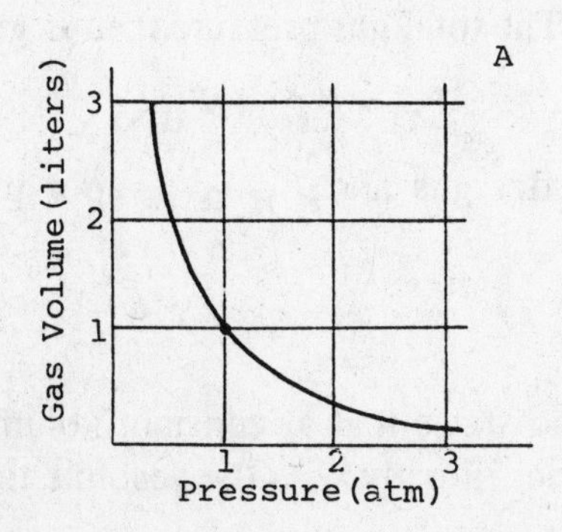

## Charles' Law

Charles' law states that at constant pressure, the volume of a given quantity of a gas varies directly with the temperature:

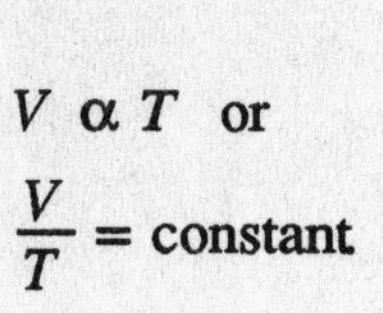

$V \alpha T$ or

$\frac{V}{T}$ = constant

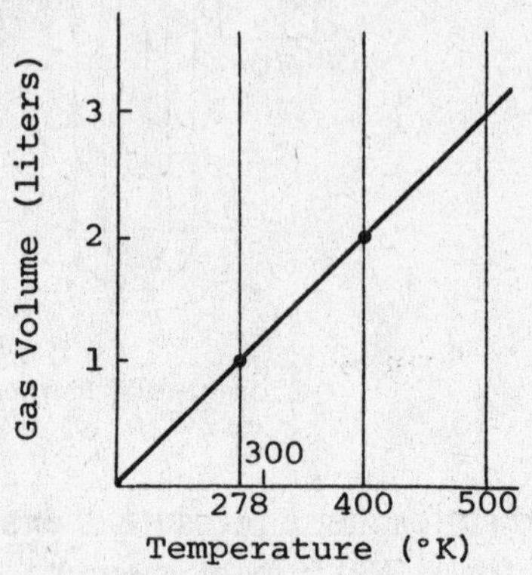


$$\frac{V_1}{T_1} = \frac{V_2}{T_2} \quad \text{or} \quad \frac{V_1}{V_2} = \frac{T_1}{T_2}$$

If Charles' law was strictly obeyed, gases would not condense when they are cooled. This means that gases behave in an ideal fashion only at relatively high temperatures and low pressures.

## Dalton's Law of Partial Pressures

The pressure exerted by each gas in a mixture is called its partial pressure. The total pressure exerted by a mixture of gases is equal to the sum of the partial pressures of the gases in the mixture. This statement, known as Dalton's law of partial pressures, can be expressed

$$P_T = Pa + Pb + Pc + \ldots$$

When a gas is collected over water (a typical laboratory method), some water vapor mixes with the gas. The total gas pressure then is given by

$$P_T = P_{gas} + P_{H_2O}$$

where $P_{gas}$ = pressure of dry gas and $P_{H_2O}$ = vapor pressure of water at the temperature of the system.

## Law of Gay-Lussac

The law of Gay-Lussac states that at constant volume, the pressure exerted by a given mass of gas varies directly with the absolute temperature:

$P \propto T$ (where volume and mass of gas are constant).

$$\frac{P_1}{T_1} = \frac{P_2}{T_2}$$

Gay-Lussac's law of combining volumes states that when reactions take place in the gaseous state, under conditions of constant temperature and pressure, the volumes of reactants and products can be expressed as ratios of small whole numbers.

## Ideal Gas Law

$$V \propto \frac{1}{P}, \; V \propto T, \; V \propto n$$

then $$V \propto \frac{nT}{P}$$

$$PV = nRT$$

The hypothetical ideal gas obeys exactly the mathematical statement of the ideal gas law. This statement is also called the equation of state of an ideal gas because it relates the variables ($P$, $V$, $n$, $T$) that specify properties of the gas. Molecules of ideal gases have no attraction for one another and have no intrinsic volume; they are "point particles." Real gases act in a less than ideal way, especially under conditions of increased pressure and/or decreased temperature. Real gas behavior approaches that of ideal gases as the gas pressure becomes very low. The ideal gas law is thus considered a "limiting law."

When using the ideal gas law, the term "R" is always a constant. If the volume is in liters, the temperature in Kelvin, and pressure is in atmospheres, $R$ has a value of 0.082 and units of liter· atm/mole Kelvin. Remember, this value of $R$ can only be used under the stated conditions.

## Combined Gas Law

The combined gas law states that for a given mass of gas, the volume is inversely proportional to the pressure and directly proportional to the absolute temperature. This law can be written

$$\frac{P_1 V_1}{T_1} = \frac{P_2 V_2}{T_2}$$

where $P_1$ is the original pressure, $V_1$ is the original volume, $T_1$ is the original absolute temperature, $P_2$ is the new pressure, $V_2$ is the new volume and $T_2$ is the new absolute temperature.

## Avogadro's Law (The Mole Concept)

Avogadro's law states that under conditions of constant temperature and pressure, equal volumes of different gases contain equal numbers of molecules.

If the initial and final pressure and temperature are the same, then the relationship between the number of molecules, $N$, and the volume, $V$, is

$$\frac{V_f}{V_i} = \frac{N_f}{N_i}$$

The laws of Boyle, Charles, Gay-Lussac, and Avogadro are all simple corollaries of the general equation of state for an ideal gas — $PV = nRT$ — under various restraining conditions (constant $T$, constant $P$, constant $V$ and constant $T$ and $P$, respectively; $n$ is assumed invariant for all).

## 2. KINETIC MOLECULAR THEORY

### Real Gases

Real gases fail to obey the ideal gas law under most conditions of temperature and pressure.

Real gases have a finite (non-zero) molecular volume; i.e., they are not true "point particles." The volume within which the molecules may not move is called the excluded volume. The real volume (volume of the container) is therefore slightly larger than the ideal volume (the volume the gas would occupy if the molecules themselves occupied no space):

$$V_{\text{real}} = V_{\text{ideal}} + nb$$

where $b$ is the excluded volume per mole and $n$ is the number of moles of gas. The ideal pressure, that is, the pressure the gas could exert in the absence of intermolecular attractive forces, is higher than the actual pressure by an amount that is directly proportional to $n^2/V^2$:

$$P_{\text{ideal}} = P_{\text{real}} + \frac{n^2 a}{V^2},$$

where $a$ is a proportionality constant that depends on the strength of the intermolecular attractions. Therefore,

$$\left(P + \frac{n^2 a}{V^2}\right)(V - nb) = nRT$$

is the Van der Waals equation of state for a real gas.

The values of the constants $a$ and $b$ depend on the particular gas and are tabulated for many real gases.

## Graham's Law of Effusion and Diffusion

Effusion is the process in which a gas escapes from one chamber of a vessel to another by passing through a very small opening or orifice.

Graham's law of effusion states that the rate of effusion is inversely proportional to the square root of the density of the gas:

$$\text{rate of effusion } \alpha \sqrt{\frac{1}{d}}\text{, and}$$

$$\frac{\text{rate of effusion (A)}}{\text{rate of effusion (B)}} = \sqrt{\frac{d_B}{d_A}} = \sqrt{\frac{M_B}{M_A}}$$

where $M$ is the molecular weight of each gas, and where the temperature is the same for both gases.

Mixing of molecules of different gases by random motion and collision until the mixture becomes homogeneous is called diffusion.

Graham's law of diffusion states that the relative rates at which gases will diffuse will be inversely proportional to the square roots of their respective densities or molecular weights:

$$\text{rate } \alpha \frac{1}{\sqrt{\text{mass}}}$$

(where, again, $T_1 = T_2$) and

$$\frac{\text{rate 1}}{\text{rate 2}} = \frac{\sqrt{M_2}}{\sqrt{M_1}} \left(\text{or } \frac{r_1}{r_2} = \frac{\sqrt{d_2}}{\sqrt{d_1}}\right)$$

## The Kinetic Molecular Theory

The kinetic molecular theory is summarized as follows:

1. Gases are composed of tiny, invisible molecules that are widely separated from one another in otherwise empty space.
2. The molecules are in constant, continuous, random and straight-line motion.
3. The molecules collide with one another, but the collisions are perfectly

elastic (that is, they result in no net loss of energy).

4. The pressure of a gas is the result of collisions between the gas molecules and the walls of the container.

5. The average kinetic energy of all the molecules collectively is directly proportional to the absolute temperature of the gas. The average kinetic energy of equal numbers of molecules of any gas is the same at the same temperature.

Any group of gas molecules at a given temperature will show a spectrum of energies. As the temperature of the gas increases, so does the average energy, as would be expected. However, at the higher temperature, the energy distribution curve is much broader, i.e., there is a larger standard deviation from the mean energy at the higher temperature.

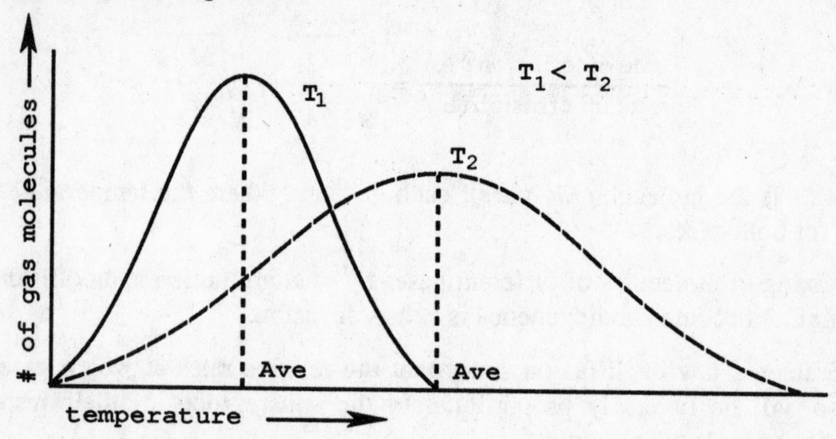


## The Boltzmann Distribution

Given a collection of $N$ molecules having total energy $E$ and distributed among $\varepsilon_i$ discrete energy levels, the fraction of molecules with energy $\varepsilon_i$ compared to $N_0$ with $\varepsilon_o$ is given by the Boltzmann distribution law

$$N_i = N_o e^{-(\varepsilon_i - \varepsilon_o)/kT}$$

where $k$ is the Boltzmann constant and $T$ is the temperature in degrees Kelvin.

The law gives the ratios of numbers of molecules with different energies as a function of $\varepsilon$ and $T$, $\varepsilon_i - \varepsilon_0$ can also be written $\Delta\varepsilon$.

$$N_i = N_o e^{-\Delta\varepsilon/kT}$$

If the energy is sealed so that $E_0 = 0$ we have

$$N_i = N_o e - Ei / kT$$

## One-Dimensional Velocity Distribution

The kinetic energy of a molecule with mass "m" and velocity "μ" moving along $x$–axis is given by $^1/_2 m\mu^2$. Making the appropriate substitution into the Boltzmann equation, and with $\mu = 0$ at $E = 0$, the number of particles with a velocity between $\mu$ and $\mu + d\mu$ becomes

$$dN - dN_o e^{-m\mu^2/2kT}$$

where $dN$ is the number of particles with velocities between $\mu$ and $\mu + d\mu$ and $dN_0$ is the number of particles with velocities between 0 and $d\mu$.

$dN_0$ is proportional to $d\mu$, and substituting, $\alpha\, d\mu$ for $dN_0$ and dividing by '$N$' gives the number of particles in '$N$' with velocities between $\mu + d\mu$ and $\mu$.

$$\frac{dN}{N} = \frac{\alpha}{N} e^{-\frac{m\mu^2}{2kT}} d\mu = f(\mu)\, d\mu$$

Normalization of the above equation yields for the constant $\alpha/N = A$

$$A\int_{-\infty}^{\infty} e^{-\frac{m\mu^2}{2kT}} d\mu = 1$$

with

$$s^2 = \frac{m\mu^2}{2kT}$$

$$ds = \left(\frac{m}{2kT}\right)^{\frac{1}{2}} d\mu$$

and

$$A\left(\frac{2kT}{m}\right)^{\frac{1}{2}} \int_{-\infty}^{\infty} e^{-s^2} ds = 1$$

The integral in terms of $s$ is a standard one and its value is $\pi^{1/2}$. Therefore

$$A = \left(\frac{m}{2\pi kT}\right)^{\frac{1}{2}}$$

By comparison with the Gauss function,

$$h = \left(\frac{m}{2kT}\right)^{\frac{1}{2}}$$

For the distribution and the mean of $\mu$ is zero. The fraction of molecules and velocities between $\mu_A$ and $\mu_B$ is then found by

$$\int_{\mu_A}^{\mu_B} f(\mu)\, d\mu$$

To evaluate the mean of μ we use the averaging technique for a continuous distribution

$$\bar{\mu} = \frac{\int_{-\infty}^{\infty} f(\mu)\, d\mu}{\int_{-\infty}^{\infty} f(\mu)\, d\mu}$$

Since the denominator in the last expression is unity, this becomes

$$\bar{\mu} = \int_{-\infty}^{\infty} \left(\frac{m}{2\pi kT}\right)^{\frac{1}{2}} e^{-\frac{m\mu^2}{2kT}} \mu\, d\mu$$

A standard table of integrals will show

$$\int_{-\infty}^{\infty} xe^{-ax^2} dx = 0$$

and therefore

$$\bar{\mu} = 0$$

The root mean square velocity can be found by a similar procedure

$$(\bar{\mu}^2) = \int_{-\infty}^{\infty} \mu^2 f(\mu)\, d\mu.$$

Again, a standard table of integrals will show that

$$\int_{-\infty}^{\infty} x^2 e^{-ax^2} dx = 2\int_{0}^{\infty} x^2 e^{-ax^2} dx = \frac{\sqrt{\pi}}{2} a^{-\frac{3}{2}}$$

and with

$$a = \frac{m}{2kT}$$

$$(\bar{\mu}^2) = \left(\frac{m}{2\pi kT}\right)^{\frac{1}{2}} \cdot \pi^{\frac{1}{2}} \cdot \left(\frac{2kT}{m}\right)^{\frac{3}{2}}$$

$$(\bar{\mu}^2) = \frac{kT}{m}$$

The root mean square speed is then

$$(\bar{\mu}^2)^{\frac{1}{2}} = \sqrt{\frac{kT}{m}}$$

Where $T$ = is the temperature in Kelvin, $M$ = the atomic weight of the molecule, and $k$ is the Boltzmann constant = $1.380662 \times 10^{-23} J/K$.

# B. LIQUIDS AND SOLIDS

## 1. KINETIC-MOLECULAR VIEW OF LIQUIDS AND SOLIDS

### Liquids

A liquid is composed of molecules that are constantly and randomly moving.

### Volume and Shape

Liquids maintain a definite volume but because of their ability to flow, their shape depends on the contour of the container holding them.

### Compression and Expansion

In a liquid the attractive forces hold the molecules close together, so that increasing the pressure has little effect on the volume. Therefore, liquids are incompressible. Changes in temperature cause only small volume changes.

### Diffusion

Liquids diffuse much more slowly than gases because of the constant interruptions in the short mean free paths between molecules.

The rates of diffusion in liquids are more rapid at higher temperatures.

### Surface Tension

The strength of the inward forces of a liquid is called the liquid's surface tension. Surface tension decreases as the temperature is raised.

### Kinetics of Liquids

Increases in temperature increase the average kinetic energy of molecules and the rapidity of their movement. If a particular molecule gains enough kinetic energy when it is near the surface of a liquid, it can overcome the attractive forces of the liquid phase and escape into the gaseous phase. This is called a change of phase (specifically, evaporation).

### Solids

Properties of solids are as follows:

1. They retain their shape and volume when transferred from one container to another.
2. They are virtually incompressible.

3. They exhibit extremely slow rates of diffusion.

In a solid, the attractive forces between the atoms, molecules, or ions are relatively strong. The particles are held in a rigid structural array, wherein they exhibit only vibrational motion.

There are two types of solids, amorphous and crystalline. Crystalline solids are species composed of structural units bounded by specific (regular) geometric patterns. They are characterized by sharp melting points.

Amorphous substances do not display geometric regularity in the solid; glass is an example of an amorphous solid. Amorphous substances have no sharp melting point, but melt over a wide range of temperatures.

When solids are heated at certain pressures, some solids vaporize directly without passing through the liquid phase. This is called sublimation. The heat required to change 1 mole of solid $A$ completely to vapor is called the molar heat of sublimation, $\Delta H_{sub}$. Note that

$$\Delta H_{sub} = \Delta H_{fus} + \Delta H_{vap}$$

## 2. PHASE DIAGRAM

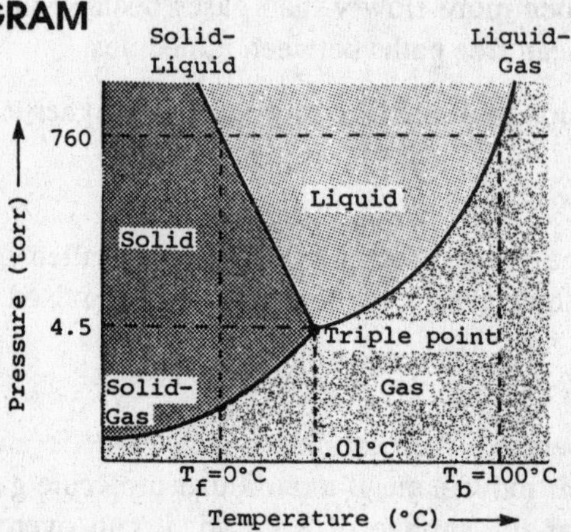


Phase diagram for water (somewhat distorted).

### Phase Equilibrium

In a closed system, when the rates of evaporation and condensation are equal, the system is in phase equilibrium.

In a closed system, when opposing changes are taking place at equal rates, the system is said to be in dynamic equilibrium. Virtually all of the equilibria considered in this review are dynamic equilibria.

## 3. CHANGES OF STATE, CRITICAL PHENOMENA

### Boiling Point and Melting Point

The boiling point of a liquid is the temperature at which the pressure of vapor escaping from the liquid equals atmospheric pressure. The normal boiling point of a liquid is the temperature at which its vapor pressure is 760mm Hg, that is, standard atmospheric pressure.

Liquids relatively strong with attractive forces have high boiling points. The melting point of a substance is the temperature at which its solid and liquid phases are in equilibrium.

### Heat of Vaporization and Heat of Fusion

The heat of vaporization of a substance is the number of calories required to convert 1g of liquid to 1g of vapor without a change in temperature.

The reverse process, changing 1g of gas into a liquid without change in temperature, requires the removal of the same amount of heat energy (the heat of condensation).

The heat needed to vaporize 1 mole of a substance is called the molar heat of vaporization, or the molar enthalpy of vaporization, $\Delta H_{vap}$, which is also represented as

$$\Delta H_{vaporization} = H_{vapor} - H_{liquid}$$

The magnitude of $\Delta H_{vap}$ provides a good measure of the strengths of the attractive forces operative in a liquid.

The number of calories needed to change 1g of a solid substance (at the melting point) to 1g of liquid (at the melting point) is called the heat of fusion.

The total amount of heat that must be removed in order to freeze 1 mole of a liquid is called its molar heat of crystallization. The molar heat of fusion, $\Delta H_{fus}$, is equal in magnitude but opposite in sign to the molar heat of crystallization and is defined as the amount of heat that must be supplied to melt 1 mole of a solid:

$$\Delta H_{fus} = H_{liquid} - H_{solid}$$

### Raoult's Law and Vapor Pressure

When the rate of evaporation equals the rate of condensation, the system is in equilibrium.

The vapor pressure is the pressure exerted by the gas molecules when they are in equilibrium with the liquid.

The vapor pressure increases with increasing temperature.

Raoult's law states that the vapor pressure of a solution at a particular temperature is equal to the mole fraction of the solvent in the liquid phase multiplied by the vapor pressure of the pure solvent at the same temperature:

$$P_{solution} = X_{solvent} P^0_{solvent}$$

and

$$P_A = X_A P^\circ_A,$$

where $P_A$ is the vapor pressure of $A$ with solute added, $P^\circ_A$ is the vapor pressure of pure $A$, and $X_A$ is the mole fraction of $A$ in the solution. The solute is assumed here to be nonvolatile (e.g., NaCl or sucrose in water).

## 4. STRUCTURE OF CRYSTALS

A crystal may be defined as a homogeneous body having the natural shape of a polyhedron. A representative portion of a crystal is called a unit cell. Just as a small swatch of fabric can show the repeating pattern that would be seen on many meters of the fabric, so can the unit cell show the pattern that is repeated throughout the crystal. The unit cell is a parallelepiped, which by variation in its dimensions we get the seven groups of crystals listed below. First, however, let us look at a unit cell for a simple compound such as sodium chloride.

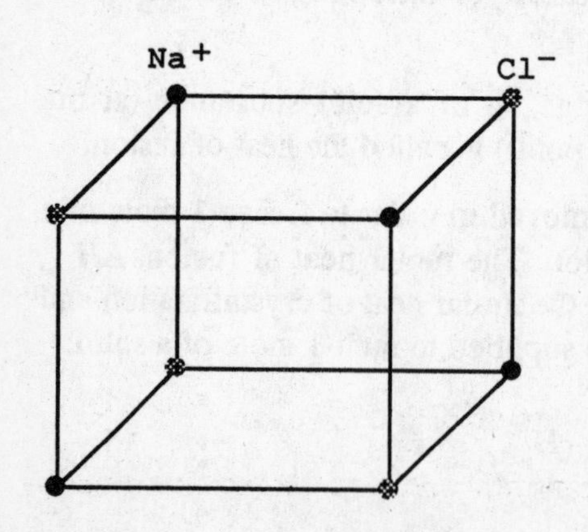


Notice the distribution of the sodium and chloride ions. If we were to graph this structure on a rectangular solids coordinate grid, we would notice that the dimensions of the unit cell in the *X*, *Y*, and *Z* planes are all equal. These are known as sides *a*, *b*, and *c*. The angles these three sides form with one another (designated as α, β, and γ) are also equal in this instance, and have a value of 90°.

Below is a more generalized unit cell that shows *a*, *b*, and *c*, and α, β, and γ, followed by a table that shows how variations in such parameters give rise to the seven groups of crystals. Keep in mind that many subgroups exist.

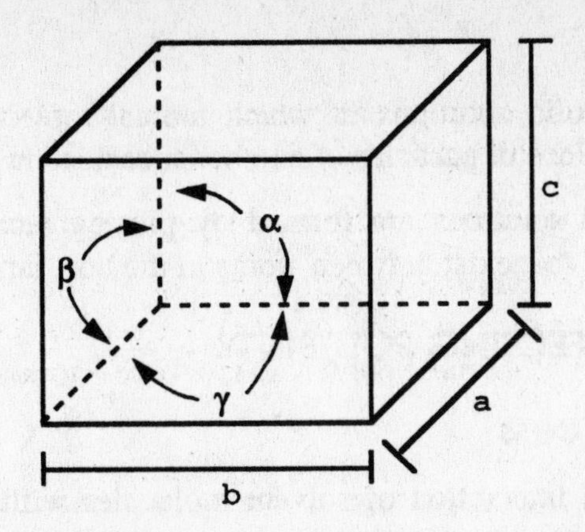


| Type | Lengths | Angles |
|---|---|---|
| Cubic | $a = b = c$ | $\alpha = \beta = \gamma = 90°$ as seen in NaCl |
| Tetragonal | $a = b \neq c$ | $\alpha = \beta = \gamma = 90°$ |
| Orthorhombic | $a \neq b \neq c$ | $\alpha = \beta = \gamma = 90°$ |
| Monoclinic | $a \neq b \neq c$ | $\alpha = \gamma = 90°, \beta \neq 90°$ |
| Triclinic | $a \neq b \neq c$ | $\alpha \neq \beta \neq \gamma \neq 90°$ |
| Hexagonal | $a = b \neq c$ | $\alpha = \beta = 90°; \gamma = 120°$ |
| Rhombohedral | $a = b = c$ | $\alpha = \beta = \gamma \neq 90°$ |

A crystal is said to be body centered if each unit cell contains at its center the same type of atoms found at its corners. A crystal is said to be face centered if the unit cell contains at the midpoints of its faces atoms of the same type that are found at its corners. Crystals that only have atoms at the corners of their lattices (e.g., wall) are called simple.

Different compounds that crystallize with the same structure are said to be isomorphisms.

When one substance can occur in two or more crystalline forms, it is said to be polymorphic.

# C. SOLUTIONS

## 1. TYPES OF SOLUTIONS

There are three types of solutions: gaseous, liquid, and solid.

The most common type of solution consists of a solute dissolved in a liquid.

The atmosphere is an example of a gaseous solution.

Solid solutions, of which many alloys (mixtures of metals) are examples, are of two types:

Substitutional solid solutions in which atoms, molecules, or ions of one substance take the place of particles of another substance in its crystalline lattice.

Interstitial solid solutions are formed by placing atoms of one kind into voids, or interstices, that exist between atoms in the host lattice.

## 2. FACTORS AFFECTING SOLUBILITY

### The Solution Process

Solvation is the interaction of solvent molecules with solute molecules or ions to form aggregates, the particles of which are loosely bonded together.

When water is used as the solvent, the process is also called aquation or hydration.

When one substance is soluble in all proportions with another substance, then the two substances are completely miscible. Ethanol and water are a familiar pair of completely miscible substances.

A saturated solution is one in which solid solute is in equilibrium with dissolved solute.

The solubility of a solute is the concentration of dissolved solute in a saturated solution of that solute.

Unsaturated solutions contain less solute than required for saturation.

Supersaturated solutions contain more solute than required for saturation. Supersaturation is a metastable state; the system will revert spontaneously to a saturated solution (stable state).

### Solubility and Temperature

The solubility of most solids in liquids usually increases with increasing temperature.

For gases in liquids, the solubility usually decreases with increasing temperature.

A positive $\Delta H°$ indicates that solubility increases with increasing temperature.

$$\log \frac{K_2}{K_1} = \frac{-\Delta H°}{2.303R}\left[\frac{1}{T_2} - \frac{1}{T_1}\right]$$

Where $K_2$ = solubility constant at $T_2$, $K_1$ = solubility constant at $T_1$ and $\Delta H°$ = enthalpy change at standard conditions. For most substances, when a hot concentrated solution is cooled, the excess solid crystallizes. The overall process of dissolving the solute and crystallizing it again is known as recrystallization, and is useful in purification of the solute.

### Effects of Pressures on Solubility

Pressure has very little effect on the solubility of liquids or solids in liquid solvents.

The solubility of gases in liquid (or solid) solvents always increases with increasing pressure.

## 3. WAYS OF EXPRESSING CONCENTRATIONS

### Concentration Units

Mole fraction is the number of moles of a particular component of a solution divided by the total number of moles of all of the substances present in the solution:

$$X_A = \frac{n_A}{n_A + n_B + n_C + \ldots}$$

$$\sum_{i=1}^{N} X_i = 1$$

Mole percent is equal to 100% × mole fraction. Weight fraction specifies the fraction of the total weight of a solution that is contributed by a particular component. Weight percent is equal to 100% × weight fraction.

Molarity ($M$) of a solution is the number of moles of solute per liter of solution:

$$\text{Molarity}\,(M) = \frac{\text{moles of solute}}{\text{liters of solution}}$$

Molality of a solution is the number of moles of solute per kilogram (1000g) of solvent.

$$\text{Molality}\,(m) = \frac{\text{moles of solute}}{\text{kilogram of solvent}}$$

When using water as a solvent, the molality ($m$) of a dilute solution is equal to the molarity ($M$).

## 4. COLLIGATIVE PROPERTIES

### Vapor Pressures of Solutions

For a solution in which a nonvolatile solute is dissolved in a solvent, the vapor pressure is due to only the vapor of the solvent above the solution. This vapor pressure is given by Raoult's law:

$$P_{\text{solution}} = X_{\text{solution}} P^{\circ}_{\text{solvent}}$$

### Colligative Properties of Solutions

Colligative property law: The freezing point, boiling point, and vapor pressure of a solution differ from those of the pure solvent by amounts which are directly proportional to the molal concentration of the solute.

The vapor pressure of an aqueous solution is always lowered by the addition of more solute, which causes the boiling point to be raised (boiling point elevation).

The freezing point is always lowered by addition of solute (freezing point depression). The freezing point depression, $\Delta T_f$, equals the negative of the molal freezing point depression constant, $K_f$, times molality ($m$):

$$\Delta T_f = -K_f(m)$$

The boiling point elevation, $\Delta T_b$, equals the molal boiling point elevation constant, $K_b$, times molality ($m$):

$$\Delta T_b = K_b(m)$$

$$\Delta T = T_{\text{solution}} - T_{\text{pure solvent}}$$

### Osmotic Pressure

Osmosis is the diffusion of a solvent through a semipermeable membrane into a more concentrated solution.

The osmotic pressure of a solution is the minimum pressure that must be applied to the solution to prevent the flow of solvent from pure solvent into the solution.

The osmotic pressure for a solution is:

$$\pi = CRT$$

where $\pi$ is the osmotic pressure, $C$ is the temperature in molality or molarity, $R$ is the gas constant, and $T$ is the temperature ($K$). (Note the formal similarity of the osmotic pressure equation to the ideal gas law, $C = n/v$.)

Solutions that have the same osmotic pressure are called isotonic solutions.

Reverse osmosis is a method for recovering pure solvent from a solution.

## 5. INTERIONIC ATTRACTIONS

$$i = \frac{(\Delta T_f)\text{ measured}}{(\Delta T_f)\text{ calculated as nonelectrolyte}}$$

$i$, the van't Hoff "factor," is defined as the ratio of the observed freezing point depression produced by a solute in solution, to the freezing point that the solution would exhibit if the solute were a nonelectrolyte. For example, since NaCl yields 2 moles of dissolved particles ($Na^+$ and $Cl^-$) in water, its van't Hoff factor is 2, and a 1 molal solution of NaCl (aq) yields a freezing point depression which is twice as large as that produced by sucrose, a nonelectrolyte.

# CHAPTER 3

# REACTIONS

## A. TYPES

### 1. FORMING AND CLEAVING COVALENT BONDS

### Definitions of Acids and Bases

### Arrhenius Theory

The Arrehenius theory states that acids are substances that ionize in water to give $H^+$ ions, and bases are substances that produce $OH^-$ ions in water.

### Bronsted-Lowry Theory

This theory defines acids as proton donors and bases as proton acceptors.

### Lewis Theory

This theory defines an acid as an electron-pair acceptor and a base as an electron-pair donor.

Whenever an Arrhenius acid and base are mixed together in equal quantities, the result will be a salt (nonmetal combined with a metal) and water. This process is termed "neutralization." For example, if hydrochloric acid is mixed with sodium hydroxide, what will the product be? Since we are mixing an Arrhenius acid and base together, we know we will get water and a salt. If we look at the equation for the reaction:

$$HCl + NaOH \rightarrow H_2O + NaCl,$$

we see that the salt formed is sodium chloride.

The hydroxides of elements that border the metal/nonmetal line on the periodic table act as acids toward strong bases and as bases toward strong acids. Such compounds are called "amphateric." For example, aluminum hydroxide is quite insoluble in water (where the number of $OH^-$ equals the number of $H^+$), but is quite soluble in either strong base or acid.

$$Al(OH)_3 + H_2O \rightarrow \text{No reaction — insoluble}$$

$$Al(OH)_3 + 3H^+ \rightarrow Al^{+++} + 3H_2O \text{ — soluble}$$

$$Al(OH)_3 + OH^- \rightarrow [Al(OH)_4]^- \text{ — soluble}$$

In the last equation the combination represented as $[Al(OH)_4]^-$ is known as a "coordination complex." This term is applied to compounds and ions in which negative or neutral polar molecules are attached to metal ions or atoms.

## 2. PRECIPITATION

A precipitation reaction is one in which soluble reactants are mixed together to form an insoluble product. For example, when a silver nitrate solution is mixed with a sodium chloride solution, silver nitrate is formed, and being insoluble, settles out of the solution. To observe this, one would see two colorless solutions poured together, immediately turning white and milky. The white silver chloride formed will settle to the bottom of the container, leaving a colorless solution of sodium nitrate above it. There are three ways to show what compounds precipitate in a gross chemical reaction:

1) $$AgNO_3 + NaCl \rightarrow NaNO_3 + AgCl\downarrow$$

An arrow pointing down designates a precipitate.

2) $$AgNO_3 + NaCl \rightarrow NaNO_3 + \underline{AgCl}$$

as does underlining the precipitate.

3) $$AgNO_{3(aq)} + NaCl_{(aq)} \rightarrow NaNO_{3(aq)} + AgCl_{(s)}$$

The nomenclature (aq) stands for "aqueous," which means the material is dissolved in water. The nomenclature (s) stands for "solid," i.e., it is not dissolved. This latter method is the preferred one to use.

Another method to show a precipitation is by writing the "net ionic equation," as opposed to the gross equations above. Recall that all of the above compounds *except* silver nitrate exist as ions when in water. Since we are working with aqueous (water) solutions, we can write:

$$Ag^+ + NO^-_3 + Na^+ + Cl^- \rightarrow Na^+ + NO^-_3 + AgCl_{(s)}$$

It will be noted that we have sodium and nitrate ions in solution on both sides of the arrow. These are called "spectator ions" and can be cancelled out to give us the net ionic equation:

$$Ag^+ + Cl^- \rightarrow AgCl_{(s)}$$

We could have also shown the precipitation with the arrow pointing down or by underlining the precipitation. It should be noted that not all molecular species are insoluble; hence, the product of a net ionic equation does not always yield a precipitate. Always look for one of the destinations used for precipitate to identify them.

## 3. OXIDATION AND REDUCTION

Oxidation is defined as a reaction in which atoms or ions undergo an increase in oxidation state. The agent that caused oxidation to occur is called the oxidizing agent and is itself reduced in the process.

Reduction is defined as a reaction in which atoms or ions undergo decrease in oxidation state. The agent that caused reduction to occur is called the reducing agent and is itself oxidized in the process.

An oxidation number can be defined as the charge that an atom would have if both of the electrons in each bond were assigned to the more electronegative element. The term "oxidation state" is used interchangeably with the term "oxidation number."

The following are the basic rules for assigning oxidation numbers:

1. The oxidation number of any element in its elemental form is zero.
2. The oxidation number of any simple ion (one atom) is equal to the charge on the ion.
3. The sum of all of the oxidation numbers of all of the atoms in a neutral compound is zero.

(More generally, the sum of the oxidation numbers of all of the atoms in a given species is equal to the net charge on that species.)

### Balancing Oxidation-Reduction Reactions Using the Oxidation Number Method

The Oxidation-Number-Change Method:

1. Assign oxidation numbers to each atom in the equation.
2. Note which atoms change oxidation number, and calculate the number of electrons transferred, per atom, during oxidation and reduction.
3. When more than one atom of an element that changes oxidation number is present in a formula, calculate the number of electrons transferred per formula unit.

4. Make the number of electrons gained equal to the number lost.
5. Once the coefficients from step 4 have been obtained, the remainder of the equation is balanced by inspection, adding $H^+$ (in acid solution), $OH^-$ (in basic solution), and $H_2O$, as required.

## Balancing Redox Equations: The Ion-Electron Method

The Ion-Electron Method:

1. Determine which of the substances present are involved in the oxidation-reduction.
2. Break the overall reaction into two half-reactions, one for the oxidation step and one for the reduction step.
3. Balance for mass (i.e., make sure there is the same number of each kind of atom on each side of the equation) for all species except H and O.
4. Add $H^+$ and $H_2O$ as required (in acidic solutions), or $OH^-$ and $H_2O$ as required (in basic solutions) to balance O first, then H.
5. Balance these reactions electrically by adding electrons to either side so that the total electric charge is the same on the left and right sides.
6. Multiply the two balanced half-reactions by the appropriate factors so that the same number of electrons is transferred in each.
7. Add these half-reactions to obtain the balanced overall reaction. (The electrons should cancel from both sides of the final equation.)

Below is a table of standard electrode potentials. Under these conditions all concentrations are 1 molar, all gases are at 1 atmosphere, and the temperature is 25°C. The first reaction on the table is the reduction of a lithium ion to a lithium:

$$Li^+ + e^- \rightarrow Li$$

with a potential listed of –3.045 volts. Were we to reverse reaction (e.g., oxidize a lithium atom to an ion):

$$Li \rightarrow Li^+ + e^-$$

the potential would be +3.045 volts. The potentials (voltages) for each of these half reactions are summed up. If the sum of the potentials are positive, the reaction is spontaneous, and will run on its own. If it is negative, the energy has to be supplied to make the reaction go.

For example, if iron metal is placed in a copper (II) sulfate solution, will the copper displace the iron spontaneously, or will no reaction occur?

We know that copper (II) sulfate, $C_4SO_4$, when in water exists as ions (e.g., $C_4^{++} + SO_4^{=}$). To reduce copper (II) ions to copper metal would require the copper ions to each gain 2 electrons, and would be written as:

$$Cu^{++} + 2e^- \rightarrow Cu^\circ.$$

Looking at our table, we can see the potential for this half-reaction is +0.337 volts. If the copper is replacing the iron, then the iron is being oxidized to ferrous (iron II) ions, e.g.,

$$Fe^\circ \rightarrow Fe^{++} + 2e^-.$$

The potential for this half reaction is +0.440 volts (remember to change the sign when reversing a reaction!). Summing the potentials for the two half-reactions we get:

$$+0.337 + 0.440 = +0.777 \text{ volts.}$$

Since the voltage is positive, it means the reaction will run by itself and give off energy. If we were to simply place an iron bar in a copper (II) sulfate solution, this energy would be given up as heat, but with a proper arrangement of the material we would be able to use this energy as electric current (with a potential of 0.777 volts under standard conditions). Such a device is called a voltaic cell, or battery. When the reaction reaches equilibrium (in this case we will have copper metal and iron (II) sulfate) we say the battery is dead. The two half-reactions above can be summed to give us the total reaction:

$$\begin{array}{l} Cu^{++} + 2e^- \rightarrow Cu^\circ \\ + Fe^\circ \rightarrow Fe^{++} + 2e \\ \hline Cu^{++} + Fe^\circ + 2e^- \rightarrow 2e^- + Fe^{++} + Cu^\circ \end{array}$$

The two electrons on each side of the arrow can be cancelled out to give us:

$$Cu^{++} + Fe^\circ \rightarrow Cu^\circ + Fe^{++}$$

Notice that in a redox equation, the number of charges on each side of the arrow are equal.

What about the reaction of aluminum chloride in water ($AlCl_3$) with the formation of chlorine gas ($Cl_2$) to give us metallic aluminum? Will that reaction go spontaneously? Again, we know that $AlCl_3$ in water exists as $Al^{+++} + Cl^-$. To reduce the aluminum ions to aluminum metal would require the reaction

$$Al^{+++} + 3e^- \rightarrow Al^\circ,$$

and from the table it can be seen that the potential for this reaction is –1.66 volts. The only donor for electrons would be the chloride ions, by the reaction:

$$2\,Cl^- \rightarrow 2e^- + Cl_2.$$

The potential for this reaction is – 1.3595 volts. Summing these potentials we get –3.02 volts. Because this value is negative, we know that the reaction will not go on its own (in other words, aluminum chloride will *not* spontaneously break down into aluminum metal and chlorine gas). However, if we were to dissolve $AlCl_3$ in water and place it in a special container called an electrolytic cell, the reaction will occur whenever a potential greater than or equal to 3.02 volts is applied. Again we can sum the two half-reactions to obtain the full reactions as follows:

$$Al^{+++} + 3e^- \rightarrow Al^\circ$$

$$2\ Cl^- \rightarrow Cl_2 + 2\ e^-$$

We first need to get the total number of electrons equal. Aluminum requires 3, but each chlorine molecule formed only gives up 2 electrons. Were we to multiply the Al reaction by the number of electrons chlorine liberates and visa versa, we obtain:

$$2 \times (Al^{+++} + 3e^- \rightarrow Al^\circ)$$

and

$$3 \times (2Cl^- \rightarrow Cl_2 + 2e^-)$$

we obtain:

$$2\ Al^{+++} + 6e^- \rightarrow 2Al^\circ$$

and

$$6\ Cl^- \rightarrow 3\ Cl_2 + 6e^-$$

which, when added yield:

$$2\ Al^{+++} + 6\ Cl^- + 6e^- \rightarrow 6e^- + 2Al^\circ + 3\ Cl_2$$

The six electrons cancel, giving:

$$2\ Al^{+++} + 6\ Cl^- \rightarrow 2Al^\circ + 3\ Cl_2.$$

### Table — Standard Electrode Potentials in Aqueous Solutions at 25°C.

| Electrode | Electrode Reaction | E°(V) |
|---|---|---|
| **Acid Solutions** | | |
| $Li \mid Li^+$ | $Li^+ + e^- \rightleftarrows Li$ | –3.045 |
| $K \mid K^+$ | $K^+ + e^- \rightleftarrows K$ | –2.925 |
| $Ba \mid Ba^{2+}$ | $Ba^{2+} + 2e^- \rightleftarrows Ba$ | –2.906 |
| $Ca \mid Ca^{2+}$ | $Ca^{2+} + 2e^- \rightleftarrows Ca$ | –2.87 |
| $Na \mid Na^+$ | $Na^+ + e^- \rightleftarrows Na$ | –2.714 |
| $La \mid La^{3+}$ | $La^{3+} + 3e^- \rightleftarrows La$ | –2.52 |
| $Mg \mid Mg^{2+}$ | $Mg^{2+} + 2e^- \rightleftarrows Mg$ | –2.363 |
| $Th \mid Th^{4+}$ | $Th^{4+} + 4e^- \rightleftarrows Th$ | –1.90 |

| | | |
|---|---|---|
| $U \mid U^{3+}$ | $U^{3+} + 3e^- \rightleftarrows U$ | −1.80 |
| $Al \mid Al^{3+}$ | $Al^{3+} + 3e^- \rightleftarrows Al$ | −1.66 |
| $Mn \mid Mn^{2+}$ | $Mn^{2+} + 2e^- \rightleftarrows Mn$ | −1.180 |
| $V \mid V^{2+}$ | $V^{2+} + 2e^- \rightleftarrows V$ | −1.18 |
| $Zn \mid Zn^{2+}$ | $Zn^{2+} + 2e^- \rightleftarrows Zn$ | −0.763 |
| $Tl \mid Tl\,I \mid I^-$ | $Tl\,I\,(s) + e^- \rightleftarrows Tl + I^-$ | −0.753 |
| $Cr \mid Cr^{3+}$ | $Cr^{3+} + 3e^- \rightleftarrows Cr$ | −0.744 |
| $Tl \mid TlBr \mid Br^-$ | $TlBr(s) + e^- \rightleftarrows Tl + Br^-$ | −0.658 |
| $Pt \mid U^{3+}, U^{4+}$ | $U^{4+} + e^- \rightleftarrows U^{3+}$ | −0.61 |
| $Fe \mid Fe^{2+}$ | $Fe^{2+} + 2e^- \rightleftarrows Fe$ | −0.440 |
| $Cd \mid Cd^{2+}$ | $Cd^{2+} + 2e^- \rightleftarrows Cd$ | −0.403 |
| $Pb \mid PbSO_4 \mid So_4^{2-}$ | $PbSO_4 + 2e^- \rightleftarrows Pb + SO_4^{2-}$ | −0.359 |
| $Tl \mid Tl^+$ | $Tl^+ + e^- \rightleftarrows Tl$ | −0.3363 |
| $Ag \mid AgI \mid I^-$ | $Ag\,I + e^- \rightleftarrows Ag + I^-$ | 0.152 |
| $Pb \mid Pb^{2+}$ | $Pb^{2+} + 2e^- \rightleftarrows Pb$ | −0.126 |
| $Pt \mid D_2 \mid D^+$ | $2D^+ + 2e^- \rightleftarrows D_2$ | −0.0034 |
| $Pt \mid H_2 \mid H^+$ | $2H^+ + 2e^- \rightleftarrows H_2$ | −0.0000 |
| $Ag \mid AgBr \mid Br^-$ | $AgBr + e^- \rightleftarrows Ag + Br^-$ | +0.071 |
| $Ag \mid AgCl \mid Cl^-$ | $AgCl + e^- \rightleftarrows Ag + Cl^-$ | +0.2225 |
| $Pt \mid Hg \mid Hg_2Cl_2 \mid Cl^-$ | $Hg_2Cl_2 + 2e^- \rightleftarrows 2Cl^- + 2Hg(l)$ | +0.2676 |
| $Cu \mid Cu^{2+}$ | $Cu^{2+} + 2e^- \rightleftarrows Cu$ | +0.337 |
| $Pt \mid I_2 \mid I^-$ | $I_2^- + 2e^- \rightleftarrows 3I^-$ | +0.536 |
| $Pt \mid O_2 \mid H_2O_2$ | $O_2 + 2H^+ + 2e^- \rightleftarrows H_2O_2$ | +0.682 |
| $Pt \mid Fe^{2+}, Fe^{3+}$ | $Fe^{3+} + e^- \rightleftarrows Fe^{2+}$ | +0.771 |
| $Ag \mid Ag^+$ | $Ag^+ + e^- \rightleftarrows Ag$ | +0.7991 |
| $Au \mid AuCl_4^-, Cl^-$ | $AuCl_4^- + 3e^- \rightleftarrows Au + 4Cl^-$ | +1.00 |
| $Pt \mid Br_2 \mid Br^-$ | $Br_2 + 2e^- \rightleftarrows 2Br^-$ | +1.065 |
| $Pt \mid Tl^+, Tl^{3+}$ | $Tl^{3+} + 2e^- \rightleftarrows Tl^+$ | +1.25 |
| $Pt \mid H^+, Cr_2O_7^{2-}, Cr^{3+}$ | $Cr_2O_7^{2-} + 14H^+ + 6e^- \rightleftarrows 2Cr^{3+} + 7H_2O$ | +1.33 |
| $Pt \mid Cl_2 \mid Cl^-$ | $Cl_2 + 2e^- \rightleftarrows 2Cl^-$ | +1.3595 |
| $Pt \mid Ce^{4+}, Ce^{3+}$ | $Ce^{4+} + e^- \rightleftarrows Ce^{3+}$ | +1.45 |
| $Au \mid Au^{3+}$ | $Au^{3+} + 3e^- \rightleftarrows Au$ | +1.50 |
| $Pt \mid Mn^{2+}, MnO_4^-$ | $MnO_4^- + 8H^+ + 5e^- \rightleftarrows Mn^{2+} + 4H_2O$ | +1.51 |
| $Au \mid Au^+$ | $Au^+ + e^- \rightleftarrows Au$ | +1.68 |
| $PbSO_4 \mid PbO_2 \mid H_2SO_4$ | $PbO_2 + SO_4 + 4H^+ + 2e^- \rightleftarrows PbSO_4 + 2H_2O$ | +1.685 |
| $Pt \mid F_2 \mid F^-$ | $F_2(g) + 2e^- \rightleftarrows 2F^-$ | +2.87 |

**Basic Solutions**

| | | |
|---|---|---|
| $Pt \mid SO_3^{2-}, SO_4^{2-}$ | $SO_4^{2-} + H_2O + 2e^- \rightleftarrows SO_3^{2-} + 2OH^-$ | −0.93 |
| $Pt \mid H_2 \mid OH^-$ | $2H_2O + 2e^- \rightleftarrows H_2 + 2OH^-$ | −0.828 |
| $Ag \mid Ag(NH_3)_2^+, NH_3(aq)$ | $Ag(NH_3)_2^+ + e^- \rightleftarrows Ag + 2NH_3\,(aq)$ | +0.373 |
| $Pt \mid O_2 \mid OH^-$ | $O_2 + 2H_2O + 4e^- \rightleftarrows 4OH^-$ | +0.401 |
| $Pt \mid MnO_2 \mid MnO_4^-$ | $MnO_4^- + 2H_2O + 3e^- \rightleftarrows MnO_2 + 4OH^-$ | +0.588 |

## Voltaic Cells

One of the most common voltaic cells is the ordinary "dry cell" used in flashlights. It is shown in the drawing below, along with the reactions occurring during the cell's discharge:

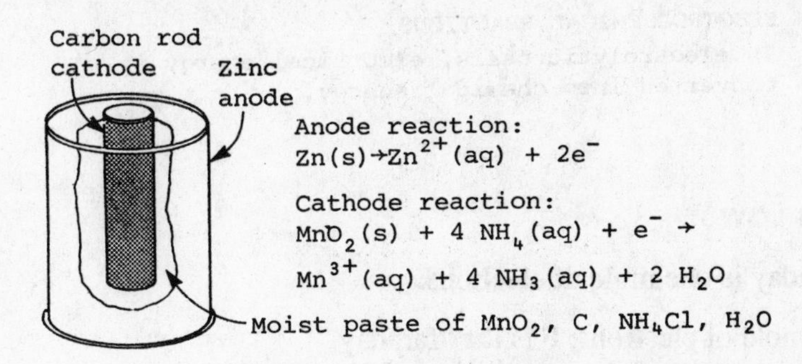


In galvanic or voltaic cells, the chemical energy is converted into electrical energy.

In galvanic cells, the anode is negative and the cathode is positive (the opposite is true in electrolytic cells).

The force with which the electrons flow from the negative electrode to the positive electrode through an external wire is called the electromotive force, or emf, and is measured in volts (V):

$$1V = \frac{1J}{\text{coul.}}$$

The greater the tendency or potential of the two half-reactions to occur spontaneously, the greater will be the emf of the cell. The emf of the cell is also called the cell potential, $E_{cell}$. The cell potential for the Zn/Cu cell can be written

$$E^0_{cell} = E^0_{Cu} - E^0_{Zn}$$

where the $E^0$s are standard reduction potentials.

The overall standard cell potential is obtained by subtracting the smaller reduction potential from the larger one. A positive emf corresponds to a negative $\Delta G$ and therefore to a spontaneous process.

## Electrolytic Cells

Reactions that do not occur spontaneously can be forced to take place by supplying energy with an external current. These reactions are called electrolytic reactions.

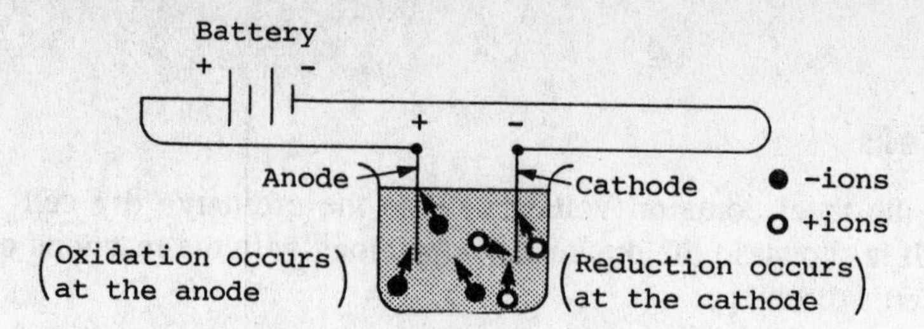


ELECTROCHEMICAL REACTIONS
In electrolytic cells, electrical energy is converted into chemical energy.

## Faraday's Law

One faraday is one mole of electrons.

(1F = 1 mole of electrons; F is the "faraday.")

1 F ~ 96,500 coul; the charge on 1 F is approximately 96,500 coulombs.

One coulomb is the amount of charge that moves past any given point in a circuit when a current of 1 ampere (amp) is supplied for one second. (Alternatively, one ampere is equivalent to 1 coulomb/second.)

Faraday's law states that during electrolysis, the passage of 1 faraday through the circuit brings about the oxidation of one equivalent weight of a substance at one electrode (anode) and reduction of one equivalent weight at the other electrode (cathode). Note that in all cells, oxidation occurs at the anode and reduction at the cathode.

## Non-Standard-State Cell Potentials

For a cell at concentrations and conditions other than standard, a potential can be calculated using the following Nernst equation:

$$E_{cell} = E^0_{cell} - \frac{.059}{n} \log Q$$

where $E^0_{cell}$ is the standard-state cell voltage, $n$ is the number of electrons exchanged in the equations for the reaction, and $Q$ is the mass action quotient (which is similar in form to an equilibrium constant).

For the cell reaction

$Zn + Cu^{2+} \rightarrow Cu + Zn^{2+}$, the term $Q = [Zn^{2+}]/\{Cu^{2+}\}$

the Nernst equation takes the form:

$$E = E^0 = \frac{.059}{n} \log \frac{\{Zn^{2+}\}}{\{Cu^{2+}\}}.$$

$$\Delta G = -nFE$$

is the Nernst equation. Also:

$$E = E^0 - \frac{RT}{nF} \ln Q$$

and

$$E = E^0 - \frac{.059}{n} \log Q$$

which is analogous to

$$\Delta G = \Delta G^\circ + RT \ln Q.$$

# B. STOICHOMETRY

## 1. RECOGNIZING THE PRESENCE OF IONIC AND MOLECULAR SPECIES

All strong acids, strong bases and soluble salts will disassociate into ions when dissolved in water:

Sulfuric acid $H2SO_4 \rightarrow 2H^+ + SO_4^=$

Cesium hydroxide $CsOH \rightarrow Cs^+ + OH^-$

Ammonium phosphate $(NH_4)_3PO_4 \rightarrow 3NH_4^+ + PO_4^=$

Weak acids and weak bases only disassociate to a very small extent. Insoluble material and covalently bonded compounds (e.g., sugars, alcohols, etc.) will not disassociate to form ions at all. A solution containing ions will conduct an electric current, whereas one without them will not. Writing net ionic equations is covered in Chapter 3, A.1., Forming and Cleaving Covalent Bonds.

## 2. BALANCING CHEMICAL EQUATIONS

When balancing chemical equations, one must make sure that there are the same number of atoms of each element on both the left and the right of the arrow.

**EXAMPLE:**

$$2\,NaOH + H_2SO_4 \rightarrow NaSO_4 + 2H_2O.$$

$$\begin{Bmatrix} \text{Na:} & 2 \text{ atoms} \\ \text{O:} & 6 \text{ atoms} \\ \text{H:} & 4 \text{ atoms} \\ \text{S:} & 1 \text{ atom} \end{Bmatrix}$$

## 3. WEIGHT AND VOLUME RELATIONSHIPS

### Calculations Based on Chemical Equations

The coefficients in a chemical equation provide the ratio in which moles of one substance react with moles of another.

**EXAMPLE:**

$$C_2H_4 + 3O_2 \rightarrow 2CO_2 + 2H_2O \text{ represents}$$

$$1 \text{ mole of } C_2H_4 + 3 \text{ moles } O_2 \rightarrow 2 \text{ moles } CO_2 + 2 \text{ moles } H_2O.$$

In this equation, the number of moles of $O_2$ consumed is always equal to three times the number of moles of $C_2H_4$ that react.

### Limiting-Reactant Calculations

The reactant that is used up first in a chemical reaction is called the limiting reactant, and the amount of product is determined (or limited) by the limiting reactant.

### Theoretical Yield and Percentage Yield

The theoretical yield of a given product is the maximum yield that can be obtained from a given reaction if the reaction goes to completion (rather than to equilibrium).

The percentage yield is a measure of the efficiency of the reaction. It is defined

$$\text{percentage yield} = \frac{\text{actual yield}}{\text{theoretical yield}} \times 100\%$$

### Percentage Composition

The percentage composition of a compound is the percentage of the total mass contributed by each element:

$$\%\text{ composition} = \frac{\text{mass of element in compound}}{\text{mass of compound}} \times 100\%$$

## Density and Molecular Weight

At "STP," standard temperature and pressure, 0°C and 760 mm of mercury pressure, 1 mole of an ideal gas occupies 22.4 liters. (The "molar volume" of the gas at STP is 22.4 *l*/mole.)

The density can be converted to molecular weight using the 22.4 liters/mole relationship:

$$MW = \text{(density) (molar volume)}$$

$$\left(\frac{g}{\text{mole}}\right) = \left(\frac{g}{l}\right)\left(\frac{l}{\text{mole}}\right)$$

## Weight-Volume Relationships

For a typical weight-volume problem, follow the following steps:

**Step 1:** Write the balanced equation for the reaction.

**Step 2:** Write the given quantities and the unknown quantities for the appropriate substances.

**Step 3:** Calculate reacting weights or number of moles (or volume, if the reaction involves only gases) for the substances whose quantities are given. Make sure that the units for each substance are identical.

**Step 4:** Use the proportion method or the factor-label method.

**Step 5:** Solve for the unknown.

## EXAMPLE:

$NaClO_3$, when heated, decomposes to NaCl and $O_2$. What volume of $O_2$ at STP results from the decomposition of 42.6 grams $NaClO_3$?

1. (using reactive masses) balanced equation:

$$213\text{ g } NaClO_3 \xrightarrow[\Delta]{} 117\text{ g NaCl} + 96\text{ g } O_2$$

$$\frac{\text{mass } O_2 \text{ produced}}{\text{mass } NaClO_3 \text{ decomposed}} = \frac{96\text{ g}}{213\text{ g}} = \frac{x}{42.6\text{ g}}$$

$$x = \frac{(96)\,(42.6)}{213} = 19.2\text{ g } O_2$$

or

2. (using moles) balanced equation:

$$2NaClO_3 \xrightarrow[\Delta]{} 2NaCl + 3O_2$$

$$\frac{\text{moles } O_2 \text{ produced}}{\text{moles } NaClO_3 \text{ decomposed}} = \frac{3}{2} = \frac{y}{(42.6\,g/106.5\,g/mole}$$

$$y = \frac{(3)\ (42.6)}{(2)\ (106.5)} = 0.6 \text{ mole } O_2$$

Finally, at STP, 1 mole (32 g) $O_2$ occupies 22.4 *l*, so

$$V_{O_2(STP)} = \left(\frac{19.2\,g}{32\,g/mole}\right)(22.4\ l/mole) = 0.0268\ l$$

or

$$V_{O_2(STP)} = (0.6 \text{ mole})\ (22.4\ l/mole) = 0.0268\ l$$

or

$$V_{O_2(STP)} = 2.68 \times 10^{-2}\ l.$$

The ideal gas law, $PV = nRT$, can be used to determine the volume of gas or the number of moles of a gas at conditions other than STP.

# C. EQUILIBRIUM

## 1. DYNAMIC EQUILIBRIUM BOTH PHYSICAL AND CHEMICAL

### The Law of Mass Action

At equilibrium, both the forward and reverse reactions take place at the same rate, and thus the concentrations of reactants and products no longer change with time.

The law of mass action states that the rate of an elementary chemical reaction is proportional to the product of the concentrations of the reacting substances, each raised to its respective stoichiometric coefficient.

For the reaction $aA + bB \rightleftarrows eE + fF$, at constant temperature,

$$K_c = \frac{[E]^e[F]^f}{[A]^a[B]^b},$$

where the [...] denotes equilibrium molar concentrations, and $K_c$ is the equilibrium constant.

The entire relationship is known as the law of mass action.

$$\frac{[E]^e[F]^f}{[A]^a[B]^b}$$

is known as the mass action expression. Note that if any of the species (*A, B, E, F*) is a pure solid or pure liquid, it does not appear in the expression for $K_c$.

For the reaction

$$N_2(g) + 3H_2(g) \rightleftarrows 2\,NH_3(g),$$

$$K_p = \frac{(P_{NH_3})^2}{P_{N_2}(P_{H_2})^3}$$

where $K_p$ is the equilibrium constant derived from partial pressures.

## Kinetics and Equilibrium

The rate of an elementary chemical reaction is proportional to the concentrations of the reactants raised to powers equal to their coefficients in the balanced equation.

For

$$aA + bB \rightleftarrows eE + fF,$$

$$\text{rate}_f = k_f[A]^a\,[B]^b,$$

$$\text{rate}_r = k_r[E]^e\,[F]^f$$

and

$$\frac{k_f}{k_r} = \frac{[E]^e[F]^f}{[A]^a[B]^b} = K_c$$

where $K_f$ and $K_r$ are rate constants for the forward and reverse reactions, respectively.

## Thermodynamics and Chemical Equilibrium

$$\Delta G = \Delta G^\circ + 2.303RT \log Q$$

The symbol $Q$ represents the mass action expression for the reaction. For gases, $Q$ is written with partial pressures. $\Delta G$ is the free energy.

At equilibrium $Q = K_{eq}$ and the products and reactants have the same total free energy, such that $\Delta G = 0$.

$$\Delta G° = -2.303RT \log K_{eq} = -RT \ln K_{eq}$$

For the equation $2NO_2(g) \rightleftarrows N_2O_4(g)$,

$$\Delta G° = -2.303RT \log \left( \frac{P_{N_2O_4}}{(P_{NO_2})^2} \right)_{eq}, \quad K_c = \frac{[N_2O_4]}{[NO_2]^2}$$

## 2. THE RELATIONSHIP BETWEEN $K_p$ AND $K_c$

$$K_p = \frac{P_E^e P_F^f}{P_A^a P_B^b} = \frac{[E]^e (RT)^e [F]^f (RT)^f}{[A]^a (RT)^a [B]^b (RT)^b}$$

and

$$K_p = \frac{[E]^e [F]^f}{[A]^a [B]^b} = (RT)^{(e+f)-(a+b)}$$

Therefore,

$$K_p = K_c (RT)^{\Delta ng}$$

where $\Delta ng$ is the change in the number of moles of gas upon going from reactants to products.

### Heterogeneous Equilibria

For heterogeneous reactions, the equilibrium constant expression does not include the concentrations of pure solids and liquids.

For the equation

$$2NaHCO_3(s) \rightleftarrows Na_2CO_3(s) + CO_2(g) + H_2O(g),$$

$$K_p = P_{CO_2(g)} P_{H_2O(g)},$$

$$K_c = [CO_2(g)]\,[H_2O(g)]$$

and

$$K_p = K_c (RT)^{\Delta ng},$$

where $\Delta ng = +2$ for the reaction.

## Le Chatelier's Principle and Chemical Equilibrium

Le Chatelier's principle states that when a system at equilibrium is disturbed by the application of a stress (change in temperature, pressure, or concentration) it reacts to minimize the stress and attain a new equilibrium position.

## Effect of Changing the Concentrations on Equilibrium

When a system at equilibrium is disturbed by adding or removing one of the substances, all the concentrations will change until a new equilibrium point is reached with the same value of $K_{eq}$.

Increase in the concentrations of reactants shifts the equilibrium to the right, thus increasing the amount of products formed. Decreasing the concentrations of reactants shifts the equilibrium to the left and thus decreases the concentrations of products formed.

## Effect of Temperature on Equilibrium

An increase in temperature causes the position of equilibrium of an exothermic reaction to be shifted to the left, while that of an endothermic reaction is shifted to the right.

## Effect of Pressure on Equilibrium

Increasing the pressure on a system at equilibrium will cause a shift in the position of equilibrium in the direction of the fewest number of moles of gaseous reactants or products.

## Effect of a Catalyst on the Position of Equilibrium

A catalyst lowers the activation energy barrier that must be overcome in order for the reaction to proceed. A catalyst merely speeds the approach to equilibrium, but does not change $K_{eq}$ (or $\Delta G°$) at all.

## Addition of an Inert Gas

If an inert gas is introduced into a reaction vessel containing other gases at equilibrium, it will cause an increase in the total pressure within the container. However, this kind of pressure increase will not affect the position of equilibrium.

## 3. EQUILIBRIUM CONSTANTS FOR REACTIONS IN SOLUTIONS

### Ionization of Water, pH

For the equation

$$H_2O + H_2O \rightleftarrows H_3O + OH^-, \; K_w = [H_3O^+]\,[OH^-]$$

$$(\text{or } K_w = [H^+]\,[OH^-]) = 1.0 \times 10^{-14} \text{ at } 25°C,$$

where $[H_3O^+]\,[OH^-]$ is the product of ionic concentrations, and $K_w$ is the ion product constant for water (or, simply the ionization constant or dissociation constant).

$$pH = -\log\,[H^+]$$

$$p\,OH = -\log\,[OH]$$

$$pK_w = pH + pOH = 14.0$$

In a neutral solution, pH = 7.0. In an acidic solution pH is less than 7.0. In basic solutions, pH is greater than 7.0. The smaller the pH, the more acidic is the solution. Note that since $K_w$ (like all equilibrium constants) varies with temperature, neutral pH is less than (or greater than) 7.0 when the temperature is higher than (or lower than) 25°C.

### Dissociation of Weak Electrolytes

For the equation

$$A^- + H_2O \rightleftarrows HA + OH^-,$$

$$K_b = \frac{[HA]\,[OH^-]}{[A^-]}$$

where $K_b$ is the base ionization constant.

$$K_a = \frac{[H^+]\,[A^-]}{[HA]}$$

where $K_a$ is the acid ionization constant.

$$K_b = \frac{K_w}{K_a}$$

for any conjugate acid/base pair, and therefore,

$$K_w = [H^+] [OH^-].$$

## Dissociation of Polyprotic Acids

For $H_2S \rightleftarrows H^+ + HS^-$,

$$K_{a_1} = \frac{[H^+] [HS^-]}{[H_2S]}$$

For $HS^- \rightleftarrows H^+ + S^{2-}$,

$$K_{a_2} = \frac{[H^+] [S^{2-}]}{[HS^-]}$$

$K_{a_1}$ is much greater than $K_{a_2}$.

Also,

$$K_a = K_{a_1} \times K_{a_2} = \frac{[H^+] [HS^-]}{[H_2S]} \times \frac{[H^+] [S^{2-}]}{[HS^-]}$$

$$= \frac{[H^+]^2 [S^{2-}]}{[H_2S]}$$

This last equation is useful only in situations where two of the three concentrations are given and you wish to calculate the third.

## Buffers

Buffer solutions are equilibrium systems that resist changes in acidity and maintain constant pH when acids or bases are added to them.

The most effective pH range for any buffer is at or near the pH where the acid and salt concentrations are equal (that is, $pK_a$).

The pH for a buffer is given by

$$\text{pH} = \text{p}K_a + \log \frac{[A^-]}{[HA]} = \text{p}K_a + \log \frac{[\text{base}]}{[\text{acid}]}$$

which is obtained very simply from the equation for weak acid equilibrium,

$$K_a = [H^+] [A^-] / [HA].$$

## Hydrolysis

Hydrolysis refers to the action of salts of weak acids or bases with water to form acidic or basic solutions.

Salts of Weak Acids and Strong Bases: Anion Hydrolysis

For $C_2H_3O_2^- + H_2O \rightleftarrows HC_2H_3O_2 + OH^-$,

$$K_h = \frac{[HC_2H_3O_2][OH^-]}{[C_2H_3O_2^-]},$$

where $K_h$ is the hydrolysis constant for the acetate ion, which is just $K_b$ for acetate.

Also,

$$K_a = \frac{[H^+][C_2H_3O_2^-]}{[HC_2H_3O_2]}$$

and

$$K_h = \frac{K_w}{K_a}$$

Salts of Strong Acids and Weak Bases: Cation Hydrolysis

For $NH_4^+ + H_2O \rightleftarrows H_3O^+ + NH_3$,

$$K_h = \frac{[H_3O^+][NH_3]}{[NH_4^+]}$$

$$(K_h = K_a \text{ for } NH_4^+)$$

Also,

$$K_h = \frac{K_w}{K_b}$$

and

$$K_b = \frac{[NH_4^+][OH^-]}{[NH_3]}$$

Hydrolysis of Salts of Polyprotic Acids

For $S^{2-} + H_2O \rightleftarrows HS^- + OH^-$,

$$K_{h_1} = \frac{K_w}{K_{a_2}} = \frac{[HS^-][OH^-]}{[S^{2-}]},$$

where $K_{a_2}$ is the acid dissociation constant for the weak acid $HS^-$.

For $HS^- + H_2O \rightleftarrows H_2S + OH^-$,

$$K_{h_2} = \frac{K_w}{K_{a_1}} = \frac{[H_2S][OH^-]}{[HS^-]}$$

where $K_{a_1}$ is the dissociation constant for the weak acid $H_2S$.

Tables giving the values of $K_A$'s and $K_B$'s for various acids and bases at different temperatures are readily available. These can be used to calculate the degree of ionization. For example, formic acid has a $K_A$ of $1.8 \times 10^{-4}$. What is the pH of a 0.0400M solution? We know that formic acid dissociates to give us a hydrogen in:

$$HCOOH \rightleftarrows HCOO^- + H^+.$$

First, we set up the equilibrium equation:

$$\frac{[H^+][HCOO^-]}{[HCOOH]} = K_A$$

We know that the equilibrium concentration of HCOOH is 0.0400 M minus the amount that dissociates to $HCOO^-$ and $H^+$. If we term this amount "$X$", then $[HCOOH] = 0.0400 - X$. We also know from stoichiometry that $[H^+] = [HCOO^-] = X$. Substituting this into the equation, we get:

$$\frac{(X)(X)}{(0.0400 - X)} = 1.8 \times 10^{-4}$$

$$\frac{X^2}{0.0400 - X} = 1.8 \times 10^{-4},\ X^2 = (0.0400)(1.8 \times 10^{-4}) - X(1.8 \times 10^{-4})$$

$$X^2 + (1.8 \times 10^{-4})X - 7.2 \times 10^{-6} = 0,$$

which can be solved as a quadratic. However, an even simpler solution is possible. Since we know from the $K_A$ very little of the acid will disassociate, we can assume $X$ is very small in terms of the 0.0400 molar concentration we subtract it from, and hence, can be ignored. This then gives us:

$$\frac{X^2}{0.0400} = 1.8 \times 10^{-4},\ X^2 = 7.2 \times 10^{-6}$$

$$X = 2.7 \times 10^{-3} = [H^+] = [HCOO^-]$$

If the $[H^+] = 2.7 \times 10^{-3}$, then the pH ($-\log [H^+]$) = 2.57. Solving the original equation by a quadratic, we get:

$$[H^+] = 2.6 \times 10^{-3} \quad \text{and} \quad pH = 2.58,$$

which shows that our initial assumption is correct in that X in the term (0.0400 – *X*) was small enough to ignore.

**Common Ion Effect** — If a large concentration of formate ions were added to the above solution (in the form of sodium formate, by the reaction $HCOONa \rightarrow HCOO^- + Na^+$), these additional formate ions would compete with the ones produced by the acid, and result in the reaction $H^+ + HCOO^- \rightarrow HCOOH$, thus resulting in an increase in the number of undisassociated formic acid molecules, and a decrease in the number of hydronium ions (represented as $H^+$). This is known as the "common ion effect."

Suppose we have one liter of the above 0.0400M HCOOH solution. What will its pH be if we add 34.0 grams of HCOONa to it? (assume no change in volume). The formula weight of HCOONa is 68.0, and gives us 0.500 moles of HCOONa. If the volume is one liter, then the concentration of HCOONa is 0.500 M. (Remember, the formula works only for *concentration*, not number of moles!) We can now use the previous formula and solve:

$$\frac{[H^+]\,[HCOO^-]}{[HCOOH]} = K_A$$

As before, [HCOOH] = 0.0400 – *X*, where "*X*" is the amount that has disassociated. $[HCOO^-]$ = the amount of formate ions from both the salt and the acid. The sodium formate was determined to be .500 molar, hence by stoichiometry so is the ion from this salt. The amount from the acid is *X*, as is the $[H^+]$. This gives us:

$$\frac{X(0.500 + X)}{(0.0400 - X)} = 1.8 \times 10^{-4}$$

which can be solved as a quadratic. Again, however, if it is assumed that *X* is very small with respect to the other numbers, it can be ignored in the subtraction and addition, giving us

$$\frac{0.500X}{.0400} = 1.8 \times 10^{-4}, \quad \text{or} \quad X = [H^+] = 1.44 \times 10^{-5}\ \text{M},$$

and the pH = 4.84. Had we solved the above equation as a quadratic, our answer would have been a pH of 4.87.

For slightly soluble compounds, a solubility constant, $K_{sp}$, is calculated in the same manner. For example, the $K_{sp}$ at 25°C for lead (II) carbonate is $1.5 \times 10^{-13}$. If 20.0 grams of $PbCO_3$ are mixed in 2.00 *l* of water, 25°C, how much of it will dissolve?

Our reaction is:

$$PbCO_3 \rightarrow Pb^{++} + CO_3^{=},$$

so we can write:

$$\frac{[Pb^{++}]\,[CO_3^{=}]}{[PbCO_3]} = K_{sp}$$

We can tell from our stoichiometry that $[Pb^{++}] = [CO_3^{=}]$, which we will assign a value of "$X$". Any material that is very insoluble, such as $PbCO_3$, is simply assigned a value of "1". Substituting in our above values yields:

$$\frac{(X)\,(X)}{1} = 1.5 \times 10^{-13}\ ,\ X^2 = 1.5 \times 10^{-13}\ ,\ X = 3.9 \times 10^{-7}\,M$$

This is the concentration of the dissolved $PbCO_3$. Since it is in 2.00 liters, we have:

$$3.9 \times 10^{-7}\ \text{moles}/l \times 2.00\ l = 7.7 \times 10^{-7}\ \text{moles}$$

total of $PbCO_3$. The formula weight of $PbCO_3$ is 267, we get

$$(7.7 \times 10^{-7}\ \text{moles}) \times (267\ \text{gm/mole}) = 2.1 \times 10^{-4}$$

grams of $PbCO_3$ in solution, or 210 micrograms have dissolved. (This is 0.0010% of the original 20.0 grams!)

# D. KINETICS

## 1. RATE OF REACTION

### Measurements of Reaction Rates

The measurement of reaction rate is based on the rate of appearance of a product or disappearance of a reactant. It is usually expressed in terms of change in concentration of one of the participants per unit time:

$$\text{rate of chemical reaction} = \frac{\text{change in concentration}}{\text{time}}$$

$$= \frac{\text{moles/liter}}{\text{sec}}$$

For the general reaction $2AB \rightarrow A_2 + B_2$,

$$\text{average rate} = \frac{[AB]_2 - [AB]_1}{t_2 - t_1} = \frac{-\Delta [AB]}{\Delta t}$$

where $[AB]_2$ means "the *AB* concentration at time $t_2$."

## 2. REACTION ORDER

### Temperature Dependence of Reaction Rate

The reaction rate increases as the temperature increases. For every 10°*K* increase in temperature, the reaction rate doubles. The rate constant is related to the temperature by the following Arrhenius equation:

$$k = A \exp\left(\frac{-E_a}{RT}\right)$$

$$\ln k = \ln A \frac{-E_a}{RT}$$

where $k$ = the rate constant

$A$ = the frequency factor or the pre-exponential factor

$E_a$ = the activation energy

$T$ = the temperature in °K.

$$\ln \frac{k_2}{k_1} = -\left(\frac{\Delta E_a}{R}\right)\left(\frac{1}{T_2} - \frac{1}{T_1}\right)$$

$$\frac{d(\ln K)}{dt} = \frac{\Delta E}{RT^2}$$

where $K$ is the equilibrium constant and $\Delta E$ is the change in energy.

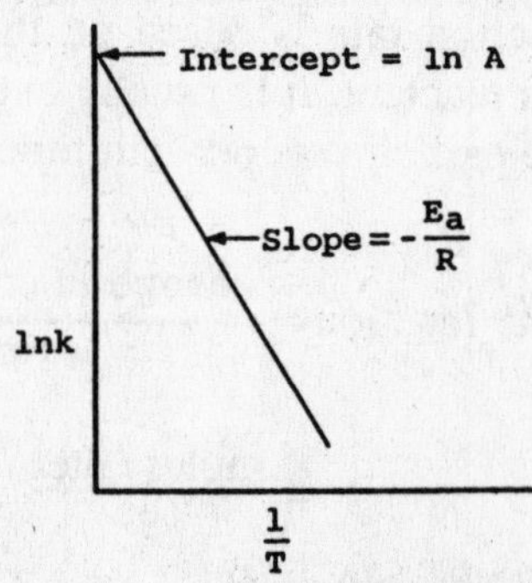


Arrhenius plot that shows the relationship between the rate constant and the temperature.

## 3. TEMPERATURE CHANGES AND EFFECT ON RATE

### Factors Affecting Reaction Rates

There are five important factors that control the rate of a chemical reaction. These are summarized below:

1. The nature of the reactants and products, i.e., the nature of the transition state formed. Some elements and compounds, because of the bonds broken or formed, react more rapidly with each other than do others.
2. The surface area exposed. Since most reactions depend on the reactants coming into contact, increasing the surface area exposed, proportionally increases the rate of reaction.
3. The concentrations. The reaction rate usually increases with increasing concentrations of the reactants.
4. The temperature. A temperature increase of 10°C above room temperature usually causes the reaction rate to double.
5. The catalyst. Catalysts speed up the rate of a reaction but do not change the equilibrium constant (i.e., it simply speeds up the rate of approach to equilibrium).

### The Arrhenius Equation: Relating Temperature and Reaction Rate

The following is the Arrhenius equation:

$$k = Ae^{-E_a/RT}$$

where $k$ is the rate constant, $A$ = the Arrhenius constant, $E_a$ = activation energy, $R$ = universal gas constant, and $T$ = temperature in Kelvin. $k$ is small when the activation energy is very large or when the temperature of the reaction mixture is low.

$$\begin{array}{ccccc} \ln k & = & \ln A & - & \dfrac{E_a}{RT} \\ \updownarrow & & \updownarrow & & \updownarrow\updownarrow \\ y & = & b & + & mx \end{array}$$

A plot of lnk versus $1/T$ gives a straight line whose slope is equal to $-E_a / R$ and whose intercept with the ordinate is ln $A$.

$$\ln\left(\frac{k_2}{k_1}\right) = \frac{-E_a}{R}\left[\frac{1}{T_2} - \frac{1}{T_1}\right]$$

$$\log\left(\frac{k_2}{k_1}\right) = \frac{-E_a}{2.303R}\left[\frac{1}{T_2} - \frac{1}{T_1}\right]$$

## 4. ACTIVATION ENERGY

The activation energy is the energy necessary to cause a reaction to occur. It is equal to the difference in energy between the transition state (or "activated complex") and the reactants.

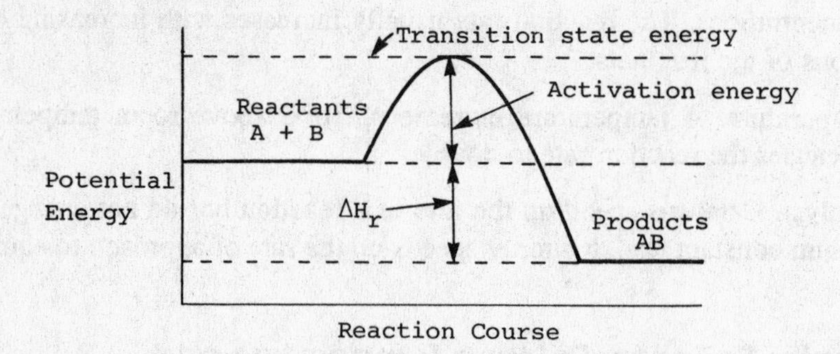


$$\Delta E = \Sigma\, E \text{ products} - \Sigma\, E \text{ reactants}$$

In an exothermic process, energy is released and $\Delta E$ of reaction is negative; in an endothermic process, energy is absorbed and $\Delta E$ is positive.

For a reversible reaction, the energy liberated in the exothermic reaction equals the energy absorbed in the endothermic reaction. (The energy of the reaction, $\Delta E$, is equal also to the difference between the activation energies of the opposing reactions, $\Delta E = Ea - Ea'$.)

A catalyst affects a chemical reaction by lowering the activation energy for both the forward and the reverse reactions, equally.

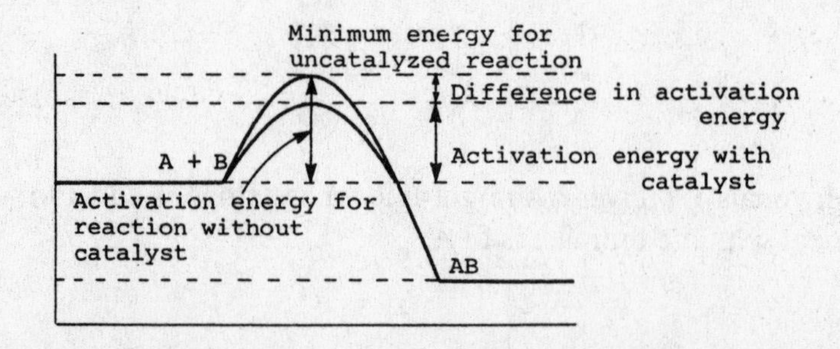

## 5. MECHANISM OF A REACTION

The mechanism of a reaction is the path that the reaction takes. The following sequence of reactions is the mechanism of the reaction

$$A_2B_5 \rightarrow A_2B_4 + {}^1/_2B_2$$

$$A_2B_5 \rightarrow AB_2 + AB_3 \quad \text{(a)}$$

$$AB_2 + AB_3 \rightarrow A_2B_5 \quad \text{(b)}$$

$$AB_2 + AB_2 \rightarrow AB_2 + {}^1/_2B_2 + AB \quad \text{(c)}$$

$$AB + A_2B_5 \rightarrow 3AB_2 \quad \text{(d)}$$

$$AB_2 + AB_2 \rightarrow A_2B_4 \quad \text{(e)}$$

Each of the above intermediate reactions is an elementary reaction. Molecular events are described by the elementary reactions.

The rate equation for a complex mechanism is the sum of the rate equations for the simple reactions composing it.

$$A + B \underset{k_{-1}}{\overset{k_1}{\rightleftarrows}} C \qquad\qquad C + B \xrightarrow{k_2} D$$

$$\frac{-dC_A}{dt} = k_1C_AC_B - k_{-1}C_C$$

$$\frac{-dC_B}{dt} = k_1C_AC_B - k_{-1}C_C + k_2C_CC_B$$

$$\frac{+dC_C}{dt} = k_1C_AC_B - k_{-1}C_C - k_2C_CC_B$$

# E. THERMODYNAMICS

## 1. STATE FUNCTIONS

### Some Commonly Used Terms in Thermodynamics

A system is that particular portion of the universe on which we wish to focus our attention.

Everything else is called the surroundings.

An adiabatic process occurs when the system is thermally isolated so that no heat enters or leaves.

An isothermal process occurs when the system is maintained at the same temperature throughout an experiment ($t_{final} = t_{initial}$).

An isopiestic (isobaric) process occurs when the system is maintained at constant pressure (i.e., $P_{final} = P_{initial}$).

The state of the system is some particular set of conditions of pressure, temperature, number of moles of each component, and their physical form (for example, gas, liquid, solid or crystalline form).

State functions depend only on the present state of the substance and not on the path by which the present state was attained. Enthalpy, energy, Gibbs free energy, and entropy are examples of state functions.

Heat capacity is the amount of heat energy required to raise the temperature of a given quantity of a substance one degree Celsius.

Specific heat is the amount of heat energy required to raise the temperature of 1g of a substance by 1.0°C.

Molar heat capacity is the heat necessary to raise the temperature of 1 mole of a substance by 1.0°C.

## 2. THE FIRST LAW OF THERMODYNAMICS

The first law of thermodynamics states that the change in internal energy is equal to the difference between the energy supplied to the system as heat and the energy removed from the system as work performed on the surroundings:

$$\Delta E = q - w$$

where $E$ represents the internal energy of the system (the total of all the energy possessed by the system). $\Delta E$ is the energy difference between the final and initial states of the system:

$$\Delta E = E_{final} - E_{initial}$$

The quantity $q$ represents the amount of heat that is added to the system as it passes from the initial to the final state, and $w$ denotes the work done by the system on the surroundings.

Heat added to a system and work done by a system are considered positive quantities (by convention).

For an ideal gas at constant temperature, $\Delta E = 0$ and $q - w = 0$ ($q = w$).

Considering only work due to expansion of a system, against constant external pressure:

$$w = P_{\text{external}} \cdot \Delta V$$

$$\Delta V = V_{\text{final}} - V_{\text{initial}}$$

## Enthalpy

The heat content of a substance is called enthalpy, $H$. A heat change in a chemical reaction is termed a difference in enthalpy, or $\Delta H$. The term "change in enthalpy" refers to the heat change during a process carried out at a constant pressure:

$\Delta H = q_p$ , $q_p$ means "heat at constant pressure."

The change in enthalpy, $\Delta H$, is defined

$$\Delta H = \Sigma H_{\text{products}} - \Sigma H_{\text{reactants}}$$

When more than one mole of a compound is reacted or formed, the molar enthalpy of the compound is multiplied by the number of moles reacted (or formed).

Enthalpy is a state function. Changes in enthalpy for exothermic and endothermic reactions are shown below:

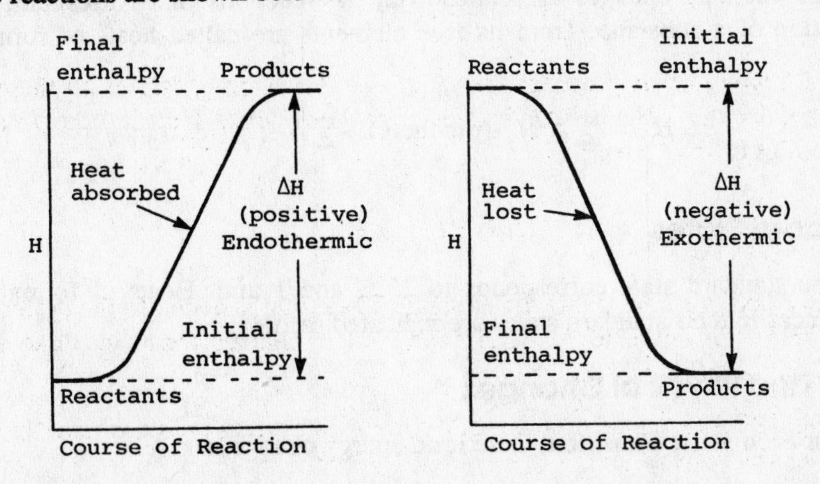


The $\Delta H$ of an endothermic reaction is positive, while that for an exothermic reaction is negative.

## Heats of Reaction

$\Delta E$ is equal to the heat absorbed or evolved by the system under conditions of constant volume:

$\Delta E = q_v$, $q_v$ means "heat at constant volume."

Since

$$H = E + PV$$

at constant pressure $\Delta H = \Delta E + P \Delta V$. Note that the term $P \Delta V$ is just the pressure-volume work ($(\Delta n)RT$) for an ideal gas at constant temperature, where $\Delta n$ is the number of moles of gaseous products minus the number of moles of gaseous reactants. Therefore,

$$\Delta H = \Delta E + \Delta nRT$$

for a reaction which involves gases. If only solid and liquid phases are present, $\Delta V$ is very small, so that $\Delta H \approx \Delta E$.

## Hess' Law of Heat Summation

Hess's law of heat summation states that when a reaction can be expressed as the algebraic sum of two or more reactions, the heat of the reaction, $\Delta H_r$, is the algebraic sum of the heats of the constituent reactions.

The enthalpy changes associated with the reactions that correspond to the formation of a substance from its free elements are called heats of formation, $\Delta H_f$.

$$\Delta H_r^0 = \sum \Delta H_f^0 \text{ (products)} - \sum \Delta H_f^0 \text{ (reactants)}$$

## Standard States

The standard state corresponds to 25°C and 1 atm. Heats of formation of substances in their standard states are indicated as $\Delta H^\circ_f$.

## Heat and Physical Changes

For each phase transition, there is an energy change, e.g.

S $\rightarrow$ liq, $\Delta H^\circ_T$ (fusion); liq $\rightarrow$ S, $\Delta H^\circ_T$ (crystallization);

S $\rightarrow$ gas, $\Delta H^\circ_T$ (sublimation); liq $\rightarrow$ gas, $\Delta H^\circ_T$ (vaporization);

$$\Delta H^\circ_T \text{ (fusion)} = - \; \Delta H^\circ_T \text{ (crystallization)}$$

Approximation of the heats of transition

For substances that are not highly joined in the liquid state, the following relationships have been noticed:

$$\Delta H^\circ_T \text{(vaporization)} \cong (88\text{J k}^{-1}\text{ mol}^{-1})T_{bp}$$

where $T_{bp}$ is the normal boiling point of the liquid. For elements,

$$\Delta H^\circ_T \text{ (fusion)} \cong (9.2\text{J k}^{-1}\text{ mol}^{-1})\, T_{mp}$$

where $T_{mp}$ is the melting point of the solid.

## Enthalpy of Heating

$$\Delta H^0 = \sum_i^{\text{phases}} \int Cp^0_i \, dT + \sum_j^{\text{transitions}} \Delta H^0_T \text{ (transition } j)$$

The following example shows how the above equation is applied.

## EXAMPLE:

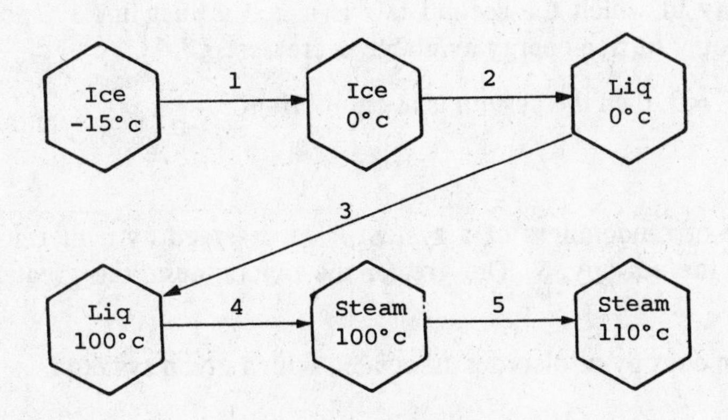


Heating ice from –15°C to steam at 110°C.

$$\Delta H^0 = \Delta H^0_{(1)} + \Delta H^0_{(2)} + \Delta H^0_{(3)} + \Delta H^0_{(4)} + \Delta H^0_{(5)}$$

$$= \int_{258\,k}^{273\,k} Cp^\circ \, dT + \Delta H^0_{273} \text{ (fusion)} + \int_{273\,k}^{373\,k} Cp^\circ \, dT$$

$$+ \Delta H^0_{373} \text{ (vaporization)} + \int_{373\,k}^{383\,k} Cp^\circ \, dT$$

The **heat capacity** of a substance is the amount of energy required to raise the temperature of one gram of the material in question, one degree Celsius.

## 3. THE SECOND LAW OF THERMODYNAMICS

The second law of thermodynamics states that in any spontaneous process there is an increase in the entropy of the universe ($\Delta S_{total} > 0$).

$$\Delta S_{universe} = \Delta S_{total} = \Delta S_{system} + \Delta S_{surroundings}$$

$$\Delta S_{surroundings} = \frac{-\Delta H_{system}}{T} \text{ at constant } P \text{ and } T$$

$$T \Delta S_{total} = -(\Delta H_{system} - T \Delta S_{system})$$

The maximum amount of useful work that can be done by any process at constant temperature and pressure is called the change in Gibbs free energy, $\Delta G$;

$$\Delta G = \Delta H - T \Delta S$$

Another way in which the second law is stated is that in any spontaneous change, the amount of free energy available decreases.

Thus, if $\Delta G = 0$, then the system is at equilibrium.

### Entropy

The degree of randomness of a system is represented by a thermodynamic quantity called the entropy, $S$. The greater the randomness, the greater the entropy.

A change in entropy or disorder associated with a given system is

$$\Delta S = S_2 - S_1$$

The entropy of the universe increases for any spontaneous process:

$$\Delta S_{universe} = (\Delta S_{system} + \Delta S_{surroundings}) \geq 0$$

When a process occurs reversibly at constant temperature, the change in entropy, $\Delta S$, is equal to the heat absorbed divided by the absolute temperature at which the change occurs:

$$\Delta S = \frac{q_{\text{reversible}}}{T}$$

## Standard Entropies and Free Energies

The entropy of a substance, compared to its entropy in a perfectly crystalline form at absolute zero, is called its absolute entropy, $S°$.

The third law of thermodynamics states that the entropy of any pure, perfect crystal at absolute zero is equal to zero.

The standard free energy of formation, $\Delta G°_f$, of a substance is defined as the change in free energy for the reaction in which one mole of a compound is formed from its elements under standard conditions:

$$\Delta G_f^0 = \Delta H_f^0 - T\,\Delta S_f^0$$

$$\Delta S_f^0 = \sum S_{f\ \text{products}}^0 - \sum S_{f\ \text{reactants}}^0$$

and

$$\Delta G_r^0 = \sum \Delta G_{f\ \text{products}}^0 - \sum \Delta G_{f\ \text{reactants}}^0$$

## Bond Energies

For diatomic molecules, the bond dissociation energy, $\Delta H°_{\text{diss}}$, is the amount of energy per mole required to break the bond and produce two atoms, with reactants and products being ideal gases in their standard states at 25°C.

The heat of formation of an atom is defined as the amount of energy required to form 1 mole of gaseous atoms from its element in its common physical state at 25°C and 1 atm pressure.

In the case of diatomic gaseous molecules of elements, the $\Delta H°_f$ of an atom is equal to one-half the value of the dissociation energy, that is,

$$-\Delta H_f^0 = \tfrac{1}{2}\,\Delta H_{\text{diss}}^0$$

where the minus sign is needed since one process liberates heat, and the other requires heat.

For the reaction HO – OH(g) →2OH(g),

$$\Delta H^\circ_{diss} = 2(\Delta H^\circ_f OH) - (\Delta H^\circ_f OH\text{–}OH).$$

The energy needed to reduce a gaseous molecule to neutral gaseous atoms, called the atomization energy, is the sum of all of the bond energies in the molecule.

For polyatomic molecules, the average bond energy, $\Delta H^\circ_{diss}$ avg., is the average energy per bond required to dissociate 1 mole of molecules into their constituent atoms.

## 4. CHANGE IN FREE ENERGY

The Gibbs free energy change ($\Delta G^\circ$) is related to the standard electrode potential, $E^\circ$ by the following equation:

$$\Delta G^\circ = -nFE^\circ$$

where $n$ is the number of electrons involved in the half reaction and $F$ is the Faraday constant, which has a value of 23,061 calories/volt, or 96,487 coulombs.

For example, calculate the Gibbs free energy change under standard conditions at 25° for the following reaction:

$$Br_2 + Pb \rightarrow PbBr_2$$

Consulting our table of half reactions we get:

$$2e^- + Br_2 \rightarrow 2Br^- = +1.065 \text{ volts}$$

$$Pb^\circ \rightarrow Pb^{++} + 2e^- = +0.126 \text{ volts}$$

This gives a total of + 1.191 volts, and a total of 2 electrons changing atoms. Putting this data into the equation yields:

$$\Delta G^\circ = -(2)(23{,}061)(1.191) = -54{,}930 \text{ calories.}$$

Notice that although a positive potential (+ 1.191 volts) designates a spontaneous reaction, a negative $\Delta G^\circ$ (– 54,930 calories) designates the same phenomena.

$\Delta G^\circ$ can also be related to the equilibrium constant, by the formula $\Delta G^\circ = -RT \ln K$ where $R$ is the ideal gas law constant (1.987 calories/mole Kelvin),* $T$ is the temperature in Kelvin, and $K$ is the equilibrium constant ($K_A$, $K_{sp}$, etc.). What is the $K$ of the above reaction?

---

* The numerical value for $R$ differs here because the units are different. $R$ can be expressed as 0.08205 $l$ atm/mole Kelvin, 1.987 cal/mole Kelvin, etc. The actual value of $R$ is still the same.

$$\Delta G° = -54{,}930,\ T = 25 + 273 = 278\ K,$$

$$\text{so} - 54{,}930 = -(1.987)\ (278)\ \ln k,\ 99.4 = \ln K,$$

$$e^{99.4} = e^{\ln K},\ 2 \times 10^{+43} = K.$$

## CHAPTER 4

# DESCRIPTIVE CHEMISTRY

## 1. HORIZONTAL, VERTICAL AND DIAGONAL RELATIONSHIPS IN THE PERIODIC TABLE

As one moves horizontally along the periodic table, the number of valence electrons increase. Each element has one more valence electron than the one to its left. The exception to this rule is the transition elements, which still contain one more electron than the element to the left of them; however, it is a non-valence electron. Each element possesses one more proton than the element to its left. To sum it up: As one moves from left to right, each atom on the periodic table increases by one electron (usually a valence electron) and one proton.

The left side of the periodic table contains the metals. These metals have a tendency to lose electrons, have high melting points, are malleable and ductile, possess a silvery lustre, and are good conductors of heat and electricity. The right hand side of the table contains the non-metals. Their properties are just the opposite of the metals. They tend to gain electrons, have varied melting points, are not malleable nor ductile, present a "dull," non-reflective appearance, and are poor conductors of heat and electricity.

If one does not include the inert gases (group VIII), reactivity decreases as one goes toward the center of the chart, and increases as one travels to the edges. The elements in rows IA and VIIA are more reactive than those in IIA and VIA; those in IIA and VIA are more reactive than those in IIIA, VA, or in the transition metals, etc.

As one travels down on the table, reactivity increases for metals and decreases for non-metals. Cesium is more reactive than lithium, and chlorine is more reactive than astatine.

The atomic radii increases as one moves down on the table. It decreases as one moves from left to right, with the exception of group VIII being the inert gases. In this case the atomic radius increases.

Electronegativity increases as one goes to the right and up on the table (again, not counting group VIII), and decreases as one goes to the left and down.

## 2. CHEMISTRY OF THE MAIN GROUPS AND TRANSITION ELEMENTS AND REPRESENTATIVES OF EACH

The groups IA elements are the most reactive of the metals. They react violently with cold water, and are so reactive that they are never found in nature in an uncombined state. Due to their high reactivity, they were not isolated and identified until the 19th century, when electric current became available for refining chemicals. Sodium is a good representative of this family. It reacts violently in water to form its hydroxide; in air it will oxidize to form its oxide. The further down the table the group IA elements are, the stronger bases they form, and the more reactive the element is.

The group IIA elements are also quite reactive, although not as reactive as those of group IA. Due to their high reactivity, they are not found in the free state in nature. They will slowly react with the oxygen in the air to form oxides, and will react slowly with hot water for hydroxide. They form moderate bases, with the strength of the base (and reactivity of the element) increasing as one goes downward.

Magnesium is a good representative of the group IIA metals. It will not react with cold water, but in boiling water, it produces its hydroxide. At normal temperatures it will react only very, very slowly with the oxygen in the air, but at an elevated temperature (e.g., a match flame), it will burst into flame.

The transition metals all react very slowly with oxygen or water, and some, not at all. Because of their low reactivity, these metals are relatively easy to isolate, and were among the first metals to be discovered by the ancients (e.g., copper and iron). Some of these metals are so unreactive (e.g., gold) that they can be found in an uncombined state in nature.

The lightest of the group IIIA elements possess non-metallic properties, while the remaining are all metals. The group IIIA elements are all mildly reactive. For example, aluminum resists combining with oxygen or reacting with water at normal temperature. It will react with hydrochloric acid.

The first element of the group IVA elements (carbon) is a non-metal, it is followed by two metalloids (silicon and germanium), and then by two metals (tin and lead). The first three elements are semi-conductors, and are fairly reactive. Were it not for the fact that life on this planet is based upon carbon, it would never have been found in the uncombined state in nature (e.g., graphite). The rest are never found free in nature, although the lower reactivities of the last two (tin and lead) allowed them to be separated and used by the ancients.

The group VA elements consists of two non-metals, two metalloids, and one metal. The more reactive (higher on the table) of these elements combine readily with oxygen and some of the more reactive metals.

The group VIA elements are quite reactive, starting with oxygen, which combines with almost all other elements. Were it not for biological processes occurring on earth, there would be no free oxygen. In fact, it is thought that due to its extreme reactivity, free oxygen was a deadly poison to early life-forms on earth. The reactivity of the group VIA elements decreases as one moves down the table. Most of the VIA elements are non-metals, the last two, however, are metalloids.

The group VIIIA elements (also known as the Halogen Family) are quite reactive, hence they are never found in the free state in nature. Fluorine is so reactive that it can combine with *all* of the other elements except helium. Reactivity decreases as one goes down the row. The first four elements have been used for killing life-forms from bacteria (germicide) to humans (war gas), due to their high degree of reactivity. Astatine, the last element in the row, is a metalloid, and only exists in radioactive forms.

The last row, group VIIA (or 0) contains the inert gases. These elements all have a filled valence shell, and hence, do not react at all in nature. They exist as monoatomic gases. When Ramsey discovered argon, the first of these gases, it was a great surprise, as their existence had not been predicted, and there was no position for them on the periodic table of the time (1895). He named it from the Greek word *argos*, meaning "lazy," since the gas was too lazy to combine with any other element. Radon, the furthest down on the table, only exists in radioactive forms. Since the 1960s, chemists have been able to make compounds with all of these elements except helium (the *only* element that has no compounds), mostly fluoride salts. These compounds are unstable, and release a lot of energy upon their decomposition. Due to the fact that they can combine with other elements (albeit under extreme circumstances), it has been suggested that they be called "noble gases," rather than "inert gases."

## 3. ORGANIC CHEMISTRY

### Alkanes

Structural Formula: $C_nH_{2n+2}$

The simplest member of the alkane family is methane ($CH_4$) which is written as:

Acidity $\quad$ H—F < H—Cl < H—Br < H—I

H—OH < H—SH < H—SeH

## Nomenclature (IUPAC System)

A) Select the longest continuous carbon chain for the parent name.

Ex.
$$CH_3-CH_2-\underset{\displaystyle CH_3}{\underset{|}{CH}}-CH_2-CH_2-CH_3$$

The parent name is hexane.

B) Number the carbons in the chain, from either end, such that the substituents are given the lowest numbers possible.

Ex.
$$\overset{1}{CH_3}-\overset{2}{CH_2}-\underset{\displaystyle CH_3}{\underset{|}{\overset{3}{CH}}}-\overset{4}{CH_2}-\overset{5}{CH_2}-\overset{6}{CH_3}$$

C) The substituents are assigned the number of the carbon to which they are attached. In the preceding example the substituent is assigned the number 3.

D) The name of the compound is now composed of the name of the parent chain preceded by the name and the number of the substituents, arranged in alphabetic order. For the same example the name is thus 3-methylhexane.

E) If a substituent occurs more than once in the molecule, the prefixes, "di–," "tri–," "tetra–," etc., are used to indicate how many times it occurs.

F) If a substituent occurs twice on the same carbon, the number of the substituent is repeated.

Ex.
$$CH_3-CH_2-\underset{\displaystyle CH_3}{\underset{|}{\overset{\displaystyle CH_3}{\overset{|}{C}}}}-CH_2-CH_3 \quad \text{3, 3–dimethylpentane}$$

$$\begin{array}{ccccccccccccc}
 & & & & & & & & & & CH_3 & & \\
 & & & & & & & & & & | & & \\
 & & & & & & & & & & CH_2 & & \\
 & & & & & & & & & & | & & \\
CH_3 & - & CH_2 & - & CH_2 & - & CH & - & CH & - & C & - & CH_2-CH_3 \\
 & & & & & & | & & | & & | & & \\
 & & & & CH_3 & - & CH & & CH_3 & - & CH_2 & & \\
 & & & & & & | & & & & | & & \\
 & & & & & & CH_3 & & & & CH_3 & &
\end{array}$$

3, 3–diethyl–5–isopropyl–4–methyloctane

## Alkenes

Alkenes (olefins) are unsaturated hydrocarbons with one or more carbon-carbon double bonds. They have the general formula, $CnH_{2n+2}$.

$$\begin{array}{c} R \quad\quad R \\ \diagdown \quad \diagup \\ C = C \\ \diagup \quad \diagdown \\ R \quad\quad R \end{array} \qquad R = \text{Alkyl or H}$$

Ex.

$$\begin{array}{c} H \quad\quad H \\ \diagdown \quad \diagup \\ C = C \\ \diagup \quad \diagdown \\ H \quad\quad H \end{array} \qquad \begin{array}{c} H \quad\quad H \\ \diagdown \quad \diagup \\ C = C \\ \diagup \quad \diagdown \\ H \quad\quad CH_3 \end{array}$$

ethylene propylene

## Nomenclature (IUPAC System)

A) Select the longest continuous chain of carbons containing the double bond. This is the parent structure and is assigned the name of the corresponding alkane with the suffix changed from "–ane" to "–ene."

B) Number the chain so that the position of the double bond is designated by the lowest possible number assigned to the first doubly bonded carbon.

Ex. $\overset{5}{CH_3} - \underset{\displaystyle \overset{|}{Br}}{\overset{4}{CH_2}} - \overset{3}{CH} = \overset{2}{CH} - \overset{1}{CH_3}$

4–bromo–2–pentene

Some common names given to families of alkenes are:

| | |
|---|---|
| $H_2C = CH - R$ | vinyl |
| $H_2C = CH - CH_2 - R$ | allyl |
| $H_3C - CH = CH - R$ | propenyl |

## Dienes

Dienes have the structural formula, $C_nH_{2n-2}$. In the IUPAC nomenclature system, dienes are named in the same manner as alkenes, except that the suffix "–ene" is replaced by "–diene," and two numbers must be used to indicate the position of the double bonds.

## Classification of Dienes

$$-\overset{|}{C}=C=\overset{|}{C}-$$ Cumulated double bonds (allenes)

$$-\overset{|}{C}=\overset{|}{C}-\underset{|}{\overset{|}{C}}=C-$$ Conjugated (alternating) double bonds

$$-\overset{|}{C}=\underset{|}{C}-(CH_2)_n-\overset{|}{\underset{|}{C}}=C-$$ Isolated (non-conjugated) double bonds

## Alkynes

Alkynes are unsaturated hydrocarbons containing triple bonds. They have the general formula, $C_nH_{2n-2}$.

$R-C\equiv C-R$ R = Alkyl or H

Ex. $H-C\equiv C-H$ Simplest Alkyne

acetylene

## Nomenclature (IUPAC System)

Alkynes are named in the same manner as alkenes, except that the suffix "–ene" is replaced with "–yne."

When both a double bond and a triple bond are present, the hydrocarbon is called an alkenyne. In this case, the double bond is given preference over the triple bond when numbering.

Ex. $CH_3-C\equiv C\ H$ $CH_3-CH=CH-C\equiv CH$

propyne 1–penten-3-yne

## Alkyl Halides

Alkyl halides are compounds in which one hydrogen atom is replaced by an atom of the halide family. An important use of alkyl halides is as intermediates in organic synthesis.

Structural formula: $C_nH_{2n+1}-X$; $X$ = Cl, Br, I, F.

## Nomenclature (IUPAC System)

| Formula | Name |
|---|---|
| $CH_3Cl$ | chloromethane |
| $CH_3CH_2Br$ | bromoethane |
| $CH_3CH_2CH_2I$ | 1–iodopropane |
| $CH_3CHICH_3$ | 2–iodopropane |
| $CH_3CH_2CH_2CH_2Cl$ | 1–chlorobutane |
| $CH_3CH_2CHBrCH_3$ | 2–bromobutane |
| $(CH_3)_3CI$ | 2–iodo–2–methylpropane |
| $CH_3CH_2CH_2CH_2CH_2Cl$ | 1–chloropentane |

## Cyclic Hydrocarbons

Cyclic hydrocarbons and cyclic alkenes are alicyclic (aliphatic cyclic) hydrocarbons.

## Nomenclature

Cyclic aliphatic hydrocarbons are named by prefixing the term "cyclo–" to the name of the corresponding open-chain hydrocarbon, having the same number of carbon atoms as the ring.

$H_2C$ $CH_2$ $H_2C$

cyclopropane

$H_2C$ — $CH_2$ / $H_2C$ — $CH_2$

cyclobutane

$CH_2$ / $H_2C$ $CH_2$ / $H_2C$ — $CH_2$

cyclopentane

Substituents on the ring are named, and their positions are indicated by numbers, the lowest combination of numbers being used.

## Aromatic Hydrocarbons

Most aromatic hydrocarbons (arenes) are derivatives of benzene. Examples of benzene derivatives are napthalene, anthracene and phenanthrene.

## Structure

Benzene has a symmetrical structure and the analysis, synthesis and molecular weight determination indicate a molecular formula of $C_6H_6$.

Napthalene structure is indicated by the oxidation of 1–nitronapthalene which shows that the substituted ring is a true benzene ring. Reduction and oxidation of the same nucleus indicates that the unsubstituted ring is a true benzene ring.

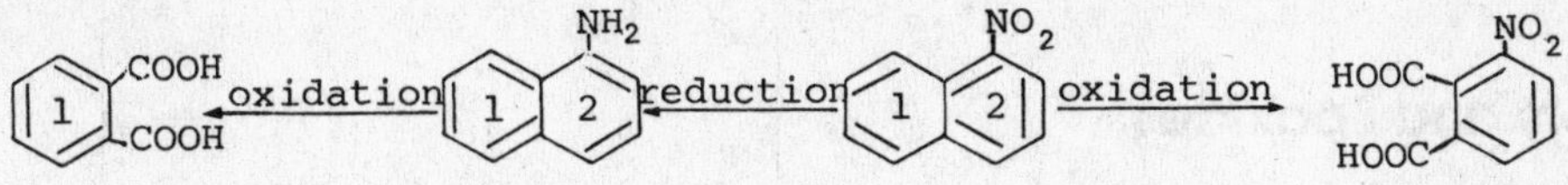

## Nomenclature (IUPAC System)

Aromatic compounds are named as derivatives of the corresponding hydrocarbon nucleus.

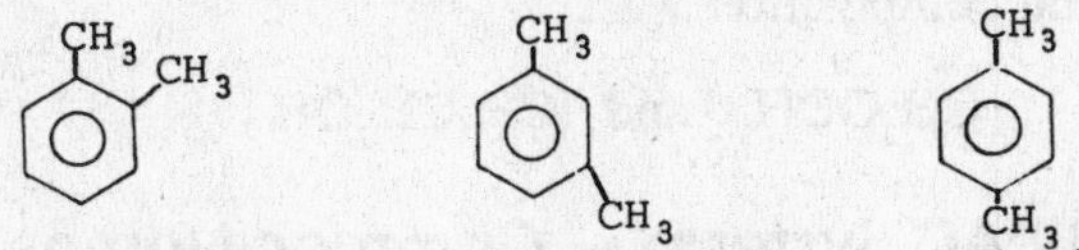

1,2-Dimethylbenzene (ortho-xylene) 1,3-Dimethylbenzene (meta-xylene) 1,4-Dimethylbenzene (para-xylene)

In the IUPAC system of nomenclature, the position of the substituent group is always indicated by numbers arranged in a certain order:

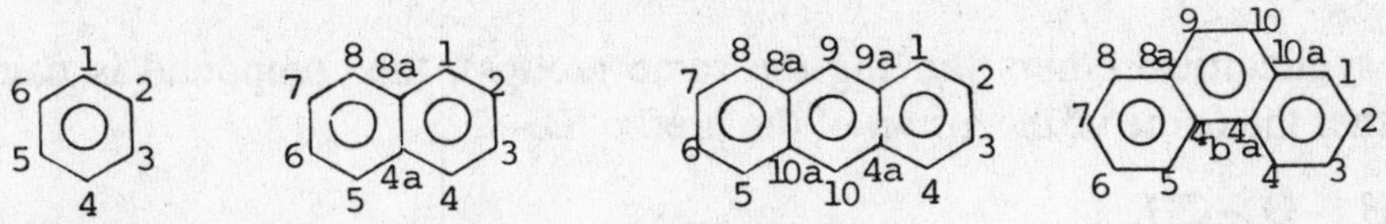

Benzene Naphthalene Anthracene Phenanthrene

## Aryl Halides

Aryl halides are compounds containing halogens attached directly to a benzene ring. The structural formula is ArX, where the aryl group, Ar, represents phenyl, napthyl, etc., and their derivatives.

## Nomenclature

Aryl halides are named by prefixing the name of the halide to the name of the aryl group. The terms meta, ortho, and para are used to indicate the positions of substituents on a disubstituted benzene ring. Numbers are also used to indicate the positions of the substituents on a benzene ring.

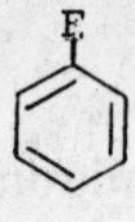

F Flouro-benzene | Cl 1-Chloronaph-thalene | Br, Cl, Cl 1-Bromo-2,4-Dichloro-benzene | Cl, Cl Ortho-Dichloro-benzene

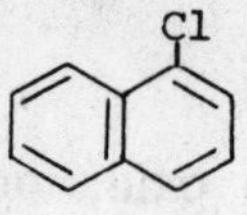

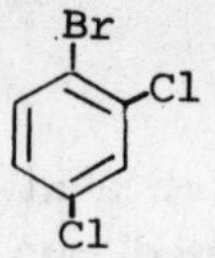

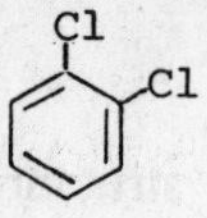

## Ethers and Epoxides

Ethers are hydrocarbon derivatives in which two alkyl or aryl groups are attached to an oxygen atom. The structural formula of an ether is $R - O - R'$, where R and R′ may or may not be the same.

## Ethers — Structure

Ethers and alcohols are metameric. They are functional isomers of alcohols with the same elemental composition

$$CH_3OCH_3 \quad \text{and} \quad CH_3CH_2OH$$

## Nomenclature (IUPAC System) — Common Names

The attached groups are named in alphabetical order, followed by the word ether.

Insert ether Structures

For symmetrical ethers (having the same groups), the compound is named using either the name of the group of the prefix "Di–."

Ex. $CH_3 - O - CH_3$

Methyl ether or Dimethyl ether

In the IUPAC system, ethers are named as alkoxyalkanes. The larger alkyl group is chosen as the stem.

Ex.

$$\begin{array}{ccccccc} & & Cl & & & & \\ & & | & & & & \\ CH_3 & - & CH & - & CH & - & CH_3 \\ & & | & & | & & \\ & & Cl & & OCH_2CH_3 & & \end{array}$$

2–Ethoxy–3,3–dichlorobutane

## Epoxides — Structure

Epoxides are cyclic ethers in which the oxygen is included in a three-membered ring.

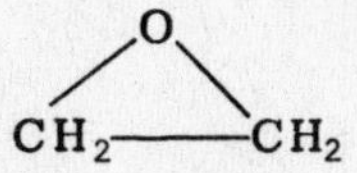

$$CH_2\overset{O}{\triangle}CH_2$$

An epoxide:
Ethylene oxide

## Alcohols and Glycols

Alcohols and hydrocarbon derivatives in which one or more hydrogen atoms have been replaced by the –OH (hydroxyl) group. They have the general formula R–OH, where R may be either alkyl or aryl.

## Nomenclature (IUPAC System)

Alcohols are named by replacing the "–e" ending of the corresponding alkane with the suffix "ol." The alcohol may also be named by adding the name of the R group to the same alcohol.

Ex. $CH_3CH_2OH$ Ethanol or methyl alcohol

Depending on what carbon atom the hydroxyl group is attached to, the alcohol is prefixed as follows:

A) Primary (–OH attached to 1° carbon) alcohols are prefixed "n–" or "1–".

B) Secondary (–OH attached to 2° carbon) alcohols are prefixed "sec–" or "2–".

C) Tertiary (–OH attached to 3° carbon) alcohols are prefixed "tert–" or "3–".

Ex. $CH_3CH_2CH_2CH_2OH$

n– or 1– butanol

$CH_3CH_2CHCH_3$ (with –OH on the third carbon)

sec– or z–butanol

$CH_3CH_2-C(CH_3)(OH)-CH_3$

tert– or 3–pantanol

## Glycols

Alcohols containing more than one hydroxyl group (polyhydroxyalcohols) are represented by the general formula $C_nH_{2n+2} - y(OH)y$. Polyhydroxyalcohols containing two hydroxyl groups are called glycols or diols.

Ex. 1,3–butanediol

$$CH_3-\underset{\underset{OH}{|}}{CH}-CH_2-CH_2-OH$$

## Carboxylic Acids

Carboxylic acids contain a carboxyl group

$$-C\begin{matrix}\nearrow O \\ \searrow OH\end{matrix}$$

bonded to either an alkyl group (RCOOH) or an aryl group (ArCOOH).

HCOOH is formic acid (methanoic acid)

$CH_3$ COOH is acetic acid (ethanoic acid)

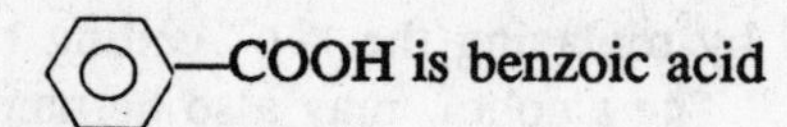

—COOH is benzoic acid

## Nomenclature (IUPAC System)

The longest chain carrying the carboxyl group is considered the parent structure and is named by replacing the "–e" ending of the corresponding alkane with "–oic acid."

| | |
|---|---|
| $CH_3CH_2CH_2CH_2COOH$ | Pentanoic acid |
| $CH_3CH_2\underset{\underset{CH_3}{\vert}}{CH}COOH$ | 2–Methylbutanoic acid |
| $C_6H_5$—$CH_2CH_2COOH$ | 3–Phenylpropanoic acid |
| $CH_3CH = CHCOH$ | 2–Butenoic acid |

The position of the substituent is indicated by a number, e.g.:

$$\overset{5}{C}-\overset{4}{C}-\overset{3}{C}-\overset{2}{C}-\overset{1}{C}OOH$$

The name of a salt of a carboxylic acid consists of the name of the cation followed by the name of the acid with the ending "–ic acid" changed to "–ate."

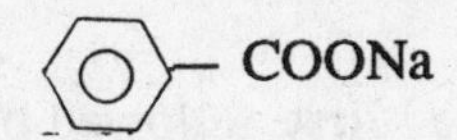

| $C_6H_5$–COONa | $(CH_3COO)_2Ca$ | $HCOONH_4$ |
|---|---|---|
| Sodium benzoate | Calcium acetate | Ammonium formate |

## Carboxylic Acid Derivatives

Carboxylic acid derivatives are compounds in which the carboxyl group has been replaced by –Cl, –OOCR, $-NH_2$, or –OR′. These derivatives are called acid chlorides, anhydrides, amides, and esters, respectively.

## Acid Chlorides — Nomenclature (IUPAC System)

When naming acid chlorides, the ending "–ic acid" in the carboxylic acid is replaced by the ending "–yl chloride."

R–C(=O)–Cl
Acid chloride

$CH_3$–C(=O)–Cl
Ethanoyl chloride
(Acetyl chloride)

$C_6H_5$–C(=O)–Cl
Benzoyl chloride

$CH_3CH_2C(=O)$–Cl
Propanoyl chloride

$CH_3CH(CH_3)CH_2CH_2C(=O)$–Cl
4-Methyl pentanoyl chloride

$O_2N$–$C_6H_4$–C(=O)–Cl
m-Nitrobenzoyl chloride

$CH_3CH{=}CHCOCl$
2-Butenoyl chloride

$C_6H_{11}$–COCl
Cyclohexane carbonyl chloride

## Carboxylic Acid Anhydrides — Nomenclature (IUPAC System)

When naming acid anhydrides, the word "acid" in the carboxylic acid is replaced by the word "anhydride."

R–C(=O)–O–C(=O)–R
Acid anhydride

$CH_3CH_2C(=O)OC(=O)CH_2CH_3$
Propionic anhydride

$C_6H_5$–C(=O)–O–C(=O)–$C_6H_5$
Benzoic anhydride

$(ClCH_2CH_2CH_2CO)_2O$
4-Chlorobutanoic anhydride

$CH_3CH_2C(=O)$–O–$C(=O)CH_3$
Acetic propanoic anhydride
(a mixed anhydride)

## Esters — Nomenclature (IUPAC System)

When naming esters, the alcohol or phenol group is named first, followed by the name of the acid with the "–ic" ending replaced by "–ate." Esters of cycloalkane carboxylic acids have the ending "carboxylate."

$R-C(=O)-OR'$
An ester

$H-C(=O)-O-CH_3$
Formic acid / Methyl alcohol
Methyl formate

$CH_3C(=O)-O-CH_2-C_6H_5$
Acetic acid / Benzyl alcohol
Benzyl acetate

$CH_3-C(=O)-OC_2H_5$
Ethyl acetate

$C_6H_5-C(=O)-OC_2H_5$
Ethyl benzoate

$C_6H_{11}-COOCH_3$
Methyl cyclohexane-carboxylate. (A)

## Amides — Nomenclature (IUPAC System)

When naming amides, the "–ic acid" of the common name (or the "–oic acid" of the IUPAC name) of the parent acid is replaced by "amide." Amides of cycloalkane carboxylic acids have the ending carbonxamide.

$CH_3-C(=O)-NH_2$
Acetamide
Ethanamide

$H_2NOCCH_2CH_2CONH_2$
1,4-Butane diamide

$o-Cl-C_6H_4-C(=O)-NH_2$
o-Chlorobenzamide

$C_4H_7-CONH_2$
Cyclobutane-carboxamide

$CH_3C(=O)-NH-CH_2CH_3$
N-Ethyl acetamide

$C_6H_5-C(=O)-N(CH_3)_2$
N,N-Dimethyl benzamide

## Arenes — Structure and Nomenclature

Arenes are compounds that contain both aromatic and aliphatic units.

The simplest of the alkyl benzenes, methyl benzene, has the common name toluene. Compounds that have longer side chains are named by adding the word "benzene" to the name of the alkyl group.

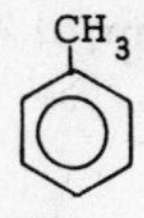

Toluene

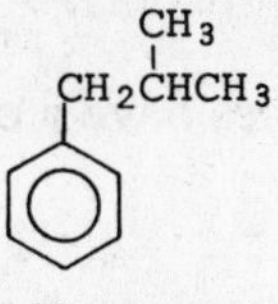

Isobutylbenzene

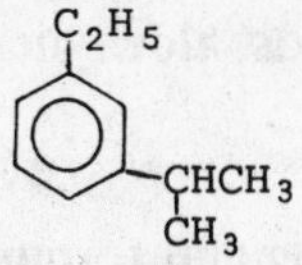

m-Ethylisopropylbenzene

**The simplest of the dialkyl benzenes, the dimethyl benzenes, have the common name xylenes, Dialkyl benzenes that contain one methyl group are named as derivatives of toluene.**

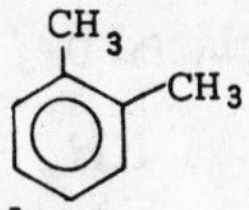

o-Xylene

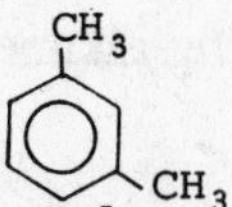

m-Xylene

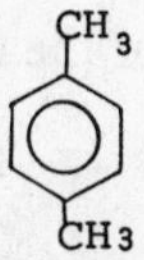

p-Xylene

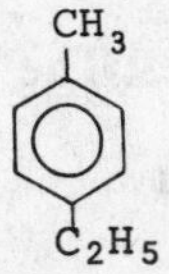

p-Ethyltoluene

**A compound that contains a complex side chain is named as a phenyl alkane ($C_6H_5$ = phenyl). Compounds that contain more than one benzene ring are named as derivatives of alkanes.**

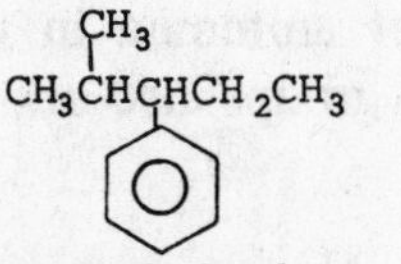

2-Methyl-3-phenylpentane

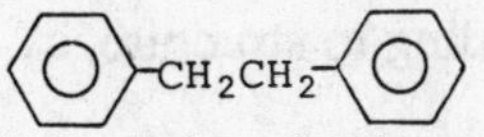

1,2-Diphenylethane

**Styrene is the name given to the simplest alkenyl benzene. Others are named as substituted alkenes. Alkynyl benzenes are named as substituted alkynes.**

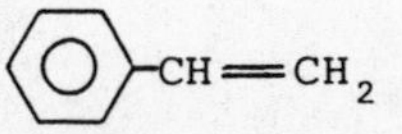

Styrene
( Vinylbenzene )
( Phenylethylene )

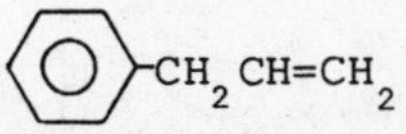

Allyl benzene
(3-Phenylpropene)

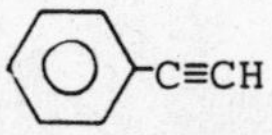

Phenylacetylene

## Aldehydes and Ketones

Carboxylic acids, aldehydes and ketones have a carboxylic group,

$$>C=O$$

in common. The general formula for aldehydes is RCHO, and that for ketones is RCOR.

## Nomenclature (IUPAC System)

A) Aldehydes: The longest continuous chain containing the carbonyl group is considered the parent structure and the "–e" ending of the corresponding alkane is replaced by "–al."

B) Ketones: The "–e" ending of the corresponding alkane is replaced by "–one."

Ex.

| H over $H-C=O$ | H over $CH_3C=O$ | $CH_3$ over $CH_3-C=O$ | $CH_3$ over $CH_3-CH_2-C=O$ |
|---|---|---|---|
| methanal (formaldehyde) | ethanal (acetaldehyde) | propanone (acetone) | butanone (methyl ethyl ketone) |

## Amines

Amines are derivatives of hydrocarbons in which a hydrogen atom has been replaced by an amino group; derivatives of ammonia in which one or more hydrogen atoms have been replaced by alkyl groups also known as amines. They are classified, according to structure, as

Primary — $R-N-H$, with H below N

Secondary — $R-N-R$, with H above N

Tertiary — $R-N-R$, with R above N

## Nomenclature (IUPAC System)

The aliphatic amine is named by listing the alkyl groups attached to the nitrogen, and following these by "–amine."

$$CH_3-NH_2 \qquad CH_3-\overset{\displaystyle CH_3 \atop |}{\underset{| \atop \displaystyle NH_2}{C}}-CH_3 \qquad C_6H_5-CH_2-\overset{\displaystyle H \atop |}{N}-CH_2CH_3$$

Methylamine tert–Butylamine Benzyl ethylamine

If an alkyl group occurs twice or three times on the nitrogen, the prefixes "di–" and "tri–" are used respectively.

Ex.

$$CH_3-NH-CH_3 \qquad CH_3-\overset{\displaystyle CH_3 \atop |}{N}-CH_3$$

dimethylamine trimethylamine

If an amino group is part of a complicated molecule, it may be named by prefixing "amino" to the name of the present chain.

Ex.

$$NH_2CH_2CH_2OH \qquad CH_3\overset{\displaystyle NH_2 \atop |}{CH}\ CH_2COOH$$

2–amino ethanol 3–aminobutanoic acid

An amino substituent that carries an alkyl group is named as an N–alkyl amino group.

Ex.

$$CH_3NH-CH_2COOH \qquad CH_3-NH\overset{\displaystyle CH_2 \atop |}{CH}(CH_2)_4CH_3$$

N–methyl amino acetic acid 2–(N–methylamino) heptane

## Phenols and Quinones — Nomenclature of Phenols

Phenols have the general formula ArOH. The –OH group in phenols is attached directly to the aromatic ring.

Phenols are named as derivatives of phenol, which is the simplest member of the family. Methyl phenols are given the name cresols. Phenols are also called "hydroxy–" compounds.

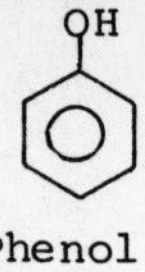

Phenol 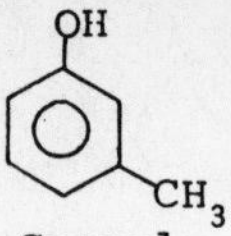 m-Cresol 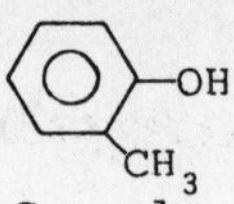 o-Cresol

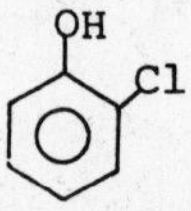

o-Chlorophenol 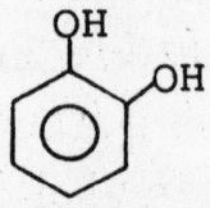 Catechol 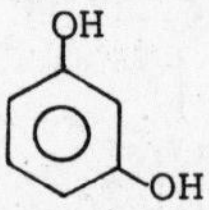 Resorcinol 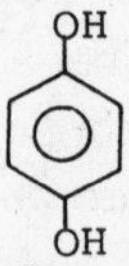 Hydroquinone

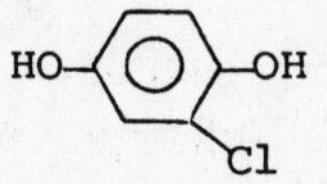

2-Chlorohydroquinone 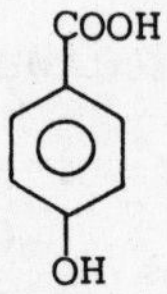 p-Hydroxy-benzoic acid Picric acid

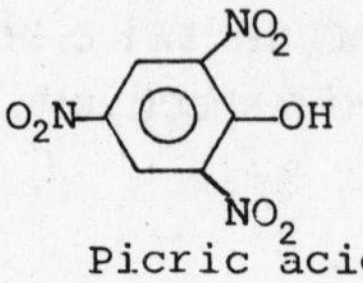

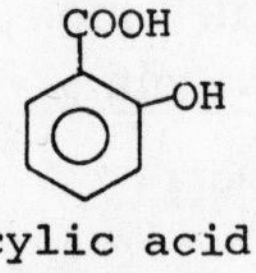

Salicylic acid 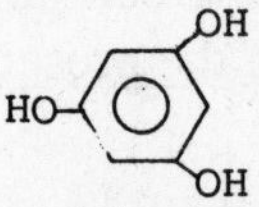 Phloroglucinol

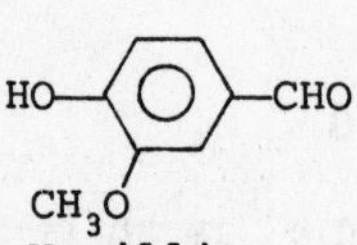

Vanillin 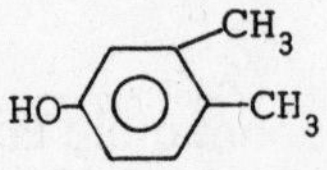 3,4-Xylenol 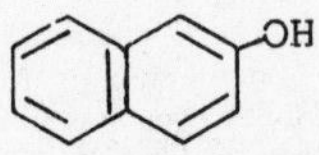 β-Naphthol (or 2-Naphthol) α-Naphthol

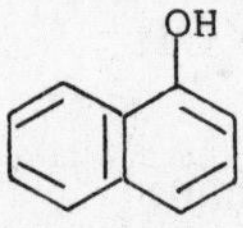

## Quinones — Nomenclature

**Quinones are cyclic, conjugated diketones named after the parent hydrocarbon.**

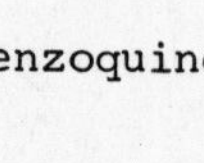

o-Benzoquinone 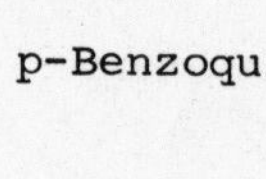 p-Benzoquinone 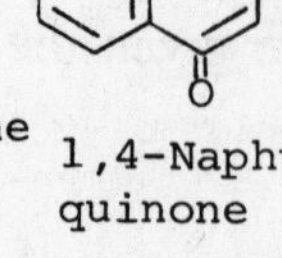 1,4-Naphthoquinone Toluquinone (2-Methyl-1,4-benzoquinone)

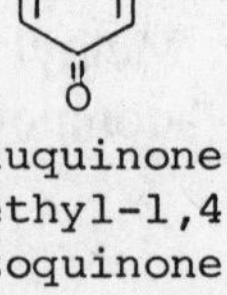

## 4. STRUCTURAL ISOMERISM

There are 2 types of butanes — normal butane and isobutane. They have the same molecular formula, $C_4H_{10}$, but have different structures. N-Butane is a straight chain hydrocarbon whereas isobutane is a branched-chain hydrocarbon. N–Butane and isobutane are structural isomers and differ in their physical and chemical properties.

```
                            H   H   H   H
                            |   |   |   |
C4H10 ≡ CH3CH2CH2CH3 ≡ H – C – C – C – C – H     n–butane
                            |   |   |   |         (straight chained)
                            H   H   H   H
```

$$C_4H_{10} \equiv CH_3CH_2CH_2CH_3$$

```
                              H   H   H
                              |   |   |
C4H10 ≡ CH3CH(CH3)CH3 ≡ H – C – C – C – H     isobutane
                              |   |   |        (branched)
                              H   |   H
                                  |
                              H – C – H
                                  |
                                  H
```

$$C_4H_{10} \equiv CH_3CH(CH_3)CH_3$$

In higher homologs of the alkane family, the number of isomers increases exponentially.

# THE ADVANCED PLACEMENT EXAMINATION IN

# CHEMISTRY

# TEST I

# THE ADVANCED PLACEMENT EXAMINATION IN
# CHEMISTRY

## ANSWER SHEET

1. (A) (B) (C) (D) (E)
2. (A) (B) (C) (D) (E)
3. (A) (B) (C) (D) (E)
4. (A) (B) (C) (D) (E)
5. (A) (B) (C) (D) (E)
6. (A) (B) (C) (D) (E)
7. (A) (B) (C) (D) (E)
8. (A) (B) (C) (D) (E)
9. (A) (B) (C) (D) (E)
10. (A) (B) (C) (D) (E)
11. (A) (B) (C) (D) (E)
12. (A) (B) (C) (D) (E)
13. (A) (B) (C) (D) (E)
14. (A) (B) (C) (D) (E)
15. (A) (B) (C) (D) (E)
16. (A) (B) (C) (D) (E)
17. (A) (B) (C) (D) (E)
18. (A) (B) (C) (D) (E)
19. (A) (B) (C) (D) (E)
20. (A) (B) (C) (D) (E)
21. (A) (B) (C) (D) (E)
22. (A) (B) (C) (D) (E)
23. (A) (B) (C) (D) (E)
24. (A) (B) (C) (D) (E)
25. (A) (B) (C) (D) (E)
26. (A) (B) (C) (D) (E)
27. (A) (B) (C) (D) (E)
28. (A) (B) (C) (D) (E)
29. (A) (B) (C) (D) (E)
30. (A) (B) (C) (D) (E)
31. (A) (B) (C) (D) (E)
32. (A) (B) (C) (D) (E)
33. (A) (B) (C) (D) (E)
34. (A) (B) (C) (D) (E)
35. (A) (B) (C) (D) (E)
36. (A) (B) (C) (D) (E)
37. (A) (B) (C) (D) (E)
38. (A) (B) (C) (D) (E)
39. (A) (B) (C) (D) (E)
40. (A) (B) (C) (D) (E)
41. (A) (B) (C) (D) (E)
42. (A) (B) (C) (D) (E)
43. (A) (B) (C) (D) (E)
44. (A) (B) (C) (D) (E)
45. (A) (B) (C) (D) (E)
46. (A) (B) (C) (D) (E)
47. (A) (B) (C) (D) (E)
48. (A) (B) (C) (D) (E)
49. (A) (B) (C) (D) (E)
50. (A) (B) (C) (D) (E)
51. (A) (B) (C) (D) (E)
52. (A) (B) (C) (D) (E)
53. (A) (B) (C) (D) (E)
54. (A) (B) (C) (D) (E)
55. (A) (B) (C) (D) (E)
56. (A) (B) (C) (D) (E)
57. (A) (B) (C) (D) (E)
58. (A) (B) (C) (D) (E)
59. (A) (B) (C) (D) (E)
60. (A) (B) (C) (D) (E)
61. (A) (B) (C) (D) (E)
62. (A) (B) (C) (D) (E)
63. (A) (B) (C) (D) (E)
64. (A) (B) (C) (D) (E)
65. (A) (B) (C) (D) (E)
66. (A) (B) (C) (D) (E)
67. (A) (B) (C) (D) (E)
68. (A) (B) (C) (D) (E)
69. (A) (B) (C) (D) (E)
70. (A) (B) (C) (D) (E)
71. (A) (B) (C) (D) (E)
72. (A) (B) (C) (D) (E)
73. (A) (B) (C) (D) (E)
74. (A) (B) (C) (D) (E)
75. (A) (B) (C) (D) (E)
76. (A) (B) (C) (D) (E)
77. (A) (B) (C) (D) (E)
78. (A) (B) (C) (D) (E)
79. (A) (B) (C) (D) (E)
80. (A) (B) (C) (D) (E)
81. (A) (B) (C) (D) (E)
82. (A) (B) (C) (D) (E)
83. (A) (B) (C) (D) (E)
84. (A) (B) (C) (D) (E)
85. (A) (B) (C) (D) (E)

# ADVANCED PLACEMENT CHEMISTRY EXAM I

## SECTION I

Time: 1 Hour 45 Minutes
85 Questions

**Note:** For all questions referring to solutions, assume that the solvent is water unless otherwise stated.

**Directions:** Each set of lettered choices below refers to the numbered statements immediately following it. Select the one lettered choice that best fits each statement. A choice may be used once, more than once, or not at all in each set.

**Questions 1 - 4**

(A) K
(B) Ca
(C) Cr
(D) Fe
(E) Cu

1. Does not form monatomic ions of $2^+$ charge.

2. Forms the oxide ion $M_2O_7^{2-}$, in which M represents the metal.

3. Is found in nature in its metallic state.

4. Is an important construction material.

**Questions 5 - 6**

(A) HF

(B) $H_2O$

(C) $NH_3$

(D) $CH_4$

(E) $NaBH_4$

5. Is an organic molecule.

6. Is a strong reducing agent.

**Questions 7 - 9**

(A) covalent bonding

(B) ionic bonding

(C) hydrogen bonding

(D) $\pi$ bonding

(E) metallic bonding

7. Involves electron transfer from one atom to another.

8. Does not occur in an aqueous solution of $CH_3CO_2Na$.

9. Is responsible for the relatively high boiling point of water.

**Questions 10 - 13**

(A) 1.0 M

(B) $5.0 \times 10^{-1}$ M

(C) $1.0 \times 10^{-1}$ M

(D) $5.0 \times 10^{-2}$ M

(E) $1.0 \times 10^{-3}$ M

What is the molarity of a solution of NaOH, if:

10. The pH is 11.00

11. 100 ml of the solution are titrated to the endpoint (using phenolphthalein as indicator) with 10.0 ml of a 1M solution of HCl.

12. 100 ml of the solution are titrated with 50 ml of a 1.0M solution of $H_2SO_4$, in the presence of phenolphthalein, to the endpoint.

13. The concentration of $H_3O^+$ ions is $2 \times 10^{-14}$ M.

**Questions 14 - 17**

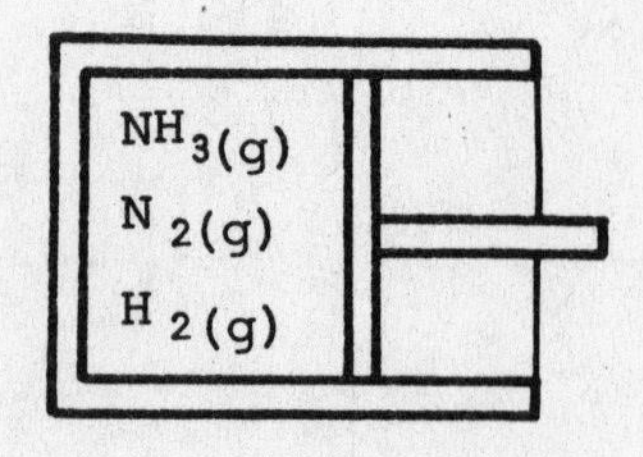

Three gases are in equilibrium in a closed chamber sealed with a piston. The equilibrium expression is:

$$2\,NH_{3(g)} \leftrightarrow N_{2(g)} + 3H_{2(g)}$$

(A) The mole fraction of $N_2$ increases

(B) The mole fraction of $N_2$ decreases

(C) The mole fraction of $N_2$ remains the same

(D) The mole fraction of $N_2$ increases and then decreases

(E) Direction of change cannot be predicted with the given information

Which of the above occurs in each of the following cases?

14. The piston is pushed into the chamber.

15. More $H_2$ is added as the piston is adjusted to maintain constant pressure.

16. The chamber is heated while the piston is held steady.

17. A catalyst is added while the piston is held steady.

18. A deuterium nucleus contains how many protons and neutrons?

| | protons | neutrons |
|---|---|---|
| (A) | 1 | 0 |
| (B) | 2 | 0 |
| (C) | 1 | 1 |
| (D) | 1 | 2 |
| (E) | 2 | 2 |

19. $\_MnO_4^- + \_H^+ + \_Fe^{2+} \rightarrow \_Fe^{3+} + \_Mn^{2+} + \_H_2O$

When the skeleton equation above is balanced and all coefficients reduced to their lowest whole-number terms, what is the coefficient for $H^+$?

(A) 4
(B) 6
(C) 8
(D) 9
(E) 10

20. Which of the following elements has the highest ionization energy?

(A) Ne
(B) Cl
(C) Si
(D) Na
(E) Li

21. Which of the following elements is the least electronegative?

(A) Al
(B) Br
(C) F
(D) Na
(E) Li

22. A sample of hydrogen gas is in a closed container, at 1.0 atmosphere pressure and 27°C. If the sample is heated to 127°C, the pressure will be approximately which of the following?

(A) 4.0 atm
(B) 1.3 atm
(C) .75 atm
(D) .67 atm
(E) .25 atm

23. Which of the following structures represents 1,1-dibromoethane?

(A) 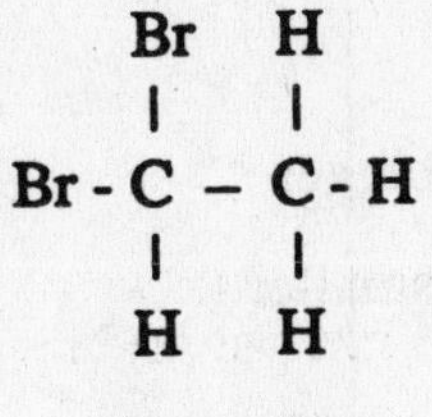
```
     Br   H
     |    |
Br - C  - C - H
     |    |
     H    H
```

(B) 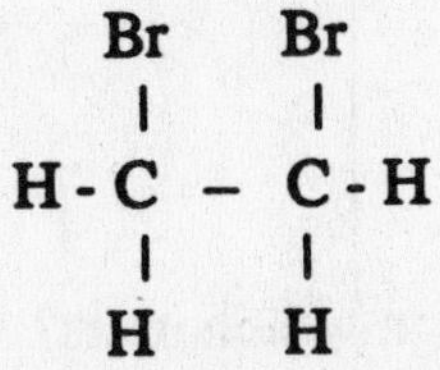
```
    Br    Br
    |     |
H - C  -  C - H
    |     |
    H     H
```

(C) $CH_3$ - $CH_2$ - Br - $CH_2$ - $CH_3$

(D) Br-C ≡ C-Br

(E) 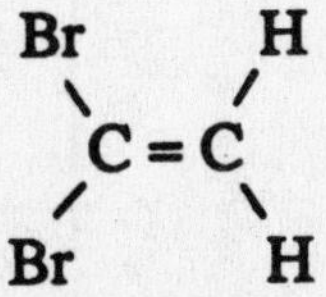
```
Br        H
  \      /
   C = C
  /      \
Br        H
```

**Questions 24 - 26**

$$A_{(aq)} + 2B_{(aq)} \rightarrow C_{(aq)}$$

The rate law for the reaction above is:

$$\text{rate} = K\,[B]^2$$

24. What is the order of the reaction with respect to B?

(A) 0
(B) 1
(C) 2
(D) 3
(E) 4

25. What will happen to the rate of the reaction if the amount of A in the solution is doubled?

(A) The rate will double
(B) The rate will halve
(C) The rate will be four times bigger
(D) The rate will be four times smaller
(E) No effect

26. The rate constant k is expressed in which units?

(A) l/mol·sec
(B) mol/l·sec
(C) $mol^2/l^2 \cdot sec$
(D) $sec^{-1}$
(E) $sec^{-2}$

27. What is the pH of a 1.0M solution of formic acid if the $K_a$ is 1.77 x $10^{-4}$?

(A) 14
(B) 11
(C) 4
(D) 2
(E) 1

28. The most probable oxidation number for the element with atomic number 53 is:

(A) -5
(B) -1
(C) +1
(D) +5
(E) +7

29. Amphoteric substances are best described as

(A) having the same number of protons and electrons but different numbers of neutrons.

(B) having the same composition but occurring in different molecular structures.

(C) being without definite shape.

(D) having both acid and base properties.

(E) having the same composition but occurring in different crystalline form.

30. Which of the following is <u>incorrect</u>?

(A) 1 liter = 1000 $cm^3$

(B) 1 meter = 100 cm

(C) 1 milliliter = 1 $cm^3$

(D) 1 liter = 1 $meter^3$

(E) 1 milliliter = $10^{-6}$ $meter^3$

31. Hydrolysis of sodium acetate yields

(A) a strong acid and a strong base.

(B) a weak acid and a weak base.

(C) a strong acid and a weak base.

(D) a weak acid and a strong base.

(E) none of the above.

32. The oxidation state of manganese in $KMnO_4$ is

(A) +1

(B) +2

(C) +3

(D) +4

(E) +7

33. Balancing the oxidation-reduction reaction

$$KMnO_4 + KCl + H_2SO_4 \rightarrow MnSO_4 + K_2SO_4 + H_2O + Cl_2$$

gives the coefficients

(A) 4,12,10,4,10,8,6

(B) 2,6,10,4,8,10,6

(C) 2,10,8,4,6,5,8

(D) 2,10,8,2,6,8,5

(E) 2,6,10,4,6,5,8

34. Which of the following is least likely to be found in nature in elemental form?

(A) hydrogen

(B) sodium

(C) carbon

(D) sulfur

(E) helium

35. Which of the following is true of an electrochemical cell?

(A) the cathode is the site of reduction

(B) the anode is negatively charged

(C) the cell voltage is independent of concentration

(D) charge is carried from one electrode to the other by metal atoms passing through the solution

(E) none of the above

36. How many grams of He occupy a volume of 11.2 liters at STP?

(A) 1g

(B) 2g

(C) 4g

(D) 6g

(E) 8g

37. What is the hydronium ion concentration of an HCl solution at pH 3?

(A) 0.001M
(B) 0.003M
(C) 0.1M
(D) 0.3M
(E) 3M

38. How many grams of Na are present in 30 grams of NaOH?

(A) 10g
(B) 15g
(C) 17g
(D) 20g
(E) 22g

39. What is the approximate density of fluorine gas at STP?

(A) 0.8g/l
(B) 1.0g/l
(C) 1.7g/l
(D) 2.0g/l
(E) 2.3g/l

**Answer questions 40 - 43 using the phase diagram below:**

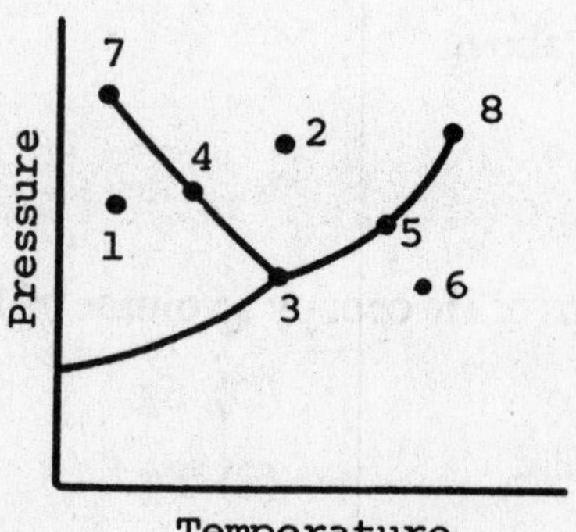

40. At which point can all three phases coexist at equilibrium?

(A) 3
(B) 4
(C) 5
(D) 7
(E) 8

41. At which point can only the solid phase exist?

(A) 1
(B) 2
(C) 3
(D) 5
(E) 6

42. Which is the critical point?

(A) 2
(B) 5
(C) 6
(D) 7
(E) 8

43. At which point can only the solid and liquid phases coexist?

(A) 3
(B) 4
(C) 5
(D) 1
(E) 2

44. How much water must be evaporated from 500 ml of 1M NaOH to make it 5M?

(A) 100 ml

(B) 200 ml

(C) 250 ml

(D) 300 ml

(E) 400 ml

45. Which of the following methods is best suited to separate a 500 ml sample of two miscible liquids whose boiling points differ by approximately 60°?

(A) distillation

(B) fractional distillation

(C) paper chromatography

(D) use of a separatory funnel

(E) none of the above

46. 20 liters of NO gas react with excess oxygen. How many liters of $NO_2$ gas are produced if the NO gas reacts completely?

(A) 5 liters

(B) 10 liters

(C) 20 liters

(D) 40 liters

(E) 50 liters

47. What is the molar concentration of $I^-$ in 1 liter of a saturated water solution of $PbI_2$ if the $K_{sp}$ of lead iodide is $1.4 \times 10^{-8}$?

(A) $3.0 \times 10^{-3}$

(B) $1.2 \times 10^{-4}$

(C) $5.9 \times 10^{-5}$

(D) $2.4 \times 10^{-3}$

(E) none of the above

48. What is the density of a diatomic gas whose gram-molecular weight is 80g?

(A) 1.9g/liter

(B) 2.8g/liter

(C) 3.6g/liter

(D) 4.3g/liter

(E) 5.0g/liter

49. How many liters of $H_2$ can be produced by the decomposition of 3 moles of $NH_3$?

(A) 4.5 liters

(B) 27 liters

(C) 67.2 liters

(D) 96 liters

(E) 101 liters

50. If a reaction causes an ion in solution to precipitate in elemental form, then this reaction is best described as being

(A) exothermic

(B) endothermic

(C) photochemical

(D) acid-base

(E) oxidation-reduction

51. Which of the following best describes the bonding usually found in Group IIIA elements?

(A) s

(B) sp

(C) $sp^2$

(D) $sp^3$

(E) $sp^3d^2$

52. The shell configuration $1s^2 2s^2 2p^6 3s^2 3p^4$ is that of

(A) O

(B) S

(C) $P^+$

(D) $Cl^-$

(E) Ar

53. The functional group $\mathrm{R{-}\overset{\overset{\displaystyle H}{|}}{C}{=}O}$ indicates a(an)

(A) alcohol

(B) aldehyde

(C) ester

(D) ether

(E) ketone

54. An atom, ${}^{B}_{A}X$, undergoes nuclear radioactive decay by emitting two beta particles and an alpha particle. The atom's new identity is given by

(A) ${}^{Y-2}_{A+2}X$

(B) ${}^{Y-2}_{A+2}Y$

(C) ${}^{B-2}_{A-4}X$

(D) ${}^{B-4}_{A}Y$

(E) ${}^{B-4}_{A}X$

55. A 0.5 molal solution could be prepared by dissolving 20g of NaOH in how much water?

(A) 0.5 liter
(B) 0.5 kg
(C) 1 liter
(D) 1 kg
(E) 2 liters

56. Which statement is true for a liquid/gas mixture in equilibrium?

(A) The equilibrium constant is dependent on temperature.
(B) The amount of the gas present at equilibrium is independent of pressure.
(C) All interchange between the liquid and gas phases has ceased.
(D) All of the above.
(E) None of the above.

57. The equilibrium expression $K_e = [CO_2]$ applies to which of the reactions below?

(A) $C + O_{2(g)} \leftrightarrow CO_{2(g)}$
(B) $CO_{(g)} + \frac{1}{2}O_{2(g)} \leftrightarrow CO_{2(g)}$
(C) $CaCO_{3(s)} \leftrightarrow CaO_{(s)} + CO_{2(g)}$
(D) $CO_{2(g)} \leftrightarrow C_{(s)} + O_{2(g)}$
(E) $CaO_{(s)} + CO_{2(g)} \leftrightarrow CaCO_{3(s)}$

(s)

58. Which of the following molecular geometries is typical of $sp^3$ bonding?

(A) tetrahedral
(B) trigonal planar
(C) linear
(D) trigonal bipyramidal
(E) octahedral

59. Which of the following molecular geometries is typical of $sp^2$ bonding?

(A) tetrahedral
(B) square planar
(C) linear
(D) trigonal planar
(E) trigonal bipyramidal

60. Which of the following molecular geometries is typical of sp bonding?

(A) tetrahedral
(B) trigonal planar
(C) linear
(D) trigonal bipyramidal
(E) octahedral

61.

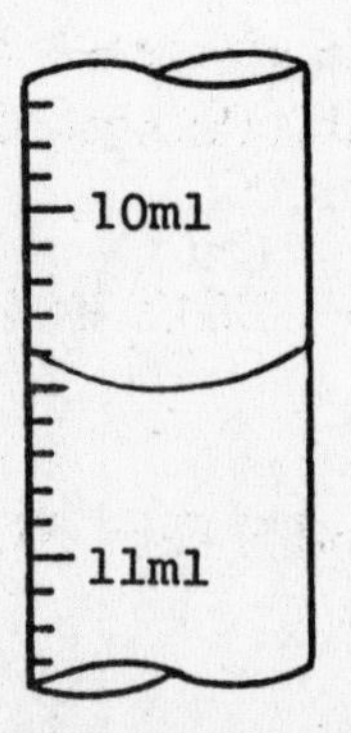


The buret above reads:

(A) 11.45 ml

(B) 10.54 ml

(C) 10.5 ml

(D) 10.40 ml

(E) 10.4 ml

62. The standard electrode in electrochemistry involves the oxidation or reduction of which of the following?

(A) gold

(B) platinum

(C) copper

(D) magnesium

(E) hydrogen

63. $Zn \rightarrow Zn^{2+} + 2e^- \quad \varepsilon° = +0.76V$

$Cr^{3+} + 3e^- \rightarrow Cr \quad \varepsilon° = -0.74V$

The anode in this cell is

(A) Zn

(B) Cr

(C) $Zn^{2+}$

(D) $Cr^{3+}$

(E) none of the above

64. The complete combustion of 1 mole of propane ($C_3H_8$) results in the liberation of 488.7 kcal. What is the heat of formation of propane?

The reaction is: $C_3H_{8(g)} + 5O_{2(g)} \rightarrow 3CO_2 + 4H_2O$

$\Delta H_f°$ (kcal/mole): $CO_2$ is -94.1 and $H_2O$ is -57.8

(A) +6.9 kcal/mole

(B) -19 kcal/mole

(C) -24.8 kcal/mole

(D) -63.6 kcal/mole

(E) -143.2 kcal/mole

65. The type of bonding in carbon tetrachloride ($CCl_4$) is

(A) ionic

(B) covalent

(C) polar covalent

(D) coordinate covalent

(E) hydrogen

66. A reaction that occurs only when heat is added is best described as:

(A) exothermic

(B) endothermic

(C) an equilibrium process

(D) spontaneous

(E) non-spontaneous

67. Neutral atoms of F (fluorine) have the same number of electrons as

(A) $B^{3-}$

(B) $N^{+}$

(C) $Ne^{-}$

(D) $Na^{-}$

(E) $Mg^{3+}$

68. An equilibrium reaction may be forced to completion by:

(A) adding a catalyst

(B) increasing the pressure

(C) increasing the temperature

(D) removing the products from the reaction mixture as they are formed

(E) decreasing the reactant concentration

69. Which of the following has the smallest mass?

(A) a hydrogen nucleus

(B) an alpha particle

(C) a neutron

(D) a helium nucleus

(E) a beta particle

70. The greatest reduction of the average kinetic energy of water molecules occurs when water is

(A) cooled as a solid.

(B) cooled as a liquid.

(C) cooled as a gas.

(D) converted from a gas to a liquid.

(E) converted from a liquid to a solid.

71. The formula $Cr(NH_3)_5SO_4Br$ represents

(A) 4 atoms
(B) 8 atoms
(C) 16 atoms
(D) 23 atoms
(E) 27 atoms

72. An equation for the electrolysis of water is

$$H_2O_{(l)} + 68.3\text{ kcal} \rightarrow H_{2(g)} + 1/2\, O_{2(g)}$$

How many liters of gaseous product are produced by the addition of 273.2 kcal of electrical energy to the reaction above?

(A) 22.4 liters
(B) 44.8 liters
(C) 134.4 liters
(D) 96.8 liters
(E) 119.2 liters

73. Which of the following structures has the IUPAC name propyl butanoate?

(A) $CH_3CH_2CH_2OCH_2CH_2CH_2CH_3$

(B) $CH_3CH_2COCH_2CH_2CH_2CH_3$

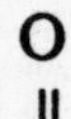

(C) $CH_3CH_2CH_2CH_2COCH_2CH_2CH_3$

O
||
(D) $CH_3CH_2CH_2COCH_2CH_2CH_3$

(E) none of the above

74. The systematic (IUPAC) name of this structure is

```
    H  H  H  OH  H  H
    |  |  |  |   |  |
H - C - C - C - C - C - C - H
    |  |  |  |   |  |
    H  H  H  H   H  H
```

(A) hexanol

(B) 3-hydroxyhexane

(C) 3-hexanol

(D) 4-hexanol

(E) isohexanol

75. The oxidation state of nitrogen in nitric acid ($HNO_3$) is

(A) +1

(B) +2

(C) +3

(D) +4

(E) +5

76. What is the volume of a sample of $1.50 \times 10^{23}$ atoms of helium at STP?

(A) 5.6 liters

(B) 11.2 liters

(C) 17.8 liters

(D) 22.4 liters

(E) none of the above

77. What is the molecular formula of a compound composed of 25.9% nitrogen and 74.1% oxygen?

(A) $NO_2$

(B) $N_2O$

(C) $N_2O_4$

(D) $N_2O_5$

(E) $N_3O_4$

78. How many milliliters of 5M NaOH are required to completely neutralize 2 liters of 3M $H_2SO_4$?

(A) 600 ml
(B) 900 ml
(C) 1200 ml
(D) 2400 ml
(E) 3333 ml

79. The valence shell of all alkaline earth metals can be designated:

(A) $Ns^1$
(B) $Ns^2$
(C) $Ns^2Np^1$
(D) $Np^1$
(E) $Nd^1$

80. The ionization energy of an element is

(A) a measure of its mass
(B) the energy required to remove an electron from the element in its gaseous state.
(C) the energy released by the element in forming an ionic bond.
(D) the energy released by the element upon receiving an additional electron.
(E) none of the above.

81. The radioactive decay of plutonium-238 ($^{238}_{94}Pu$) produces an alpha particle and a new atom. That new atom is:

(A) $^{234}_{92}Pu$
(B) $^{234}_{92}U$
(C) $^{234}_{92}Cm$
(D) $^{242}_{96}Pu$
(E) $^{242}_{96}Cm$

82. When a beta particle is emitted from an atomic nucleus,

(A) the atomic number increases by one.

(B) the atomic number decreases by one.

(C) the atomic mass increases by one.

(D) the atomic mass decreases by one.

(E) none of the above.

**Questions 83 - 85**

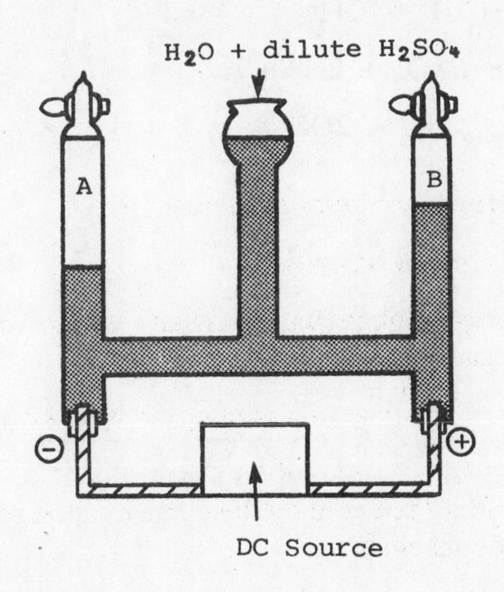


83. What type of reaction is occurring with this setup?

(A) single replacement

(B) double replacement

(C) electrolysis

(D) oxidation

(E) reduction

84. What will be found at point A?

(A) $O_2$

(B) $H_2$

(C) $H_2O$ vapor

(D) $H_2SO_4$ vapor

(E) free electrons

85. Which of the following is the anode reaction?

(A) $2H_2O \rightarrow H_3O^+ + OH^-$

(B) $H_2O \rightarrow 1/2\ O_2 + H_2$

(C) $H_2O \rightarrow H^+ + OH^-$

(D) $H_2O \rightarrow 1/2\ O_2 + 2H^+ + 2e^-$

(E) $2H_2O \xrightarrow{+2e^-} H_2 + 2OH^-$

# ADVANCED PLACEMENT CHEMISTRY EXAM I

## SECTION II

**Directions:** The percentages given for each individual part indicates the scoring for this section. Spend approximately 35 minutes on Parts A and B together and 40 minutes on Part C.

The methods and steps used to solve a problem must be shown clearly. Partial credit will be given if work is shown. Pay attention to significant figures.

### Part A
(30 Percent)

1) Consider a 0.05M solution of sodium acetate

(a) Demonstrate the relationship between the hydrolysis constant $K_h$ and $K_a$ (acid dissociation constant).

(b) What would the pH of the solution be if we added 0.06M acetic acid?

(c) What would happen if we added 0.001M hydrochloric acid?

(d) Determine the pH of the solution described in (c).

given: $K_a = 1.76 \times 10^{-5}$

## Part B

(30 Percent)

**Solve ONE of the two problems presented. Only the first problem answered will be scored.**

2) For the reaction A → B + C the data below were obtained:

Table 2

| t(in sec.) | 0 | 900 | 1800 |
|---|---|---|---|
| conc. of A at 25°C | 50.8 | 19.7 | 7.62 |
| conc. of A at 35°C | 80.0 | 26.7 | 8.9 |

(a) Determine the order of the reaction.

(b) Calculate K at 25°C.

(c) How long would it take for half of the original material to decompose (at 25°C)?

(d) Find the activation energy of the reaction.

3) Table 3

| | Standard Heat of Formation (kJ/mol) | Absolute entropies at 25°C and 1 atm. (J/mol x K) |
|---|---|---|
| C(s) | 0 | 5.69 |
| $CO_2$(g) | -393.5 | 213.6 |
| $H_2$(g) | 0 | 130.6 |
| $H_2O$(l) | -285.85 | 70 |
| $O_2$(g) | 0 | 205.0 |
| $CH_4$(g) | -74.9 | 186.2 |

Using table 3, determine:

(a) the heat of combustion of $CH_4$.

(b) the balanced equation for the formation of $CH_4$ from its elements.

(c) the standard entropy change ΔS° for the formation of $CH_4$ at 25°C.

(d) the standard free energy of formation ΔG° for $CH_4$ at 25°C.

## Part C
(40 Percent)

**Choose THREE of the following five topics. Only the first three answers will be scored. Your answers will be scored based on their accuracy, relevance of the details chosen, and appropriateness of the descriptive material used. Be as specific as possible and use illustrative examples and equations where helpful.**

4) (a) Define vapor pressure and graph its relationship with temperature.

(b) Define heat of combustion and explain what must be known in order to obtain its value, once the identity of the reactants and of the products in a combustion reaction are known.

(c) Explain the relationship between the freezing point of a solution and the molality of a solution.

(d) Define critical pressure and temperature.

5) Show the titration curves (pH x % neutralized acid) for the cases below.

(a) a strong acid and a strong base

(b) a strong acid and a weak base

6) Define covalent radii and describe how they vary on the periodic table.

7) What are the conditions under which gases deviate from ideal behavior?

8) Describe some of the most important chemical properties of halogens.

# ADVANCED PLACEMENT CHEMISTRY EXAM I

## ANSWER KEY

| | | | | | |
|---|---|---|---|---|---|
| 1. | A | 29. | D | 57. | C |
| 2. | C | 30. | D | 58. | A |
| 3. | E | 31. | D | 59. | D |
| 4. | D | 32. | E | 60. | C |
| 5. | D | 33. | D | 61. | B |
| 6. | E | 34. | B | 62. | E |
| 7. | B | 35. | A | 63. | A |
| 8. | E | 36. | B | 64. | C |
| 9. | C | 37. | A | 65. | C |
| 10. | E | 38. | C | 66. | B |
| 11. | C | 39. | C | 67. | E |
| 12. | A | 40. | A | 68. | D |
| 13. | B | 41. | A | 69. | E |
| 14. | B | 42. | E | 70. | D |
| 15. | B | 43. | B | 71. | E |
| 16. | E | 44. | E | 72. | C |
| 17. | C | 45. | A | 73. | D |
| 18. | C | 46. | C | 74. | C |
| 19. | C | 47. | A | 75. | E |
| 20. | A | 48. | C | 76. | A |
| 21. | D | 49. | E | 77. | D |
| 22. | B | 50. | E | 78. | D |
| 23. | A | 51. | C | 79. | B |
| 24. | C | 52. | B | 80. | B |
| 25. | E | 53. | B | 81. | B |
| 26. | A | 54. | E | 82. | A |
| 27. | D | 55. | D | 83. | C |
| 28. | B | 56. | A | 84. | B |
| | | | | 85. | D |

# ADVANCED PLACEMENT CHEMISTRY EXAM I

# DETAILED EXPLANATIONS OF ANSWERS

## SECTION I

1. (A)

Almost all transition elements form monatomic ions of $2^+$ charge, including Cr, Fe and Cn. Calcium also forms $2^+$ cations by losing both of its 4s valence electrons. The valence shell of potassium (K) contains only one electron ($4s^1$). After losing this electron, potassium achieves the noble gas configuration of argon. It is then very difficult to remove another electron, hence $K^{2+}$ ions are never found.

2. (C)

$Cr_2O_7^{2-}$ (dichromate ion) is a commonly encountered oxidizing agent.

3. (E)

All of these metals except copper are easily oxidized to their cationic form and hence are not found in their metallic state in nature.

4. (D)

Iron (Fe) is an important construction material.

5. (D)

Organic chemistry is the study of carbon-based molecules. Organic molecules can be defined as those which contain carbon.

6. (E)

Sodium borohydride ($NaBH_4$) and other boron compounds are used in the reduction of various organic molecules. Since boron is less electronegative than hydrogen, boron compounds can, in effect, add $H^-$ to other molecules.

7. (B)

In ionic bonding one or more electrons are completely donated by an electropositive element such as sodium to an electronegative element such as chlorine. The attraction between the resulting ions is electrostatic. A bond is considered ionic if the two bonding elements differ in electronegativity by more than 1.7.

8. (E)

In the sodium acetate molecule ($CH_3CO_2Na$) covalent, ionic, and $\pi$ bonding are represented. The C-H, C-C, and C-O bonds are all covalent. In addition, the C-O bonds have double bond character, which means that they use $\pi$ bonding in addition to the usual bonding. The acid group ($-CO_2^-$) and the sodium atom are bound ionically. The acid group takes on a negative charge and the sodium atom takes on a positive charge ($Na^+$). Hydrogen bonding exists in all aqueous solutions: among water molecules and between molecules of the solute and the water. Metallic bonding is not present in this model.

9. (C)

By comparison with other Group VI B compounds, we would expect water to boil at -100°C. Water does not boil until its temperature reaches +100°C because the water molecules must gain enough kinetic energy to overcome hydrogen bonding. This bonding consists of an attraction between the oxygen atom of one water molecule and the hydrogen atoms of neighboring molecules.

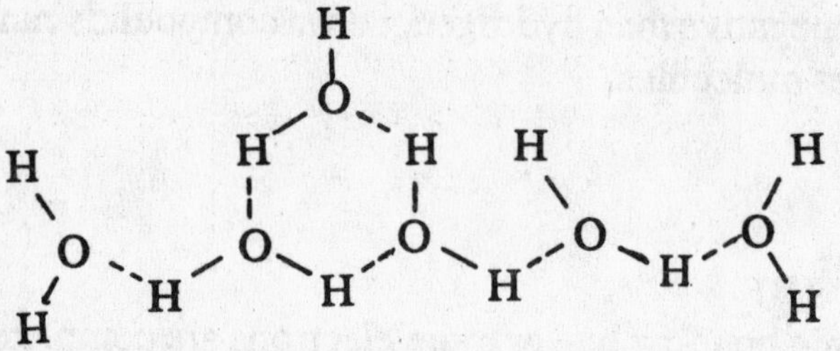

– covalent bonds
-- hydrogen bonds

This attraction is in part electrostatic, since the oxygen of a water molecule carries a partial negative charge and the hydrogens carry a partial positive charge.

10. (E)

pH is defined as:

$$pH \equiv -\log [H^+]$$

in which $[H^+]$ is the concentration of free protons. At a pH of 11.00, $[H^+] = 10^{-11}$. We can find $[OH^-]$ by substituting this value in the expression for the dissociation constant of water:

$$K_{dissociation} = 10^{-14} = [H^+]\,[OH^-]$$
$$10^{-14} = 10^{-11}\,[OH^-]$$
$$[OH^-] = 10^{-3}\text{ M}$$

11. (C)

At the equivalence point (the endpoint) [Acid] = [Base]. We added 10 ml of 1.0M HCl, or:

(.010 liter) (1.0 mol/liter) = .01 mol HCl

Thus the original 100 ml of basic solution contains .01 ml NaOH. The concentration is:

$$\frac{.01 \text{ mol}}{.100 \text{ liter}} = 0.1 \text{ M NaOH or } 1.0 \times 10^{-1} \text{ M NaOH}$$

We could also use the equation:

(concentration acid) (volume acid) =

(concentration base) (volume base)

12. (A)

Since $H_2SO_4$ is a diprotic acid, $[H^+] = 2 \times [H_2SO_4]$. Using the answer from questions #11,

$$(2 \times 1.0\text{M})(50.0 \text{ ml}) = [\text{NaOH}](100 \text{ ml})$$

$$\frac{(2 \times 1.0\text{M})(50.0 \text{ ml})}{(100 \text{ ml})} = 1.0\text{M NaOH}$$

13. (B)

We can substitute in the expression for $K_w$:

$$K_w = 10^{-14} = [H^+][OH^-]$$

$$10^{-14} = [2 \times 10^{-14}][OH^-]$$

$$[OH^-] = \frac{10^{-14}}{2 \times 10^{-14}} = 0.5\text{M or } 5.0 \times 10^{-1} \text{ M}$$

14. (B)

Le Chatelier's Principle applies to both questions 14 and 15. The principle states that a system perturbed from equilibrium will react in the opposite direction so as to restore equilibrium. In the first case, pushing the piston into the chamber increases the pressure of the reaction mixture. The system reacts so as to bring the pressure back down to a new equilibrium point. In the equilibrium expression for the reaction there are two moles of gas on the left side for every four moles on the right. In order to reduce the pressure, then, the equilibrium must shift to the left. Thus, the mole fraction of $N_2$ decreases.

15. (B)

In this problem the equilibrium is perturbed by an increase in the mole fraction of $H_2$. In order to correct this, the reaction must shift to the left until a new equilibrium point is reached. This results in a decrease in the amount of $N_2$ present.

16. (E)

Several competing effects may arise from heating the chamber. One important unknown is whether the reaction is exothermic or endothermic. For example, if the reaction is endothermic, we might write:

$$2NH_{3(g)} + \text{heat} \quad \leftrightarrow \quad N_{2(g)} + 3H_{2(g)}$$

Adding heat would drive the reaction forward in the same way that adding more $NH_3$ would. Since we do not know the heat of reaction we cannot predict the effect of heating the chamber.

17. (C)

Catalysts affect the rates of reactions by reducing their activation energies. They do <u>not</u> affect equilibrium constants or the energy changes (free energy, enthalpy, etc.) associated with reactions.

18. (C)

Deuterium is an isotope of hydrogen with one extra neutron. Hydrogen usually has only one proton and no neutrons in its nucleus, hence deuterium has one proton and one neutron. Tritium is an isotope with one proton and two neutrons.

19. (C)

The completely balanced reaction is:

$$MnO_4^- + 8H^+ + 5Fe^{2+} \rightarrow 5Fe^{3+} + Mn^{2+} + 4H_2O$$

20. (A)

Ionization energy (I.E.) is the energy required to remove an electron from an elemental atom. The lowest I.E. would occur for the alkali metals (Li, Na, etc.), since by removing an electron these atoms will have a highly stable noble gas configuration. The highest I.E. would occur for noble gases (e.g. Ne), since the noble gas configuration would be destroyed by removing an electron.

21. (D)

Electronegativity is a measure of the attraction of an atom for electrons. It increases as we move up and to the right on the periodic table (noble gases are not assigned electronegativity values). Sodium is the farthest down-left on the table of the choices given.

22. (B)

We can use the ideal gas law, PV=nRT. Since V and n are constant, an increase in T will cause a proportional increase in P. Remember that we must use the absolute (Kelvin) temperature scale. The temperature increases from 300°K to 400°K, thus the pressure will also increase by a third from 1.0 atm to 1.3 atm. We might express this in the equation below:

$$\frac{T_1}{T_2} = \frac{P_1}{P_2}$$

23. (A)

Dibromoethane is ethane, $CH_3CH_3$, with two hydrogen atoms replaced by bromine. The numbers "1,1" indicate that both bromine atoms are on the first carbon.

Thus

```
     Br  H
     |   |
Br - C - C - H
     |   |
     H   H
```

is the correct structure.

24. (C)

The rate law includes the concentration of B raised to the second power, thus it is second order in B. The rate law includes no factor dependent on A and so it is zero order in A. The rate law is second order overall.

25. (E)

Since A is not a factor in the rate law, changes in the concentration of A have no effect on the rate of reaction.

26. (A)

The rate is expressed in mol/l·sec and [B] in mol/l. We can write:

$$mol/l \cdot sec = K\,(mol/l)^2$$

which rearranges to:

$$K = l/mol \cdot sec$$

27. (D)

The acid dissociation constant, $K_a$, applies to this reaction:

$$HA_{(aq)} \leftrightarrow H^+_{(aq)} + A^-_{(aq)}$$

where

$$K_a = \frac{[H^+][A^-]}{[HA]}$$

We know that $K_a = 1.77 \times 10^{-4}$ and [HA] = 1.0M (minus a negligible amount).

$$1.77 \times 10^{-4} = \frac{[H^+][A^-]}{1.0}$$

$$1.77 \times 10^{-4} = [H^+][A^-]$$

since $[H^+] = [A^-]$, $\quad 1.77 \times 10^{-4} = [H^+]^2$

$$1.33 \times 10^{-2} = [H^+]$$

since $pH = -\log[H^+]$, $\quad pH = 1.88 \cong 2$

28. (B)

53 is one short of 54, the noble gas configuration of Xenon. Element 53 would be far more likely to gain one electron, forming an ion of charge -1, than to lose electrons. Element 53 is iodine, a halogen.

29. (D)

An amphoteric substance has both acid and base properties. Isotopes of an element have the same number of protons and electrons but different numbers of neutrons.[1] Isomers of a compound are indicated by the same molecular formulas but different structures.[2] Amorphous substances are designated as having no definite shape.[3] Allotropes of a substance have the same composition but have different crystalline structures.[4]

(1) (for example ${}^{12}_{6}C$ and ${}^{13}_{6}C$)

(2) (for example 1-propanol and 2-propanol)

(3) (for example the product obtained when liquid sulfur is poured in water)

(4) (for example rhombic and monoclinic sulfur)

30. (D)

The correct expression would be

$$1 \text{ liter} = 1000 \text{ cm}^3 \times \left[\frac{1\text{m}}{100 \text{ cm}}\right]^3$$

$$1 \text{ liter} = 1000 \text{ cm}^3 \times \frac{1\text{m}^3}{1 \times 10^6 \text{ cm}^3}$$

$$1 \text{ liter} = 1 \times 10^{-3} \text{ m}^3$$

31. (D)

The hydrolysis reaction for sodium acetate proceeds as follows:

$$CH_3\overset{O}{\overset{\parallel}{C}}ONa + H_2O \rightarrow CH_3\overset{O}{\overset{\parallel}{C}}OH + NaOH$$

The products of the reaction are a weak acid (acetic acid) and a strong base (sodium hydroxide).

32. (E)

The oxidation states of the atoms of a neutral compound must add up to equal zero. For $KMnO_4$, the oxidation state of K must be +1 since it is in Group IA and the oxidation state of O must be -2 since it is in Group VIA. Thus we have:

$$1 + Mn + 4(-2) = 0 \quad \text{and} \quad Mn = +7$$

33. (D)

The oxidation state of the manganese atom changes from +7 to +2 and the oxidation state of the chlorine atom changes from -1 to 0. Thus

$$Mn^{+7} + 5e^- \rightarrow Mn^{2+}$$

$$Cl^{-1} \rightarrow Cl^{\circ} + e^-$$

Multiplying the chlorine half cell by five balances the electrons gained with the electrons lost and gives

$$Mn^{+7} + 5Cl^{-1} \rightarrow 5Cl^{\circ} + Mn^{2+}$$

Using this in the original equation:

$$KMnO_4 + 5KCl + H_2SO_4 \rightarrow MnSO_4 + K_2SO_4 + H_2O + \tfrac{5}{2}\,Cl_2$$

Note that 5Cl is identical to $\frac{5}{2}$ $Cl_2$ stoichiometrically. The remaining coefficients are easily obtained. Since there are 6K on the left side we place a 3 in front of $K_2SO_4$. This gives us 4S on the right side so we place a 4 in front of $H_2SO_4$. We now have 20 O on the left side so we place a 4 in front of $H_2O$. This gives us the balanced equation:

$$KMnO_4 + 5KCl + 4H_2SO_4 \rightarrow MnSO_4 + 3K_2SO_4 + 4H_2O + \frac{5}{2} Cl_2$$

Multiplying by two to remove the fraction before $Cl_2$ gives us the coefficients 2, 10, 8, 2, 6, 8, 5.

34. (B)

Elemental sodium will not be found in nature since almost anything is capable of oxidizing it to $Na^+$. Elemental hydrogen is found as $H_2$ gas, carbon is found as diamond or graphite, sulfur is found as an elemental solid ($S_8$) and helium is a monatomic gas.

35. (A)

The cathode of an electrochemical cell is defined as the site of reduction and the anode is defined as the site of oxidation.

36. (B)

Since one mole of an ideal gas occupies a volume of 22.4 liters at STP we have

$$11.2 \text{ liters} \times \frac{1 \text{ mole of He}}{22.4 \text{ liters}} = 0.5 \text{ moles of He}$$

Converting to grams of He

$$0.5 \text{ moles of He} \times \frac{4\text{g of He}}{1 \text{ mole of He}} = 2\text{g of He}$$

37. (A)

The pH of a solution is defined as

$$pH = -\log [H^+]$$

in which $[H^+]$ is the hydronium ion concentration. Rearranging this equation, we obtain

$$[H^+] = 10^{-pH}$$

Substituting the given pH, we have $[H^+] = 10^{-3} = 0.001$.

38.(C)

We find that there are 23g of sodium (its atomic weight) in 40g of NaOH (its molecular weight). Using this as a conversion factor

$$30\text{g of NaOH} \times \frac{23\text{g of Na}}{40\text{g of NaOH}} = 17\text{g of Na}$$

Another method for determining this quantity is to convert 30g of NaOH to moles

$$30\text{g of NaOH} \times \frac{1\text{ mole of NaOH}}{40\text{g of NaOH}} = 0.75\text{ mole of NaOH}$$

and calculating the weight of sodium corresponding to 0.75 mole

$$0.75\text{ mole of Na} \times \frac{23\text{g of Na}}{1\text{ mole of Na}} = 17\text{g of Na}$$

since one mole of Na is present in one mole of NaOH.

39. (C)

Since one mole of an ideal gas occupies a volume of 22.4 liters at STP and fluorine is a diatomic gas (gases with the endings -gen or -ine are usually diatomic) we have

$$\frac{38\text{g of } F_2}{1 \text{ mole of } F_2} \times \frac{1 \text{ mole of } F_2}{22.4 \text{ liters}} = \frac{1.7\text{g}}{\text{liter}}$$

40. (A)

All three phases (solid, liquid and gas) may coexist at a single pressure/temperature combination known as the triple point. This point occurs at the intersection of the solid-liquid, solid-gas and liquid-gas equilibrium curves as illustrated by point 3.

41. (A)

Examining a labeled phase diagram we see that the solid phase can only exist at point 1.

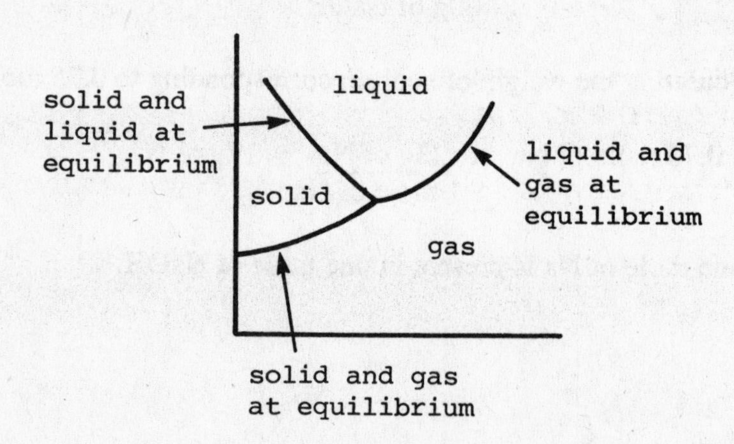

42. (E)

The critical point is the point above which a gas cannot change into a liquid. This means that a liquid cannot exist above this point, but at and below this point a liquid can exist. The temperature at the critical point is called the critical temperature and the pressure is called the critical pressure. The critical point in the phase diagram shown is the point 8, since above it a gas cannot be liquefied.

43. (B)

Referring to the phase diagram previously given we see that the solid and liquid phases coexist on the line that contains point 4.

44. (E)

Using the relationship

$$M_1V_1 = M_2V_2$$

where M is the molarity and V is the volume we have

$$(1M)\ (500\ ml) = (5M)\ (x)$$

and $x = 100$ ml

This value gives us the volume of the final solution obtained by evaporating 500 - 100 or 400 ml of water from the initial solution.

Note that the product MV gives the number of moles of solute

$$MV = \left(\frac{moles}{liter}\right) x\ liter = moles$$

45. (A)

A simple distillation is sufficient to separate this mixture into its components. Fractional distillation would be the method of choice if the boiling points of the components were similar. Paper chromatography separates compounds by virtue of their differing amounts of interaction with the solvent used and has no use in separating this mixture. In addition, only small quantities of testing material are reasonable for paper chromatography. A separatory funnel is used to extract a compound from one immiscible liquid to another and is not suited for this purpose.

46. (C)

The reaction in question is

$$NO + \tfrac{1}{2}O_2 \rightarrow NO_2$$

or using the given coefficients

$$20\ NO + 10\ O_2 \rightarrow 20\ NO_2$$

Note that the unit of the coefficients used is liters, not moles. This does not affect the calculation since moles and liters are directly related in the case of gases (1 mole of a gas occupies 22.4 liters at STP).

47. (A)

The solubility product of lead (II) iodide is given by

$$K_{sp} = [Pb^{2+}]\,[I^-]^2$$

where $[Pb^{2+}]$ and $[I^-]$ are the concentrations of lead ion and iodide in solution, respectively. We know that $K_{sp} = 1.4 \times 10^{-8}$ and that the concentration of iodide in solution is twice that of the lead ion from the dissociation:

$$PbI_2 \rightarrow Pb^{2+} + 2I^-$$

Setting $[Pb^{2+}] = x$, we know that $[I^-] = 2x$. Thus

$$K_{sp} = [Pb^{2+}][I^-]^2 = (x)(2x)^2 = 1.4 \times 10^{-8}$$

Solving for x gives

$$4x^3 = 1.4 \times 10^{-8}$$

and $$x = 1.5 \times 10^{-3}$$

Recalling that $[I^-] = 2x$ we have

$$[I^-] = 2(1.5 \times 10^{-3}) = 3 \times 10^{-3}$$

48. (C)

Recalling that density $= \rho = \frac{m}{v}$ gives

$$\rho = \frac{80 \text{ g/mole}}{22.4 \text{ l/mole}} = 3.6 \text{g/liter}$$

49. (E)

The equation for the reaction is:

$$2NH_3 \rightarrow N_2 + 3H_2$$

Multiplying each coefficient by $\frac{3}{2}$ gives

$$3NH_3 \rightarrow \frac{3}{2} N_2 + \frac{9}{2} H_2$$

Thus 3 moles of $NH_3$ decompose to produce 4.5 moles of $H_2$. Converting to liters (since 1 mole = 22.4 l):

$$4.5 \text{ moles } H_2 \times \frac{22.4 \text{ l of } H_2}{1 \text{ mole } H_2} \approx 101 \text{ l of } H_2$$

50. (E)

The transition from ionic to elemental form involves the transfer of electrons. Such reactions are called oxidation-reduction reactions.

51. (C)

Boron is one of the elements in Group IIIA and has the valence electron configuration $2s^22p^1$. The boron atom hybridizes from its ground state as follows

2p1 – – energy    2p11 – → 2p –

2x⇅    2s1    $sp^2$ 111

The hybrid orbitals are composed of one original s orbital and two original p orbitals. Thus the hybridization is $sp^2$.

52. (B)

The given electronic configuration indicates an atom with 16 electrons. Sulfur, atomic number 16, has 16 electrons. Oxygen, atomic number 8, has 8 electrons, $P^+$, atomic number 15, has 14 electrons, $Cl^-$, atomic number 17, has 18 electrons and Ar, atomic number 18, has 18 electrons.

53. (B)

The functional group $R - \overset{\overset{H}{|}}{C} = O$ represents an aldehyde. An alcohol is indicated by R - OH, an ester by $R - \overset{\overset{O}{\|}}{C}O - R$, an ether by R - O- R and a ketone by $R - \overset{\overset{O}{\|}}{C} - R$.

54. (E)

An alpha particle is a helium nucleus, $^4_2He$, and a beta particle is an electron. Beta particles are produced in the nucleus by the decomposition of a neutron into a proton and an electron. Thus if a beta particle is emitted, the atomic number increases by 1 while the mass number remains unchanged. For this radioactive decay, we have:

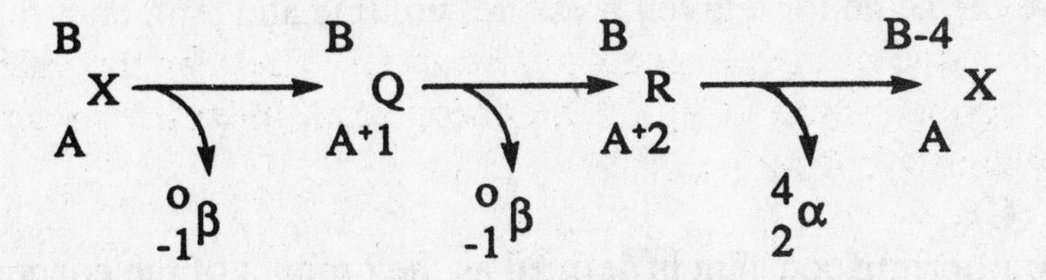


Different letters were chosen to represent the nuclei since the number of protons in the nucleus determines the atomic identity. The initial and final nuclei are represented by the same letter since they are isotopes of the same element.

55. (D)

The molality of a solution (m) is defined as the number of moles of solute dissolved in one kilogram of solvent. The number of moles of NaOH to be used is determined to be:

$$20\text{g of NaOH} \times \frac{1 \text{ mole of NaOH}}{40\text{g of NaOH}} = 0.5 \text{ mole of NaOH}$$

Thus:

$$0.5\text{m} = \frac{0.5 \text{ mole of NaOH}}{x \text{ kilograms of water}}$$

Rearranging

$$x = \frac{0.5}{0.5} = 1 \text{ kg of water}$$

56. (A)

The equilibrium constant is dependent only on temperature but the amount of each substance present at equilibrium is dependent on pressure, volume and temperature. There is still an interchange between the phases, but the same number of molecules leave and enter both phases so the equilibrium concentrations and equilibrium constant are the same for a given pressure, volume and temperature.

57. (C)

The equilibrium constant is defined as the product of the concentrations of the gaseous products raised to the power of their coefficients divided by the product of the gaseous reactant concentrations raised to the power of their coefficients. Only gaseous reactants and products are included in $K_e$ since the concentrations of liquids and solids participating in the reaction are assumed to be large (as compared to those of the gases) and relatively constant. The expressions of the equilibrium constants for the reactions given are:

(A) $K_e = \frac{[CO_2]}{[O_2]}$

(B) $K_e = \frac{[CO_2]}{[CO][O_2]^{1/2}}$

(C) $K_e = [CO_2]$

(D) $K_e = \frac{[O_2]}{[CO_2]}$

(E) $K_e = \frac{1}{[CO_2]}$

58. (A)

In $sp^3$ hybridization, four bonding orbitals are positioned as far from each other as possible. This requirement gives us a tetrahedral geometry:

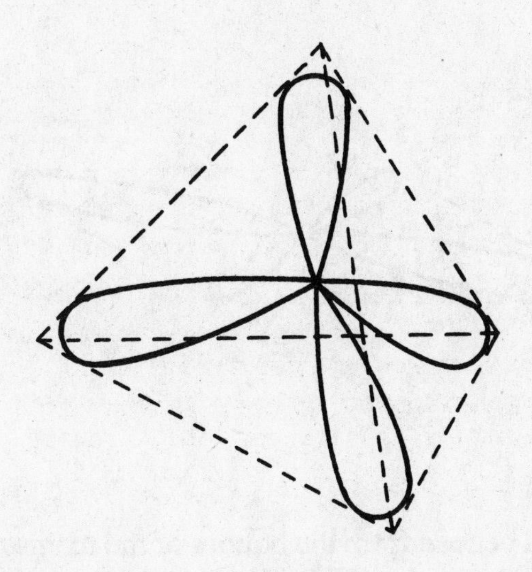

59. (D)

In $sp^2$ hybridization, three bonding orbitals are positioned as far from each other as possible. This requirement yields trigonal-planar geometry:

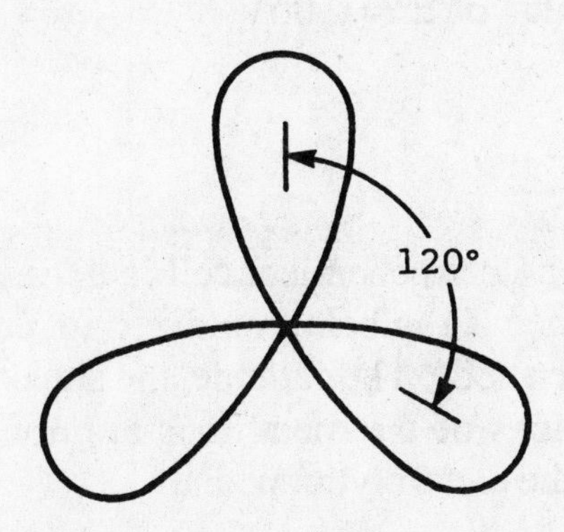

60. (C)

In sp hybridization we are once again separating the orbitals as much as possible. The two sp orbitals point in opposite directions along a line. Note that hybridization does not always result in maximum separation of orbitals. $dsp^2$ hybridization results in a square, planar geometry:

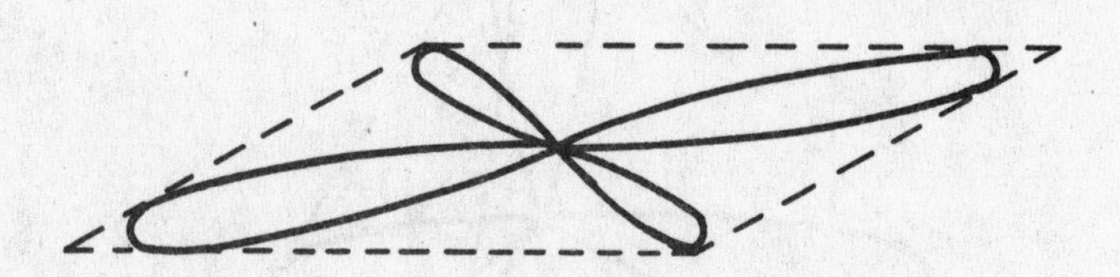

61. (B)

The buret should be read from the bottom of the meniscus (the curve in the liquid surface). Tenths of milliliters can be read off the scale and hundredths should be estimated. Notice that the volume measured increases going <u>down</u> the scale.

62. (E)

The hydrogen electrode has been chosen as the standard electrode with an assigned value of $E° = 0.00V$.

63. (A)

The anode of any electrochemical cell is defined to be the site of oxidation. Thus, since Zn is being oxidized to $Zn^{2+}$ in this cell it is determined to be the anode. The cathode, the site of reduction, is Cr in this cell. The solutions of the metal ions are not the anode nor the cathode but rather the electrolytic media.

64. (C)

Recalling that the change in enthalpy for a reaction is given by the sum of the heats of formation of the reactants subtracted from the sum of the heats of formation of the products, we have

$$\Delta H^\circ = \sum_{\text{products}} \Delta H_f - \sum_{\text{reactants}} \Delta H_f \quad \text{or}$$

$$-488.7 = 3(-94.1) + 4(-57.8) - (x)$$

$$x = 3(-94.1) + 4(57.8) + 488.7$$

$$x = -24.8 \text{ kcal/mole}$$

65. (C)

The type of bonding in an element or a compound may be determined by the difference in electronegativities of the atoms engaged in the bond. The bonding is covalent if the electronegativity difference is less than 0.5, polar covalent if from 0.5 to 1.7 and ionic if greater than 1.7. Coordinate covalent bonding is characterized by one atom supplying two electrons to the bond while the other supplies none. Hydrogen bonding occurs between protons of one molecule and highly electronegative (especially oxygen) atoms of another molecule. The electronegativities of carbon and chlorine are 2.5 and 3.0, respectively. The electronegativity difference is 0.5, making the bonding polar covalent.

66. (B)

An endothermic reaction is one in which heat may be considered one of the "reactants." An exothermic reaction releases heat upon formation of the products. An equilibrium reaction may be either exothermic or endothermic. The same holds true for spontaneity; spontaneity can only be determined if one also knows the entropy change ($\Delta S$) for the reaction.

67. (E)

Neutral fluorine atoms have 9 electrons as determined by their atomic number. Magnesium atoms have 12 electrons so $Mg^{3+}$ has 9 electrons. Boron has 5 electrons so $B^{3-}$ has 8 electrons (the same as oxygen). Nitrogen has 7 electrons so $N^{+}$ has 6 electrons (the same as carbon). Neon has 10 electrons so $Ne^{-}$ has 11 electrons (the same as sodium). Sodium has 11 electrons so $Na^{-}$ has 12 electrons (the same as magnesium).

68. (D)

Le Chatelier's Principle may be used to predict equilibrium reactions: If a stress is placed on a system in equilibrium, the equilibrium shifts so as to counteract that stress. Hence, increasing the reactant concentration favors formation of the products while decreasing the reactant concentration favors formation of the reactants. The same holds true for altering the product concentrations. Increasing the temperature favors the reaction that absorbs heat while decreasing the temperature favors the reaction that releases heat. Increasing the pressure favors the reaction that decreases the volume of a closed system while decreasing the pressure favors the reaction resulting in an increased volume (moles of gaseous product produced are the only things counted since liquids and solids occupy a relatively small volume in comparison). However, temperature and pressure dependencies cannot be inferred from this question. The addition of a catalyst alters the reaction rate but not the position of equilibrium. The only way completion can be obtained is that we remove the products as they are formed. Now the state of the reaction becomes nonequilibrium, but it tries to come in equilibrium state once again. This leads to formation of more products which in turn leads to completion of the given reaction.

69. (E)

A beta particle is a fast electron of mass $9.11 \times 10^{-28}$g while a proton and a neutron both have a mass of $1.67 \times 10^{-24}$g. A hydrogen nucleus is a proton, and an alpha particle is a helium nucleus (two protons and two neutrons). Thus the electron (beta particle) has the smallest mass of the choices given.

70. (D)

Molecules in the gaseous state have the greatest kinetic energy. The difference in energy between the liquid and gas phases is greater than the difference in energy between the solid and liquid phases. This may readily be seen by the energy changes occurring in water; the heat of fusion of water is 80 calories/gram while the heat of vaporization is 540 calories/gram.

71. (E)

$Cr(NH_3)_5SO_4Br$ represents 27 atoms. They are: 1 x Cr, 5 x N, 5 x 3 x H, 1 x S, 4 x O and 1 x Br.

72. (C)

1.5 moles of gaseous product (one mole of $H_2$ and 0.5 mole of $O_2$) are produced for every 68.3 kcal of energy added to water. Since 4 x 68.3 kcal = 273.2 kcal, 4 x 1.5 or 6 moles of gaseous product are formed. Since molar volume is given as 22.4 liters at STP, 6 x 22.4 liters = 134.4 liters of gaseous product are evolved.

73. (D)

The first word of the ester name is the name of the group attached to oxygen (propyl in our case). The second word is the name of the parent carboxylic acid with the suffix -ic replaced by -ate (butanoate in our case). Thus, we are looking for an ester with a propyl group attached to the oxygen of butanoic acid. This structure is

$$CH_3CH_2CH_2\overset{\overset{\displaystyle O}{||}}{C}OCH_2CH_2CH_3$$

$CH_3CH_2CH_2OCH_2CH_2CH_2CH_3$ has the assigned name butyl propyl ether

$$\overset{\displaystyle O}{\|}$$
$CH_3CH_2COCH_2CH_2CH_2CH_3$ has the assigned name butyl propanoate

$$\overset{\displaystyle O}{\|}$$
$CH_3CH_2CH_2CH_2COCH_2CH_2CH_3$ has the assigned name propyl pentanoate

74. (C)

Alcohols are named by replacing the -e of the corresponding hydrocarbon name by the suffix -ol. The position of the hydroxy substituent is numbered from the shorter end of the chain. Thus the structure is named 3-hexanol. It is a hexanol because the parent hydrocarbon has six carbons and the prefix 3- (not 4-) is used to indicate the location of the hydroxy group on the third carbon.

75. (E)

The oxidation states of the element comprising a neutral compound must have a sum of zero. Thus, nitrogen in $HNO_3$ has an oxidation state of +5, since hydrogen and oxygen generally have oxidation states of +1 and -2, respectively.

i.e. $(1) + (x) + (-6) = 0 \quad x = 5$

76. (A)

A mole is defined to be $6.02 \times 10^{23}$ atoms, molecules, ions, particles, etc. Thus, $1.5 \times 10^{23}$ atoms represents 0.25 mole. Recalling that the molar volume of any ideal gas at STP is 22.4 liters, we may calculate the volume of $1.5 \times 10^{23}$ atoms to be

$$(0.25)\,(22.4) = 5.6 \text{ liters}$$

77. (D)

A 100g sample of this gas contains 25.9g of nitrogen and 74.1g of oxygen. Dividing each of these weights by their respective atomic weights gives us the molar ratio of N to O for the gas. This gives

$$N_{\frac{25.9}{14}} \quad O_{\frac{74.1}{16}} \quad = \quad N_{1.85}\,O_{4.63}$$

Dividing both subscripts by the smallest subscript gives

$$N_{\frac{1.85}{1.85}} \quad O_{\frac{4.63}{1.85}} \quad = \quad N_1O_{2.5}$$

Doubling both subscripts so as to have whole numbers gives us $N_2O_5$.

78. (D)

We must first remember that the concentration of hydronium ions ($H_3O^+$) in 3M $H_2SO_4$ is 6M, since sulfuric acid is diprotic (it releases two $H^+$s). In two liters of 3M $H_2SO_4$, there are (2 x 3 mol/liter x 2 liters) = 12 moles of $H_3O^+$. To neutralize this, we need 12 moles of $OH^-$. Thus the volume of 5M NaOH required is

$$\frac{12\text{ mol}}{5\text{ mol/liter}} = 2.4\text{ liters} = 2400\text{ ml}$$

79. (B)

The alkaline earth metals are those in the second column of the periodic table. They have two s electrons in their valence shell, designated $Ns^1$ and $Ns^2$, in which N is the principal quantum number. (N=2 for Be, N=3 for Mg, etc.).

80. (B)

The ionization energy is defined as the energy required to remove the most loosely bound electron from an element in the gaseous state. The energy released by an element in forming an ionic solid with another element is the lattice energy of that ionic compound. The electronegativity of an element gives the relative strength with which the atoms of that element attract valence electrons in a chemical bond.

81. (B)

Plutonium-238 has a mass of 238 and an atomic number of 94. The atomic mass tells us the number of protons and neutrons in the nucleus while the atomic number tells us the number of protons. An alpha particle ($^4_2\alpha$) is a helium nucleus composed of 2 neutrons and 2 protons (atomic mass of 4). Hence, upon emitting an alpha particle, the atomic number decreases by 2 and the atomic mass decreases by 4. This gives us $^{234}_{92}X$. Examining the periodic table we find that element 92 is uranium. Thus our new atom is $^{234}_{92}U$. $^{234}_{92}Pu$ and $^{234}_{92}Cm$ are impossible since the atomic number of plutonium is 94 and that of curium is 96. $^{242}_{96}Pu$ and $^{242}_{96}Cm$ are impossible since these nuclei could only be produced by fusion of $^{238}_{94}Pu$ with an alpha particle. In addition, $^{242}_{96}Pu$ is incorrectly named.

The reaction (decay) is: $^{238}_{94}Pu \rightarrow {}^{234}_{92}U + {}^{4}_{2}\alpha$

82. (A)

A beta particle is a high-energy electron. In beta decay, a neutron in the nucleus emits an electron and becomes a proton. Thus, the mass number of the nucleus remains the same since there is no change in the number of nucleons (neutrons or protons). The nuclear charge, however, increases by one with the addition of a proton. The atomic number thus increases by one.

83. (C)

The electrical source indicates that an electrolysis is occurring. Both oxidation and reduction are taking place in the reaction.

84. (B)

The reaction in question is

$$2H_2O \rightarrow 2H_2 + O_2$$

where oxygen is being oxidized at the anode

$$O^{2-} \rightarrow \frac{1}{2} O_2 + 2e^-$$

and hydrogen is being reduced at the cathode

$$2H^+ + 2e^- \rightarrow H_2$$

The side of the apparatus labelled A is the cathode (reduction site) since it is connected to the negative terminal of the source. Therefore $H_2$ will be found at A.

85. (D)

The oxidation reaction is

$$H_2O \rightarrow \frac{1}{2} O_2 + 2H^+ + 2e^-$$

Reaction E is a reduction while reactions A and C are not electrochemical in nature. Reaction B is the overall reaction for the cell.

# SECTION II

## PART A

1. a) The hydrolysis of a salt that results from the combination of a strong base and weak acid is represented below:

$A^- + H_2O \leftrightarrow HA + OH^-$

By definition the hydrolysis constant Kh is:

$$Kh = \frac{[HA]\,[OH^-]}{[A^-]}$$

If we multiply by $\frac{[H^+]}{[H^+]}$ we will obtain:

$$Kh = \frac{[HA]\,[OH^-]}{[A^-]} \times \frac{[H^+]}{[H^+]} = \frac{[H^+]\,[HA]\,[OH^-]}{[H^+]\,[A^-]} \times \frac{[HA]^{-1}}{[HA]^{-1}}$$

$$= \frac{[OH^-]\,[H^+]}{\frac{[H^+]\,[A^-]}{[HA]}}$$

Notice $[OH^-]\,[H^+]$ is Kw (ionic constant of water) and $\frac{[H^+]\,[A^-]}{[HA]}$ is Ka.

By substituting the values above, it is possible to conclude:

$$Kh = \frac{Kw}{Ka}$$

b) $Ka = \frac{[H^+]\,[Ac^-]}{[HAc]}$

$[Ac^-]$ = concentration of salt present = 0.05M
[HAc] = 0.06M

$$1.76 \times 10^{-5} = \frac{[H^+]\,[0.05]}{[0.06]}$$

$$[H^+] = 1.76 \times 10^{-5} \times \frac{0.06}{0.05} = 2.11 \times 10^{-5}$$

$$pH = -\log [H^+] = -\log [2.11 \times 10^{-5}] = 4.67$$

c) All H+ from HCl would react with $Ac^-$ present to produce HAc, which is a weak acid. Therefore, $Ac^-$ would decrease and [HAc] would increase. In other words, $[H^+]$ would not change. This situation is an example of a buffer solution, in which the pH is kept constant.

d) [HAc] = 0.06 + 0.001 = 0.061
[Ac] = 0.05 - 0.001 = 0.049

$$Ka = \frac{[H^+]\,[Ac^-]}{[HAc]}$$

$$1.76 \times 10^{-5} = \frac{[H^+]\,[Ac^-]}{[HAc]}$$

$$1.76 \times 10^{-5} = \frac{[H^+] \times 0.049}{0.061}$$

$$[H^+] = 1.41 \times 10^{-5}$$

$$pH = -\log 1.41 \times 10^{-5} = 5 - 0.15 = 4.85$$

As predicted in (c), the pH does not change substantially.

## PART B

2. a) To find the order of the reaction H is necessary to assume a certain order and see whether the reaction is of the order that is assumed.

Zero order:

$$\frac{-dc_A}{dt} = K \qquad\qquad \frac{-dc_A}{dt} = K\,dt$$

$$-\int_{c_{A_O}}^{c_A} dc_A = K \int_{t_o}^{t} dt$$

$$K = \frac{c_{AO} - c_A}{t - t_o}$$

Let's find the reaction rate constant at two different intervals of time and call them $K_1$ and $K_2$.

$$K_1 = \frac{50.8 - 19.7}{900 - 0} = \frac{31.1}{900} = 0.0345$$

$$K_2 = \frac{19.7 - 7.62}{1800 - 900} = \frac{12.08}{900} = 0.034$$

Since K1 ≠ K2 we can conclude that the reaction is not a zero-order reaction.

First order:

$$\frac{dc_a}{dt} = -Kc_A \qquad\qquad \frac{dc_A}{c_A} = -Kdt$$

$$\int_{cA_o}^{c_A} \frac{dc_A}{c_A} \qquad \int_{t_o}^{t} -Kdt$$

$$\frac{\ln \frac{cA_o}{c_A}}{t-t_o} = K$$

Again calculating the rate constant at different intervals, we obtain:

$$K = \frac{\ln \frac{c_A}{c_{Ao}}}{t - t_o} \qquad K = \frac{\ln \frac{c_{Ao}}{c_A}}{t - t_o}$$

$$K_1 = \frac{\ln \frac{50.8}{19.7}}{900-0} = \frac{\ln 2.57}{900}$$

$$K_2 = \frac{\ln \frac{19.7}{7.62}}{900-0} = \frac{\ln 2.57}{900}$$

Since $K_1 = K_2$, this reaction is a first-order reaction.

b) $K = \frac{\ln 2.57}{900} \qquad 1 \times 10^{-3}$

c) If $c_A = 1/2\, c_{Ao}$ then $(t_o - t)$ is the time that we are looking for.

$$\frac{\ln \frac{c_{Ao}}{c_A}}{t - t_o} = K$$

$$t - t_o = \frac{1}{K} \ln \frac{c_{Ao}}{c_A}$$

For $c_A = \frac{1}{2} c_{Ao}$

$$t - t_o = \frac{1}{K} \ln \frac{c_{Ao}}{1/2\, c_{Ao}}$$

$$t - t_0 = \frac{1}{K} \ln 2$$

As stated in (a) and (b), $K = \ln 2.57 \cong 1 \times 10^{-3}$

$$t - t_o = \frac{\ln 2}{1 \times 10^{-3}}$$

$$t - t_o = 693 \text{ per}$$

d) In order to determine the activation energy Ea, it is necessary to know the rate constant at two different temperatures.

$$K_2 \text{ (at } 35°C) = \frac{\ln 80/26.7}{900} = \frac{\ln 3}{900} = 1.22 \times 10^{-3}$$

$$K_1 \text{ (at } 25°C) = 1 \times 10^{-3}$$

K, Ea, and temperature are related by the Arrhenius equation:

$$K - Ae^{-(Ea/RT)}$$

in which A is a proportionality constant and R is the gas constant.

$$K_1 = Ae^{-(Ea/Rt1)}$$

$$K_2 = Ae^{-(Ea/Rt2)}$$

$$\frac{K_1}{K_2} = \frac{Ae^{-(Ea/RT1)}}{Ae^{-(Ea/Rt1)}}$$

The proportionality constants cancel out, and the natural logwithms of both sides of the equation are determined:

$$\ln \left(\frac{K_1}{K_2}\right) = \ln \frac{e^{-(Ea/RT1)}}{e^{-(Ea/RT2)}} = \ln e^{-(Ea/RT1)} - \ln e^{-(Ea/RT2)}$$

$$\ln \left(\frac{K_1}{K_2}\right) = \frac{Ea}{RT_2} - \frac{Ea}{RT_1} = \frac{Ea}{R} \left(\frac{1}{T_2} - \frac{1}{T_1}\right)$$

Since ln x = 2.303 log x, the following can be done:

$$2.303 \log (K_1/K_2)\, R = Ea$$
$$\frac{1}{T_2} - \frac{1}{T_1}$$

T1 = 25 + 273 = 298K

T2 = 35+ 273 = 308K

R = 8.324 J.mol-1 + K-1

Substituting these values for the appropriate variables in the expression for Ea, we get:

$$Ea = \frac{2.303 \log \frac{(1 \times 10\text{-}3)}{(1.22 \times 10\text{-}3)}\ (8.314\ J\ mol\text{-}1\ K\text{-}1)}{\left(\frac{1}{308K} - \frac{1}{298K}\right)}$$

Ea = 11.52 kJ/mol or 2741.35 cal/mol.

3. The equation for the combustion of $CH_4$ is as follows:

$$CH_4 + 2O_2 \rightarrow CO_2 + 2H_2O$$

a) $\Lambda H_{comb} = \Sigma H_{prod} - \Sigma H_{react} = (H_{CO_2} + H_{H_2O}) - (H_{CH_4} + H_{O_2})$

$= -393.5 + 2(-285.5) - (-74.9) = -889.6$ KJ/mol x K

b) $CH_4 \rightarrow 1C(S) + 2H_2\ (8) \rightarrow CH_4\ (8)$

c) $\Lambda S = \Sigma S_{prod} - \Sigma S_{react} = (S_{CH_4}) - (S_C + S_{H_2})$

$\Lambda S = (1)(186.2J/mol \times K) - (1)\ (5.69J/mol \times K) - (2)(130.6J/mol \times K) = 80.7\ J/mol \times K$

d) $\Delta G = \Delta H - T\Delta S$

$= -889.6 kJ/mol - (298K) (.0807 KJ/mol \times K)$
$= 913.6 kJ/mol$

## PART C

4. a) The pressure exerted by a vapor in equilibrium with its liquid phase is called the vapor pressure of the liquid. Vapor pressure increases with temperature. If the logarithm of vapor pressure versus the reciprocal of temperature is plotted, a straight line should be obtained.

b) The heat of combustion of a substance is defined as the change in the enthalpy (quantity of heat liberated) of a system due to the combustion (oxidation) of 1 mole of the substance.

According to Hess' law the heat of combustion depends only on the final and initial states of the reaction.

Given reactants and products, it is possible to calculate the heat of combustion by obtaining the difference between the enthalpy values of reactants and products.

c) According to Raoult's law, in a dilute solution, a change in the freezing point of a solution is directly proportional to the molality of the solution.

d) The critical temperature of a substance is the temperature above which the substance can exist only as a gas, regardless of how much pressure is applied to it. The critical pressure of a substance is the pressure that must be applied to a substance at its critical temperature in order for the substance's liquid phase to be in equilibrium with its gas phase.

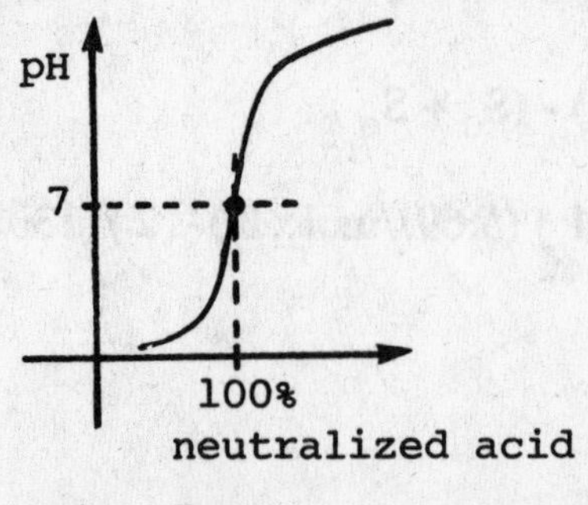

Fig. 5-a

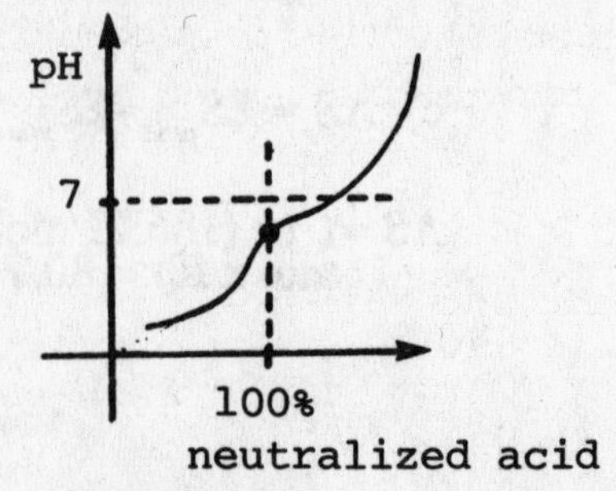


Fig. 5-b

5. In figure 5-a, the pH changes suddenly to 7 when all acid has been consumed. In 5-b the pH is less than 7 when all acid has been consumed because the hydrolisis of the resulting salt gives an acid solution.

6. Covalent radii are defined as half the distance between two nuclei of identical atoms when they are joined by a single covalent bond.

In general the covalent radius decreases from left to right across a period. This change can be explained by increasing nuclear charges of the atoms within a period. Proceeding down a group, successive elements have larger covalent vadii as a result of greater numbers of electron shells.

7. The ideal behavior equations are a good approximation in situations in which intermolecular attraction is not significant. This happens at low pressures (intermolecular distances are large) and high temperatures (at which the great deal of kinetic energy that is present prevents the forces of intermolecular attraction from operating). But at high pressures and low temperatures, at which intermolecular attraction, volume and the sizes of molecules become significant, then deviation from ideal behavior occurs.

8. 1) Their valence shells have the electron configuration $ns_2np_5$.

2) They generally form one single covalent bond with less electronegative non-metals, although they can form several bonds to more electronegative non-metals.

3) In a compound with a metal, a halogen usually accepts one electron to form a stable anion.

4) They form diatonic molecules in which the atoms are bound together.

# THE ADVANCED PLACEMENT EXAMINATION IN

# CHEMISTRY

# TEST II

# THE ADVANCED PLACEMENT EXAMINATION IN
# CHEMISTRY

## ANSWER SHEET

1. (A) (B) (C) (D) (E)
2. (A) (B) (C) (D) (E)
3. (A) (B) (C) (D) (E)
4. (A) (B) (C) (D) (E)
5. (A) (B) (C) (D) (E)
6. (A) (B) (C) (D) (E)
7. (A) (B) (C) (D) (E)
8. (A) (B) (C) (D) (E)
9. (A) (B) (C) (D) (E)
10. (A) (B) (C) (D) (E)
11. (A) (B) (C) (D) (E)
12. (A) (B) (C) (D) (E)
13. (A) (B) (C) (D) (E)
14. (A) (B) (C) (D) (E)
15. (A) (B) (C) (D) (E)
16. (A) (B) (C) (D) (E)
17. (A) (B) (C) (D) (E)
18. (A) (B) (C) (D) (E)
19. (A) (B) (C) (D) (E)
20. (A) (B) (C) (D) (E)
21. (A) (B) (C) (D) (E)
22. (A) (B) (C) (D) (E)
23. (A) (B) (C) (D) (E)
24. (A) (B) (C) (D) (E)
25. (A) (B) (C) (D) (E)
26. (A) (B) (C) (D) (E)
27. (A) (B) (C) (D) (E)
28. (A) (B) (C) (D) (E)
29. (A) (B) (C) (D) (E)
30. (A) (B) (C) (D) (E)
31. (A) (B) (C) (D) (E)
32. (A) (B) (C) (D) (E)
33. (A) (B) (C) (D) (E)
34. (A) (B) (C) (D) (E)
35. (A) (B) (C) (D) (E)
36. (A) (B) (C) (D) (E)
37. (A) (B) (C) (D) (E)
38. (A) (B) (C) (D) (E)
39. (A) (B) (C) (D) (E)
40. (A) (B) (C) (D) (E)
41. (A) (B) (C) (D) (E)
42. (A) (B) (C) (D) (E)
43. (A) (B) (C) (D) (E)
44. (A) (B) (C) (D) (E)
45. (A) (B) (C) (D) (E)
46. (A) (B) (C) (D) (E)
47. (A) (B) (C) (D) (E)
48. (A) (B) (C) (D) (E)
49. (A) (B) (C) (D) (E)
50. (A) (B) (C) (D) (E)
51. (A) (B) (C) (D) (E)
52. (A) (B) (C) (D) (E)
53. (A) (B) (C) (D) (E)
54. (A) (B) (C) (D) (E)
55. (A) (B) (C) (D) (E)
56. (A) (B) (C) (D) (E)
57. (A) (B) (C) (D) (E)
58. (A) (B) (C) (D) (E)
59. (A) (B) (C) (D) (E)
60. (A) (B) (C) (D) (E)
61. (A) (B) (C) (D) (E)
62. (A) (B) (C) (D) (E)
63. (A) (B) (C) (D) (E)
64. (A) (B) (C) (D) (E)
65. (A) (B) (C) (D) (E)
66. (A) (B) (C) (D) (E)
67. (A) (B) (C) (D) (E)
68. (A) (B) (C) (D) (E)
69. (A) (B) (C) (D) (E)
70. (A) (B) (C) (D) (E)
71. (A) (B) (C) (D) (E)
72. (A) (B) (C) (D) (E)
73. (A) (B) (C) (D) (E)
74. (A) (B) (C) (D) (E)
75. (A) (B) (C) (D) (E)
76. (A) (B) (C) (D) (E)
77. (A) (B) (C) (D) (E)
78. (A) (B) (C) (D) (E)
79. (A) (B) (C) (D) (E)
80. (A) (B) (C) (D) (E)
81. (A) (B) (C) (D) (E)
82. (A) (B) (C) (D) (E)
83. (A) (B) (C) (D) (E)
84. (A) (B) (C) (D) (E)
85. (A) (B) (C) (D) (E)

# ADVANCED PLACEMENT CHEMISTRY EXAM II

## SECTION I

Time: 1 Hour 45 Minutes
85 Questions

**Note**: For all questions referring to solutions, assume that the solvent is water unless otherwise stated.

**Directions**: Each set of lettered choices below refers to the numbered statements immediately following it. Select the one lettered choice that best fits each statement. A choice may be used once, more than once, or not at all in each set.

### Questions 1 – 3

(A) 7.02
(B) 7.13
(C) 7.41
(D) 7.66
(E) 7.88

1. $H^+$ ion concentration in solution= .0000000389 grams/liter

2. $H^+$ ion concentration in solution= .0000000221 grams/liter

3. $H^+$ ion concentration in solution= .0000000751 grams/liter

**Questions 4 - 5**

(A) Molecules are moving least rapidly and are closest together.

(B) Water is in this state at 12 degrees Centigrade.

(C) Mercury is in this state at room temperature.

(D) Molecules are moving most rapidly.

(E) Molecules maintain a definite volume but shape depends upon the contours of the container holding them.

4. Gas

5. Solid

**Questions 6 – 9**

(A) sodium chlorate

(B) sodium chloride

(C) sodium chlorite

(D) sodium hypochlorite

(E) sodium perchlorate

6. NaC1

7. NaC10

8. $NaC10_2$

9. $NaC10_3$

**Questions 10 – 13**

(A) alcohols

(B) alkanes

(C) alkenes

(D) alkynes

(E) amines

10. $C_2H_5OH$

11. $C_2H_6$

12. $C_2H_2$

13. $C_2H_4$

14. Suppose that the reaction rate of an inorganic reaction mixture at 35 degrees Centigrade is double the reaction rate at an earlier temperature setting. All other environmental factors are held constant. This earlier temperature was most likely:

(A) 0°

(B) 10°

(C) 25°

(D) 40°

(E) 45°

15. $HCOO^- + H_2O \rightarrow HCOOH + OH^-$

$$Rate = K [HCOOH] [OH^-]$$

What is the order of this reaction?

(A) 1

(B) 2

(C) 4

(D) 5

(E) 6

16. Select the indicator that changes color at a pH of 1.

(A) congo red
(B) malachite green
(C) methyl violet
(D) phenophthalein
(E) thymol blue

17. Select the characteristic that is not a standard condition for comparing gas volumes:

(A) Pressure -- 31 inches
(B) Pressure -- 760 torr
(C) Temperature -- 0 degrees Centigrade
(D) Temperature -- 32 degrees Fahrenheit
(E) Temperature -- 273 degrees Kelvin

18. A recorded Fahrenheit value in lab is 122 degrees. Its corresponding Kelvin temperature value is:

(A) 32
(B) 52
(C) 152
(D) 273
(E) 323

19. Consider the balanced equation:

$$2H_2 + O_2 \rightarrow 2H_2O$$

What volume of oxygen in liters must be available at STP to allow six grams of hydrogen to react and form water?

(A) 2
(B) 11.2
(C) 16
(D) 22.4
(E) 33.6

20. Select the element with an atomic number of 19 and one electron in its valence shell:

(A) calcium
(B) chlorine
(C) hydrogen
(D) potassium
(E) sodium

21. A certain element commonly has 8 protons, 8 neutrons, and 8 electrons. Select the combination of particles that denote an isotope of this particular atom.

(A) 4 protons, 4 neutrons, 4 electrons
(B) 8 protons, 8 neutrons, 4 electrons
(C) 8 protons, 10 neutrons, 8 electrons
(D) 10 protons, 8 neutrons, 8 electrons
(E) 10 protons, 10 neutrons, 8 electrons

22. A probable compound formed from calcium and oxygen has the formula:

(A) $CaO$
(B) $Ca_2O_2$
(C) $Ca_2O$
(D) $Ca_2O_3$
(E) $Ca_3O_2$

23. Select the incorrect statement about alpha, beta, and gamma rays of radiation.

(A) All affect a photographic plate.

(B) Alpha rays possess charged particles.

(C) Beta rays move most rapidly.

(D) Beta rays lack charged particles.

(E) Gamma rays display high frequency waves.

24. The valence of cobalt is:

(A) 1

(B) 2

(C) 3

(D) 4

(E) 6

25. Among the choices given, the atom with the largest size is:

(A) bromine

(B) chlorine

(C) fluorine

(D) helium

(E) iodine

26. The final stable, disintegration product in the decay of uranium, $_{92}U^{238}$, is:

(A) $_{83}Bi^{210}$

(B) $_{82}Pb^{206}$

(C) $_{82}Pb^{210}$

(D) $_{84}Po^{210}$

(E) $_{90}Th^{234}$

27. The correct ranking of alkali metals from most reactive to least reactive is:

(A) Be–Mg–Co–Sr–Ba
(B) Cs–Rb–K–Na–Li
(C) F–Cl–Br–I
(D) I–Br–Cl–F
(E) Li–Na–K–Rb–Cs

28. Select the most unreactive element:

(A) Cl
(B) H
(C) Na
(D) S
(E) Xe

29. Each of the following is a statement of Dalton's laws <u>except</u>:

(A) Any gas in a mixture exerts its partial pressure.
(B) Atoms are permanent and cannot be decomposed.
(C) Each gas's pressure depends on other gases in a mixture.
(D) Gases can exist in a mixture.
(E) Substances are composed of atoms.

30. Consider the balanced equation:

$$2KClO_3 \rightarrow 2KCl + 3O_2$$

If 72 grams of oxygen gas are produced, the amount of potassium chlorate required in grams is:

(A) 112
(B) 224
(C) 336
(D) 448
(E) 1020

31. Select the organic compound:

(A) $C_3H_8$

(B) $CO_2$

(C) CO

(D) HCl

(E) NaCl

32. An atom has an atomic mass of 45 and an atomic number of 21. Select the correct statement about its atomic structure:

(A) The number of electrons is 24.

(B) The number of neutrons is 21.

(C) The number of protons is 24.

(D) The number of electrons and neutrons is equal.

(E) The number of protons and neutrons is unequal.

33. Boron is bombarded by alpha particles. Complete the products formed in the following equation by transmutation.

$$_5B^{11} + {}_2He^4$$

(A) $_6C^{12} + {}_0n^1$

(B) $_3L^6 + {}_0n^1$

(C) $_7N^{14} + {}_0n^1$

(D) $_{15}P^{31} + {}_0n^1$

(E) $_{14}Si^{28} + {}_1H^1$

34. The least bond energy in Kcals per mole is found with:

(A) C–C

(B) H–Br

(C) H–Cl

(D) H–F

(E) O–H

35. The bond angles between carbon and hydrogen in methane are best labeled as:

(A) covalent
(B) ionic
(C) linear
(D) tetrahedral
(E) trihybrid

36. Select the light metal with a low melting point:

(A) copper
(B) iron
(C) lithium
(D) phosphorous
(E) sulfur

37. $HCl + NaOH \rightarrow NaCl + H_2O$ is an example of a reaction classified as:

(A) decomposition
(B) double replacement
(C) reversible
(D) single replacement
(E) synthesis

38. Consider this reaction under standard lab conditions:

$$FeS + 2HCl \rightarrow FeCl_2 + H_2S$$

If 22 grams of iron sulfide are completely reacted to form products, the volume of hydrogen sulfide gas produced is:

(A) 5.6
(B) 11.2
(C) 22.4
(D) 44.4
(E) 88.0

39. Compute the quantity in grams of sucrose ($C_{12}H_{22}O_{11}$) required to make a 1M strength solution of 500 ml.

(A) 85.5
(B) 171
(C) 342
(D) 684
(E) 982

40. Consider the following balanced equation:

$$2K + 2HCl \rightarrow 2KCl + H_2$$

The respective oxidation numbers for K, H, and Cl before and after reaction:

(A) go from 0, –1, +1 to –1, +1, 0
(B) go from 0, +1, –1 to +1, –1, 0
(C) go from 1, –1, 0 to –1, +1, –1
(D) go from 1, –1, 0 to –1, –1, 0
(E) go from 0, 0, 1 to 1, 1, –1

41. Enzymes, which are organic catalysts, always partly consist of:

(A) carbohydrates
(B) lipids
(C) nucleic acids
(D) proteins
(E) steroids

42. Al correct ranking of elements in order of decreasing electronegativity is:

(A) Al–F–O–Cs–Na
(B) Cs–Na–Al–O–F
(C) F–O–Al–Na–Cs
(D) Na–F–Al–O–Cs
(E) O–Cs–Al–F–Na

43. The critical pressure of a substance is a value that is necessary to:

(A) convert a gas to a solid at its critical temperature

(B) convert a liquid to a solid at its critical temperature

(C) freeze a liquid at its critical temperature

(D) liquefy a gas at its critical temperature

(E) vaporize a liquid at its critical temperature

**Questions 44 – 46** refer to the valence electron dot formulas in the figure below. The letters merely identify the different atoms. They do not stand for actual known elements.

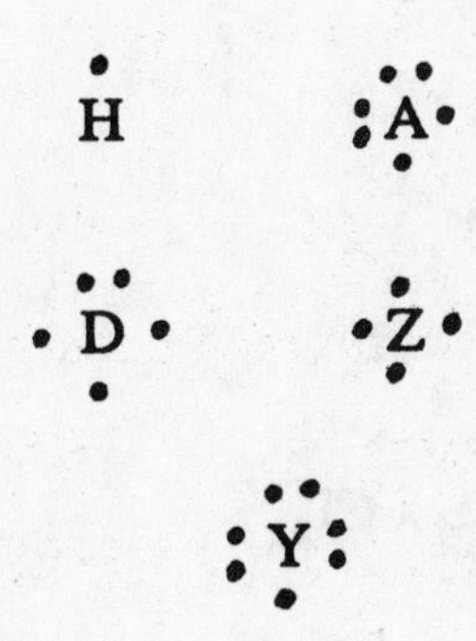


44. The most active nonmetal is:

(A) A

(B) D

(C) H

(D) Y

(E) Z

45. A likely bonding association is:

(A) $HA_2$
(B) $HD_4$
(C) $DH_5$
(D) $ZH_4$
(E) YZ

46. Element D has a valence of:

(A) 1
(B) 3
(C) 4
(D) 5
(E) 7

47. For the reaction $PCl_5 \rightarrow PCl_3 + Cl_2$, the rate of the reaction is proportional by mole amounts to:

(A) $Cl_2 \text{ x } PCl_3$

(B) $PCl_5$

(C) $\frac{Cl_2 \text{ x } PCl_3}{PCl_5}$

(D) $\frac{PCl_5}{Cl_2 \text{ x } PCl_3}$

(E) $PCl_5 \text{ x } PCl_3 \text{ x } Cl_2$

48. Complete ionization of an aluminum hydroxide particle yields:

(A) $Al^+$, $OH^-$
(B) $Al^+$, $2OH^-$
(C) $Al^+$, $3OH^-$
(D) $2Al^+$, $3OH^-$
(E) $2Al^+$, $OH^-$

49. $H_2 + S \leftrightarrow H_2S$ + energy

In this reversible reaction, select the factor that will shift the equilibrium to the right:

(A) adding heat

(B) adding $H_2S$

(C) blocking hydrogen gas reaction

(D) removing hydrogen sulfide gas

(E) removing sulfur

50. The valence of tellurium is:

(A) 1

(B) 2

(C) 3

(D) 4

(E) 5

51. Select the aqueous conductor:

(A) alcohol

(B) distilled $H_2O$

(C) glucose

(D) hydrochloric acid

(E) sucrose

52. $HC_2H_3O_2 \leftrightarrow H^+ + C_2H_3O_2^-$

Consider the above equation.

.5 moles/liter of acetic acid dissociates into hydrogen and acetate ions. The equilibrium concentration of the hydrogen ions is $2.9 \times 10^{-3}$ moles/liter. The ionization constant for the acid is

(A) $1.7 \times 10^{-5}$
(B) $8 \times 10^{-2}$
(C) $4 \times 10^{-1}$
(D) $4 \times 10^{-2}$
(E) $4 \times 10^{2}$

53. The pOH of a .0001 M KOH solution is:

(A) 1
(B) 2
(C) 4
(D) 7
(E) 11

54. Carbon's valence shell electron configuration can be symbolized as:

(A) $s^1, p^1$
(B) $s^1, p^2$
(C) $s^2, p^2$
(D) $s^2, p^4$
(E) $s^4, p^2$

55. Carbon monoxide directly interferes with the human body function of:

(A) gamma globulins
(B) growth hormone
(C) hemoglobin
(D) insulin
(E) iodine

56. Select the substance that is molecular <u>and</u> a compound:

(A) gold

(B) hydrogen gas

(C) methane

(D) oxygen gas

(E) sodium chloride

57. $\_\_H_2S + \_\_O_2 \rightarrow \_\_H_2O + \_\_SO_2$

Balancing this equation yields the following coefficients from left to right:

(A) 1-1-2-2

(B) 2-3-2-2

(C) 2-2-2-3

(D) 3-2-2-2

(E) 3-2-3-2

58. The volume in milliliters of a .3M solution of NaOH needed to neutralize 3 liters of a .01M HCl solution is:

(A) .1

(B) 1

(C) 10

(D) 100

(E) 1000

59. (A)

```
        O
        ||
    H - C
        |
    H - C - OH
        |
   HO - C - H
        |
    H - C - OH
        |
    H - C - OH
        |
    H - C - OH
        |
        H
```

(B)

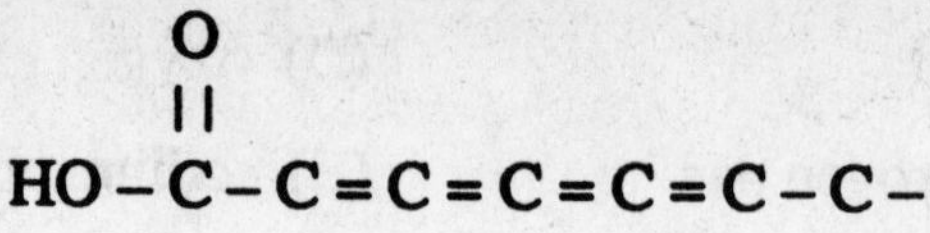

```
         O
         ||
HO - C - C = C = C = C = C - C -
```

(C)

```
       O   H   H   H   H
       ||  |   |   |   |
HO  -  C - C - C - C - C
           |   |   |   |
           H   H   H   H
```

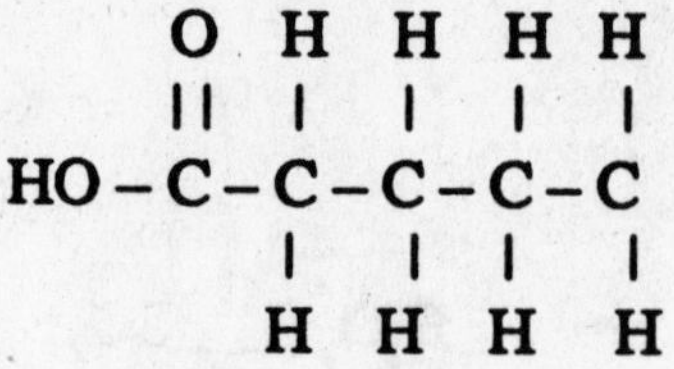

(D)

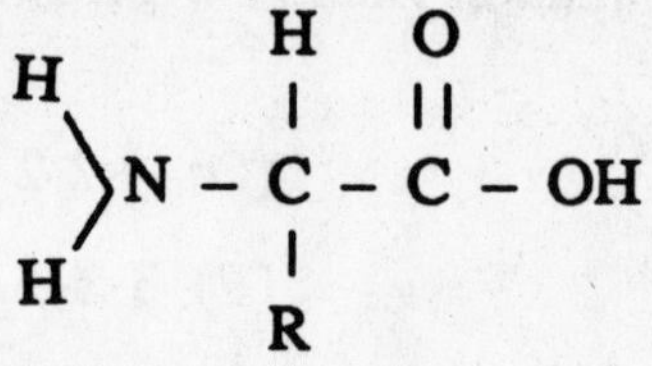

```
           H   O
 H         |   ||
  \ N  -   C - C - OH
 /         |
 H         R
```

(E)

```
      H       H
  H     \  C  /    H
    \  /     \   /
     C         C
    /           \
   H             H
```

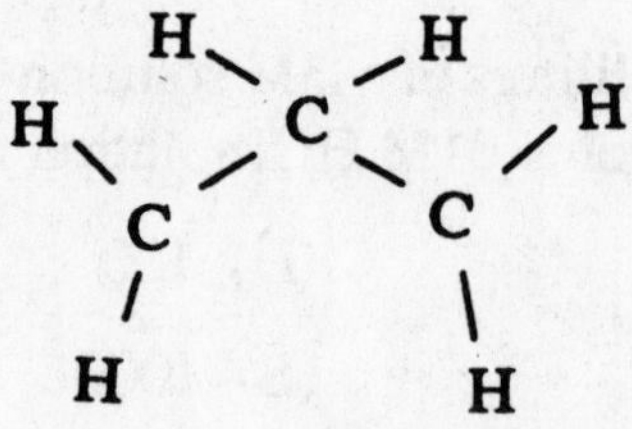

The following refers to the diagram above. The outline for a molecular sub-unit for a saturated fat molecule is depicted at:

(A) A

(B) B

(C) C

(D) D

(E) E

60. An ice cube is placed in an open glass of water at room temperature. Describe the resultant effect on its energy content and entropy.

| | Energy | Entropy |
|---|---|---|
| (A) | decrease | decrease |
| (B) | increase | increase |
| (C) | increase | decrease |
| (D) | increase | remains constant |
| (E) | remains constant | increase |

61. $NH_4^+$ is the radical named:

(A) ammonium

(B) hydrate

(C) hydroxyl

(D) nitrate

(E) nitrite

62. Solid zinc oxide, ZnO, has a heat of formation of about –84 kilocalories per mole. Select the correct statement for a reaction producing 162 grams of zinc oxide.

(A) 42 kilocalories are absorbed

(B) 81 kilocalories are absorbed

(C) 81 kilocalories are released

(D) 168 kilocalories are absorbed

(E) 168 kilocalories are released

63. One mole of a substance dissolved in 1000 grams of water elevates the boiling point by .52°C and depresses the freezing point by 1.86°C. 23 grams of an alcohol, $C_2H_5OH$, is dissolved in a kilogram of water. At standard conditions, water's new boiling and freezing points are respectively (in ° C):

(A) 100.26°, –.93°
(B) .26°, –1.86°
(C) .52°, –.93°
(D) 100.52°, –1.86°
(E) 101.04°, –1.86°

64. Select the correct rule about solubility of substances:

(A) All ammonium salts are insoluble.
(B) All nitrates are insoluble.
(C) All silver salts, except $AgNO_3$ are insoluble.
(D) All sodium salts are insoluble.
(E) Sulfides of sodium, potassium, and magnesium are insoluble.

65. A salt formed by a neutralization reaction of a strong acid and weak base is:

(A) HCl
(B) NaCl
(C) $Na_2CO_3$
(D) $NH_4Cl$
(E) $NH_4CN$

66. Substances are neither created nor destroyed, but simply changed from one form to another. This is the law of:

(A) change of matter
(B) conservation of energy

(C) conservation of matter

(D) multiple proportions

(E) thermodynamics (second law)

67. $FeS_2$ or "fool's gold" is also known as

(A) hematite

(B) lodestone

(C) magnetite

(D) pyrite

(E) siderite

68. Members of a common horizontal row of the periodic table should have the same:

(A) atomic number

(B) atomic mass

(C) electron number in the outer shell

(D) number of energy shells

(E) valence

69. The most common isotope of hydrogen has an atomic number and mass, respectively, of:

(A) 1,0

(B) 1,1

(C) 1,2

(D) 2,1

(E) 2,2

70. 500 ml of a gas experiences a pressure change from 760 mm of mercury pressure to a barometric pressure of 800. If all other laboratory factors are held constant, its new volume in ml is:

(A) 400
(B) 425
(C) 475
(D) 525
(E) 595

71. An atom's electron number is 11 while the number of neutrons is 12. Its atomic mass is:

(A) 11/12
(B) 11
(C) 12
(D) 23
(E) 132

**Questions 72 and 73 refer to the figure below.**

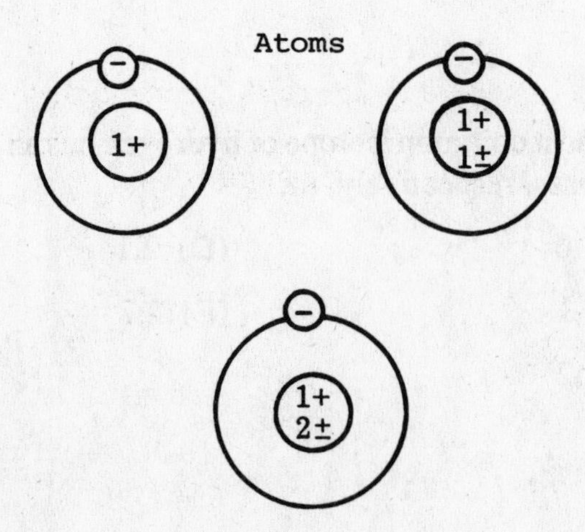

72. The relationship between these three atoms is that they are:

(A) isobars
(B) isomers
(C) isometric
(D) isotonic
(E) isotopes

73. All three atoms represent the element:

(A) C
(B) H
(C) He
(D) N
(E) O

74. 44.8 liters of a gas are collected in a lab under standard conditions. The number of molecules in this volume is:

(A) 112,000
(B) $6.02 \times 10^{23}$
(C) $12.04 \times 10^{23}$
(D) $18.06 \times 10^{23}$
(E) 1,112,000

75. 600 ml of a gas experiences a temperature change from 30 degrees C to 90 degrees C with other laboratory variables held constant. Its new volume in ml is:

(A) 666.34
(B) 697.34
(C) 743.34
(D) 893.34
(E) 1200.03

76. The Law of Entropy states that:

(A) energy is neither created nor destroyed, but changed from one form to another

(B) gas pressures are determined independently in a mixture

(C) heat flows to a more concentrated medium

(D) matter is neither created nor destroyed

(E) systems tend toward increasing disorder

77. Select the incorrect statement about the element neon:

(A) its chemical symbol is Ne.

(B) it is a noble gas.

(C) it is a reactive substance.

(D) neon has two energy shells.

(E) neon is found in family eight of the periodic table.

78. One faraday of electricity is passed through an HCl electrolyte solution. Select the correct electrode result.

(A) 1 gram of chloride ions is deposited at the anode.

(B) 1 gram of hydrogen ions is deposited at the cathode.

(C) 5 grams of hydrogen ions are deposited at the anode.

(D) 35 grams of chloride ions are deposited at the anode.

(E) 36.5 grams of chloride ions are deposited at the cathode.

79. A strip of zinc is dipped in a solution of copper sulfate. Select the correct occurring half-reaction.

(A) $Co^{++}$ + 2 electrons → Co, reduction

(B) Cu + 2 electrons → $Cu^{++}$, reduction

(C) Cu → $Cu^{++}$ + 2 electrons, reduction

(D) Zn → $Zn^{++}$ + 2 electrons, oxidation

(E) Zn + 2 electrons → $Zn^{++}$, oxidation

80. The approximate quantity of copper, $Cu^{++}$, in grams, needed to furnish $6.02 \times 10^{23}$ electrons in an electrolytic cell is:

(A) 1
(B) 15.88
(C) 31.77
(D) 63.54
(E) 95.37

81. A subatomic nucleon is the:

(A) hyperon
(B) lepton
(C) neutrino
(D) positron
(E) proton

82.

```
     H   H   H   H   H
     |   |   |   |   |
 H - C - C - C - C - C - H
     |   |   |   |   |
     H   H   H   H   H
```

This question refers to the structural formula of pentane. Its molecular weight is:

(A) 6
(B) 12
(C) 60
(D) 72
(E) 98

83. Atmospheric pressure is measured by a:

(A) barometer
(B) spectrophotometer
(C) sphygmomanometer
(D) thermocouple
(E) thermometer

84. A solution has a density of 1.2 in grams/ml. Pure water is added to it. A probable new solution density is:

(A) .9
(B) 1.0
(C) 1.1
(D) 1.2
(E) 1.5

85. The molecule represented in the structural formula below is:

(A) ammonia
(B) methane
(C) nitrogen
(D) urea
(E) uric acid

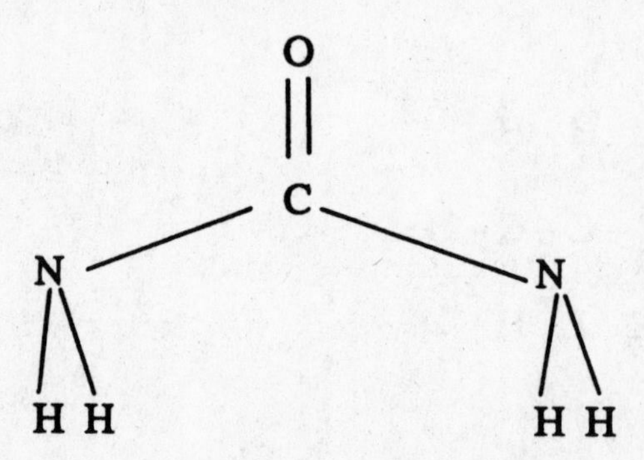

# ADVANCED PLACEMENT CHEMISTRY EXAM II

## SECTION II

**Directions**: The percentages given for each individual part indicates the scoring for this section. Spend approximately 35 minutes on Parts A and B together and 40 minutes on Part C.

The methods and steps used to solve a problem must be shown clearly. Partial credit will be given if work is shown. Pay attention to significant figures.

### Part A
(30 Percent)

Solve the steps in the following problem series.

Table: Heats of Formation

| Compound | H (Kcal/mole) |
|---|---|
| $H_2O$ (gas) | – 57.8 |
| $H_2S$ | – 5.3 |
| $SO_2$ | – 71.0 |

1. ____$SO_2$ + ____$H_2S$ → ____S + ____$H_2O$

(a) Balance the above-stated reaction with the correct coefficients.

$SO_2$ ________

$H_2S$ ________

S ________

$H_2O$ ________

Assume the coefficients in (a) represent actual mole amounts of each involved substance.

(b) Determine $\Delta H$ for this reaction.

(c) Assuming that 64 grams of sulfur dioxide react with 68 grams of hydrogen sulfide, calculate the volume of water vapor formed if collected under standard conditions.

(d) Calculate the changed volume of produced water vapor collected if standard conditions change to 790 mm mercury of pressure and 20 degrees Centigrade.

## Part B
(30 Percent)

**Solve ONE of the two problems presented. Only the first problem answered will be scored.**

2. ____Mg + $O_2$ → ____MgO

(a) Balance the above-stated reaction with the correct coefficients.

____Mg

____$O_2$

____MgO

(b) Write the change in oxidation states for the elements involved.

| element | reactant | product |
|---|---|---|
| Mg | ____ | ____ |
| O | ____ | ____ |

(c) Calculate the total mole amounts of each substance involved in the above reaction.

____Mg

____$O_2$

____MgO

(d) Balance the equation below and write an expression for the equilibrium constants $K_c$ and $K_p$:

$$NaHCO_3 \leftrightarrow NaCO_3(s) + CO_2(g) + H_2O(g)$$

At 125°C, the value of the equilibrium constant $K_c$ = $2.34 \times 10^{-4}$ mole$^2$/l$^2$. Find the partial pressures of $CO_2(g)$ and $H_2O(g)$. How does sodium bicarbonate serve as a fire extinguisher?

3. Complete the following four-step problem:

The molecular formula for sulfuric acid is $H_2SO_4$

(a) Write a balanced equation for its reaction with sodium hydroxide.

(b) 2.45 grams of sulfuric acid are dissolved in 2 liters of solution. Calculate the normality of that solution.

(c) Calculate the molarity of the solution described in (b).

(d) Calculate the pH of the solution described in (b), assuming complete dissociation of the acid.

## Part C
(40 percent)

Choose **THREE** of the following five topics. Only the first three answers will be scored. Your answers will be scored based on their accuracy, relevance of the details chosen, and appropriateness of the descriptive material used. Be as specific as possible and use illustrative examples and equations where helpful.

4. Compare and contrast elements, compounds, and mixtures. Emphasize possible similarities among the three classes along with the differences.

5. Outline the pattern of organization and predictive value of the periodic table. Characteristics to include are:

   (a) Mendeleev's original work

   (b) metals/nonmetals

   (c) families

   (d) periods

   (e) valences

   (f) element activity/atom size

   (g) atomic number/atomic mass

6. Describe four broad patterns of chemical equations summarized by reactions. Write two balanced-equation examples for each.

7. Compare compound categories of acids, bases, and salts. Use formulas in the comparison. Describe distinguishing characteristics of each. Include common examples encountered on each compound type.

8. Describe buffers and their mode of action. Include common examples of their action, such as:

   (a) homeostasis of body fluids

   (b) pH influence on environmental waterways

# ADVANCED PLACEMENT CHEMISTRY EXAM II

## ANSWER KEY

| | | | | | |
|---|---|---|---|---|---|
| 1. | C | 29. | C | 57. | B |
| 2. | D | 30. | C | 58. | D |
| 3. | B | 31. | A | 59. | C |
| 4. | D | 32 | E | 60. | B |
| 5. | A | 33. | C | 61. | A |
| 6. | B | 34. | A | 62. | E |
| 7. | D | 35. | D | 63. | A |
| 8. | C | 36. | C | 64. | C |
| 9. | A | 37. | B | 65. | D |
| 10. | A | 38. | A | 66. | C |
| 11. | B | 39. | B | 67. | D |
| 12. | D | 40. | B | 68. | D |
| 13. | C | 41. | D | 69. | B |
| 14. | C | 42. | C | 70. | C |
| 15. | B | 43. | D | 71. | D |
| 16. | C | 44. | D | 72. | E |
| 17. | A | 45. | D | 73. | B |
| 18. | E | 46. | B | 74. | C |
| 19. | E | 47. | B | 75. | C |
| 20. | D | 48. | C | 76. | E |
| 21. | C | 49. | D | 77. | C |
| 22. | A | 50. | B | 78. | B |
| 23. | D | 51. | D | 79. | D |
| 24. | B | 52. | A | 80. | C |
| 25. | E | 53. | C | 81. | E |
| 26. | B | 54. | C | 82. | D |
| 27. | B | 55. | C | 83. | A |
| 28. | E | 56. | C | 84. | C |
| | | | | 85. | D |

# ADVANCED PLACEMENT CHEMISTRY EXAM II

# DETAILED EXPLANATIONS OF ANSWERS

## SECTION I

1. (C) 2. (D) 3. (B)

pH = –log of the $H^+$ ion concentration in moles (mole = 1 gram) per liter. Another way to express this is $\frac{1}{\log [H^+]}$

For solution A the value translates into $3.89 \times 10^{-8}$

$$\frac{1}{3.89 \times 10^{-8}} = \frac{10^8}{3.89}$$

From the log table, the log of 3.89 is .5899. In division, logs are subtracted: 8 – .5899 = 7.41. For solution B:

$$\frac{10^8}{2.21} = 8 - .3444 \text{ (log of 2.21)} = 7.66.$$

For solution C: $10^8 = 8 - .8756$ (log of 7.51) = 7.13.

4. (D) 5. (A)

The two choices are general principles comparing solids and liquids for rapidity of particle movement and intermolecular distance. Liquid water does not freeze and become a solid until 0° C. Mercury, unlike most metals, is a liquid at room temperature.

6. (B) 7. (D) 8. (C) 9. (A)

The most common oxygen-containing salt ends in the suffix "ate": $NaClO_3$ = sodium chlorate. Chlorine's oxidation state is +5, as sodium's is +1, and oxygen's total is –6 (3 x –2). +5 and +1 balance the –6 oxidation state. The next lowest chlorine oxidation state ends in "ite": sodium chlorite, $NaClO_2$. Chlorine's oxidation state here is +3, along with sodium's +1 and oxygen's total –4. The next lowest chlorine oxidation state uses the "hypo. . .ite" prefix-suffix: NaClO → Na is +1, Cl is +1 and O is –2. The next lowest state uses the "ide" suffix: NaCl, sodium chloride. Chlorine's oxidation state is –1 to sodium's +1.

10. (A) 11. (B) 12. (D) 13. (C)

Alkanes have the general molecular formula $C_nH_{2n+2}$, such as butane, $C_2H_6$. Alkenes conform to $C_nH_{2n}$, for example butene, $C_2H_4$. Alkynes, $C_nH_{2n-2}$, include butyne, $C_2H_2$. Alcohols, such as butyl alcohol, include an OH in the carbon chain. Amines are nitrogen derivatives, and are not represented.

14. (C)

Reaction rates of inorganic substances usually double with every 10° increase in temperature. Therefore, if 35°C represents the new, doubled rate, then the original temperature must be 10° less: 35° – 10° = 25°C.

15. (B)

Two reactants are involved in this reaction, each with a coefficient of one. The order is found by adding their understood coefficients, one plus one.

16. (C)

Other color indicator changes are: congo red 4, malachite green 12, phenophthalein 9, thymol blue 3.

17. (A)

All three standard temperature choices express identical intensities from three different measurement scales. At the standard atmospheric pressure of sea level, a mercury barometer reveals an elevated mercury column of 760mm (torr) or 29.92 inches.

18. (E)

By temperature conversion, C = 5/9 (F – 32). C = 5/9 (122 – 32) = 50. A Centigrade value is converted to Kelvin by adding the constant 273: 50 + 273 = 323.

19. (E)

Two moles (4 grams) of hydrogen gas react with one mole (32 grams) of oxygen gas in this balanced reaction. Six grams of $H_2$ are 3 moles. By simple proportion, 1.5 moles of $O_2$ must react. One mole occupies 22.41 STP. 1.5 moles occupy 33.6 liters by simple proportion.

20. (D)

With atomic number 19 (19 electrons), the electrons will fill up the energy levels by 2 (K) – 8 (L) – 8 (M) – 1 (N). This array places potassium in family one of the periodic table.

21. (C)

Isotopes of an element have the same atomic number but their atomic mass varies due to a varying number of neutrons. This question compares two isotopes of oxygen.

22. (A)

Two electrons from calcium satisfy oxygen's requirement for eight valence electrons to obey the octet rule. They thus combine in a one-to-one ratio.

23. (D)

Beta rays consist of negatively charged particles moving close to the speed of light. Alpha rays are positive, and penetrating gamma rays are high frequency, near X-rays. All affect a photographic plate, a trait that led to their initial detection.

24. (B)

Cobalt is in family two of the periodic table and hence has two valence electrons. All elements of a given numbered family have that common valence electron number.

25. (E)

Iodine is at the bottom of the halogens in family seven of the periodic table. Elements that appear at the bottom of any family have more energy levels and are larger in size.

26. (B)

All of the choices are formed throughout the various steps of radioactive $_{92}U^{238}$ decay. $_{82}Pb^{206}$ is thus associated with $_{92}U^{238}$ in nature. The relative amounts of the two, along with knowledge of half-life periods, can be used to calculate the age of a geological structure harboring these two atoms.

27. (B)

The alkali metals are in family one of the period table. They are progressively more active as metals, losing their one valence electron as they descend through the vertical array.

28. (E)

Xe, xenon, is found in family eight of the periodic table, the noble gases. All other choices are active metals or nonmetals of other families.

29. (C)

Dalton's laws of partial pressures state that gases in a mixture exert their individual pressures independently.

30. (C)

By the balanced equation, two moles (224 grams) of $KClO_3$ yield 3 moles of $O_2$ (48 grams). By simple proportion:

$$\frac{(2KClO_3-)}{3O_2} = \frac{224}{48} = \frac{X}{72} \qquad \text{so, } X = 336.$$

31. (A)

This is the formula for propane. Its structural formula is:

```
    H   H   H
    |   |   |
H - C - C - C - H
    |   |   |
    H   H   H
```

Note that its carbon skeleton, indicative of organic compounds, is lacking in the other molecules.

32. (E)

With an atomic number of 21, the electron and proton numbers are each 21. For a mass of 45, 24 neutrons must exist with the 21 protons.

33. (C)

An alpha particle consists of two protons and two neutrons as in a helium nucleus, $_2He^4$. In the bombardment, boron incorporates the two protons for an atomic number increase from 5 to 7. These two protons plus one captured neutron yield an atomic mass increase from 11 to 14, thus $_7N^{14}$ as in nitrogen. The second neutron remains free.

34. (A)

Carbon is neither a strong metal nor a strong nonmetal and has less electron-attracting power than the strong nonmetals bromine, chlorine, fluorine, and oxygen, in their covalent bond formation.

35. (D)

Methane, $CH_4$, is a symmetrical molecule in terms of the direction of its four covalent bonds. Each C–H bond is at an approximate 109 degree bond angle, oriented toward the corner of an imaginary tetrahedron, (a four-sided figure).

36. (C)

Metals in family one and two of the periodic table have these characteristics. Note lithium's location in the periodic table.

37. (B)

The respective cations, $H^+$ and $Na^+$, and anions, $Cl^-$ and $OH^-$, swap places in the change from reactants to products.

38. (A)

One mole of FeS, 88 grams (56 + 32) forms one mole of $H_2S$ gas in the balanced equation. By simple proportion, one-quarter mole of the given 22 grams, thus yields one-quarter mole of $H_2S$. One mole of a gas fills 22.4 liters. .25 mole x 22.4 liters = 5.6 liters.

39. (B)

One mole of sucrose is 342 grams:

| | | | |
|---|---|---|---|
| C | 12 x 12 (atomic weight) | = | 144 |
| H | 22 x 1 | = | 22 |
| O | 11 x 16 | = | 176 |

These total to 342. 342 grams in one liter makes a 1M strength. In one-half liter, 500 ml, this measured amount is also halved.

40. (B)

Among reactants, uncombined potassium is 0. Hydrogen is +1 metallic behavior and chlorine is –1 (nonmetal) while combined in a compound. Among products, potassium is now combined, +1 (metal). Chlorine remains –1 as combined. Liberated hydrogen is uncombined as a liberated gas, 0.

41. (D)

Enzymes are proteins which act as catalysts for biochemical reactions.

42. (C)

Electronegativity is electron-attracting power. Fluorine, in the upper right-hand corner of the periodic table, is the most active nonmetal and has the highest electronegativity. Oxygen follows as very active. Aluminum is a metal, which means less electronegativity: it actually donates electrons. Sodium and cesium are metals, too, and cesium has the least electronegativity.

43. (D)

For each gas, a temperature is reached where the kinetic energy of the molecules is so great that no pressure, however large, can liquefy the gas. Any pressure is insufficient to compress gas molecules back to the liquid state where molecules are closer together. This temperature is the critical temperature. The accompanying pressure is the critical pressure.

44. (D)

Nonmetals tend to accept electrons to obey the octet rule. They have five or more electrons in their valence shell. Atom Y has seven, being very active and close to fulfillment.

45. (D)

Z has four electrons. Four hydrogens, each with one electron, can share and fulfill Z with its remaining four. Hydrogen is also satisfied, gaining a second electron for fulfillment of its only energy shell. Four covalent bonds are formed.

46. (B)

With 5 valence electrons, D is in need of 3 more for octet fulfillment.

47. (B)

For the general reaction

$$A \rightarrow B + C$$

The rate is written as

$$\text{rate} \propto [A]^n$$

where n is the experimentally determined order of the reaction. The only answer which conforms with the rate as written above is (B), i.e. rate $[PCl_5]$. In this case, $n = 1$.

48. (C)

Aluminum hydroxide's formula is $Al(OH)_3$. Thus, one aluminum ion and three hydroxyl ions are yielded by dissociation.

49. (D)

The left to right reaction is exothermic, therefore, adding heat drives the reaction equilibrium to the left. From the equilibrium constant

$$K_{eq} = \frac{[H_2S]}{[H_2][S]}$$

it is clear that an increase in the concentration of $H_2S$ increases the value of this ratio, i.e. the equilibrium is disturbed. To return to the equilibrium constant value, $H_2S$ decomposes, so that the reaction is shifted to the left. Blocking $H_2$ removes reactant, inhibiting formation to the right side. Removing S has the same effect. Removing $H_2S$, however, lowers the value of the equilibrium constant. To restore it, more $H_2S$ is produced, i.e., the reaction shifts to the right. The above analysis is the application of Le Chatelier's principle.

50. (B)

Tellurium, Te, is in family six of the periodic table. As with other nonmetals in this family, e.g. oxygen and sulfur, it has six valence electrons and thus needs two more for a stable outer 8. Thus, it can form two covalent bonds.

51. (D)

Hydrochloric acid, HCl, almost completely ionizes in solution, therefore allows passage of electricity; i.e. it acts as a conductor. The other four choices do not ionize.

52. (A)

For a reaction $HA \leftrightarrow H^+ + A^-$, given initial concentrations: $[x] \leftrightarrow 0 + 0$ and equilibrium concentrations: $[x-y] \leftrightarrow [y]+[y]$, then the ionization constant Ka is

$$Ka = \frac{[y]\,[y]}{[x-y]}$$

$$Ka = \frac{[y]^2}{[x-y]}$$

If x >> y (x is much greater than y), then

$$Ka \sim \frac{[y]^2}{x}$$

In our example, $x = 0.5$ moles/l and $y = 2.9 \times 10^{-3}$

$$\therefore K_a = \frac{\left(2.9 \times 10^{-3}\right)^2}{0.5 - 0.0029} \approx \frac{8.41 \times 10^{-6}}{0.5}$$

$$K_a \approx 16.82 \times 10^{-6} \approx 1.7 \text{x} 10^{-5}$$

53. (C)

pOH is minus the log of a solution's hydroxyl ion concentration. A .0001M KOH solution furnishes $10^{-4}$ hydroxyl ions in moles per liter. Molarity and normality are the same for KOH, as it yields one hydroxyl ion per KOH unit. $pOH = -\log [OH-]$ $\therefore$ $pOH = -\log 10^{-4} = 4$.

54. (C)

Carbon has four valence electrons. Two occupy the smaller s orbital and the two remaining fill the larger p orbital.

55. (C)

Carbon monoxide has a greater affinity for hemoglobin in red blood cells than oxygen does. By tying it up, less hemoglobin is available to combine with oxygen and thus supply body cells. This causes asphyxiation.

56. (C)

Methane, $CH_4$, has two elements with atoms bonded covalently. NaCl, sodium chloride, is a compound but employs ionic bonding. The other three are not compounds, consisting of only one element.

57. (B)

Three molecules of $O_2$ (6 atoms) supply 2 water molecules (2 oxygen atoms) and 2 sulfur dioxide molecules (4 atoms). Four hydrogen atoms from $H_2S$ (2 x 2) also supply the four hydrogens in the water produced (2 x 2).

58. (D)

The acid and base react by the equation:

$$HCl + NaOH \rightarrow HCl + H_2O.$$

Each compound donates single ions to form an NaCl unit. By this assumption: molarity of an acid x volume of an acid = (normality of a base) x (volume of a base).

By substitution:

$$.01 \times 3 = .3 \times V_2$$

$$V_2 = \frac{.01 \times 3 \text{ liters}}{.3} = .1 \text{ Liter}$$

.1 Liter = 100 ml.

59. (C)

Illustration A shows the molecular formula of glucose. (D) shows an amino acid's formula. (E) is cyclopropane. Both (B) and (C) are outlines of fatty acids. (B), however, is not saturated with hydrogen atoms, covalently bonded to its carbon chain. This is because of the double C–to–C bonds, limiting availability for hydrogen covalent bonding. Carbon's valence is four. The carbon atoms in illustration (C) are single bonded, leaving more bonding sites for hydrogen atoms. It is relatively saturated with these atoms. Fatty acids are the molecular sub-units of fat molecules.

60. (B)

The ice cube will melt by gaining heat from the water. The temperature of the water drops below room temperature; therefore heat flows from the surroundings into the water until room temperature is attained. The resultant effect is an increase in energy for the ice and water system. When the ice melts, it changes from an ordered to a disordered system. Entropy is a measure of disorder (randomness). The higher the disorder, the higher the entropy. In this case, energy and entropy both increase.

61. (A)

Hydroxyl is $OH^-$. Nitrate is $NO_3^-$ and nitrite is $NO_2^-$.

62. (E)

Zinc oxide's formula weight is 81 (65 + 16). Therefore, 162 grams is twice this weight or two moles. If one mole liberates 84 calories, two liberates twice that amount of energy. The minus sign indicates energy liberation.

63. (A)

The alcohol's molecular weight is 46: 24 ($C_2$) + 6 (6H) + 16(O). Twenty-three grams is one-half mole. It, therefore, changes boiling and freezing points by one-half the stated increments.

64. (C)

The other statements are the opposite of what is correct due to substances' ability to ionize (soluble) or not ionize (insoluble) among the molecules of the solvent water. Silver salts form insoluble precipitates except for the ionizing silver nitrate, $AgNO_3$.

65. (D)

HCl is an acid. NaCl is formed from a strong acid and strong base. $Na_2CO_3$ is a product of a weak acid and strong base. $NH_4CN$ is produced from two weak compounds. A strong acid, HCl, and weak base $NH_4OH$, react to produce $NH_4Cl$.

66. (C)

This is a law applied every time an equation is balanced.

67. (D)

Hematite is $Fe_2O_3$. Magnetite or lodestone is $Fe_3O_4$. Siderite is $FeCO_3$. All choices given are the most common iron ores in the earth's crust.

68. (D)

The number of energy shells stays constant with the valence electrons tending to increase from left to right. Atomic numbers and mass change for each element.

69. (B)

Hydrogen's most common atomic form has one proton and one electron.

70. (C)

Boyle's law predicts an inverse pressure-volume relationship. A pressure increase means a volume decrease. The multiplied fraction is thus less than one:

$$500 \times \frac{760}{800} = 475$$

71. (D)

Atomic mass is the sum of protons and neutrons. In a neutral atom, the number of positive protons equals the electron number. Hence, 11 is added to 12.

72. (E) 73. (B)

All of these atoms vary in neutron number, thus changing the atomic mass. This is a definition of an element's isotopes, hydrogen's in this case.

74. (C)

22.4 liters of a gas is occupied by one mole. This mole contains $6.02 \times 10^{23}$ molecules, Avogadro's number. The given volume in this problem is twice that.

75. (C)

Charles' law predicts a direct H proportion relationship between volume and temperature. The Centigrade temperatures are first converted to Kelvin by adding the constant 273. The multiplied fraction is greater than one, reflecting volume increase:

$$600 \times \frac{363}{303} = 743.34$$

76. (E)

This is a strict statement of the law, predicting that the universe is gradually approaching randomness. Chemical reactions are a part of this.

77. (C)

Neon, Ne, is in family eight of the periodic table. The outer of its two electron shells is filled with 8, obeying the octet rule.

78. (B)

One faraday of current deposits one gram equivalent of a substance at its attracting electrode. Hydrogen has a mass of one. As a positive ion, it will travel to the cathode (negative electrode), during electrolysis.

79. (D)

Zinc, a more active metal than copper, will replace it in the sulfate salt. It will alter its state from a neutral atom, Zn, to a cation $Zn^{++}$ as it combines with the minus two sulfate radical. To become positive two, it loses electrons which is oxidation.

80. (C)

Copper's atomic weight is 63.54. One gram equivalent of a substance furnishes $6.02 \times 10^{23}$ electrons if each atom yields one electron. Copper, plus two, yields two electrons per atom. Thus, only one-half of its gram equivalent is required. $0.5 \times 63.54 = 31.77$.

81. (E)

Protons and neutrons, found in an atom's nucleus, are sometimes collectively termed nucleons.

82. (D)

By counting the carbon and hydrogen atoms, the molecular formula is $C_5H_{12}$.

C: 5 x 12; H: 12 x 1. The sum is 72.

83. (A)

Air pressure is measured by the height of a column of mercury in a tube.

84. (C)

Pure water has a density of one and will thus dilute the solution, but not to a density less than that of water itself.

85. (D)

Other choices have the following formulas: methane – $CH_4$, nitrogen –N, ammonia – $NH_3$. Urea is $CO(NH_2)_2$. The illustration shows its structural formula.

# SECTION II

## PART A

1. (a) $SO_2 + 2H_2S \rightarrow 3S + 2H_2O$

An $SO_2$ molecule must produce two water molecules. Its two oxygens yield two water molecules with one oxygen atom each. Two water molecules, with four hydrogen atoms, require two $H_2S$ molecules with their four hydrogen atoms, 2 x 2. One $SO_2$ molecule and two $H_2S$ molecules in turn yield three sulfur atoms.

(b) Heat of formation is the energy involved in the formation of one mole of a compound. Use the table and balanced equation. Note that the energy values are in kilocalories per mole. From the equation, one mole of $SO_2$ plus two moles of $H_2S$ yield three moles of sulfur plus two moles of water. From the table, energy amounts are applied to this. Negative values mean energy liberated. Positive values refer to energy absorbed. Elements have 0 values.

| $\Delta$ H (products) | $\Delta$ H (reactants) |
|---|---|
| ($2H_2O$) –57.8Kc x 2 moles | ($SO_2$) –71Kc x 1 mol = –71 Kc |
| = –115.6 Kc | ($2H_2S$) –5.3Kc x 2 moles = –10.6Kc |
| | –81.6 |

$\Delta$ H = $\Delta$ H products – $\Delta$ H reactants
= –115.6 – (–81.6)
= – 34 kilocalories

(c) The given gram values reflect the mole amounts of reactants in the equation:

$SO_2$ [32 + (2x16) ] = 62 x 1 mole = 62 grams

$H_2S$ [2(1) + (32) ] = 34 x 2 moles = 68 grams

From this equation, one mole of $SO_2$ reacts with 2 moles of $H_2S$ to yield two moles of $H_2O$. One mole of any gas at STP occupies 22.4 liters. Two moles occupy 44.8 liters.

(d) Boyle's law predicts an inversely proportional change between gas volume and pressure. Since pressure increases from 760 (standard) to 790, volume will decrease. The multiplied fraction is, therefore, less than one, $\frac{760}{790}$.

Charles' law predicts a directly proportional change between gas volume and Kelvin temperature. As temperature increases so will volume. Thus, the fraction is greater than one. Centigrade temperatures are converted to Kelvin by adding the constant 273.

$$\text{Thus, } \frac{20°C}{0°C \text{ (standard)}} = \frac{293°K}{273°K}$$

New volume equals: =

$$44.8 \times \frac{760}{790} \times \frac{293}{273} = 46.26 \text{ liters}$$

## PART B

**Solve ONE of the two problems presented. Only the first problem answered will be scored.**

2. (a) $2 Mg + O_2 \rightarrow 2 MgOa$

A diatomic oxygen atom yields two MgO molecules with one oxygen atom each. For reactants, two magnesium atoms are needed to yield the two MgO molecules.

(b)

| element | reactant | product |
|---|---|---|
| Mg | O | +2 |
| O | O | –2 |

Uncombined magnesium lacks a signed oxidation state until it combines in a compound. Combined with –21 oxygen (in need of two negative electrons), it donates two electrons to become +2. Formerly uncombined oxygen changes to –2 as it receives two electrons.

(c) From the balanced equation, two moles of Mg combine with one mole of $O_2$ to produce two moles of MgO.

two moles of Mg = 2 x 24 (atomic mass) = 48 grams
one mole of $O_2$ = 16 x 2 atoms = 32 grams
two moles of MgO = 2 x (24 + 16) = 2 x 40 = 80 grams

(d) $2NaHCO_3 \leftrightarrow NaCO_3(s) + CO_2(g) + H_2O(g)$.

The above is the balanced equation.

According to the Law of Mass action:

$$K_c' = \frac{[Na_2CO_3(s)]\,[CO_2(g)]\,[H_2O(g)]}{[NaHCO_3(s)]^2}$$

The above is an example of a heterogeneous equilibrium between gases $CO_2$, $H_2O$ and solids $NaHCO_3$, $Na_2CO_3$.

In heterogeneous equilibrium, concentrations of pure solids and liquids are taken as constants and can be incorporated into an overall equilibrium constant $K_c$

$$\therefore K_c = \frac{K'[NaHCO_{3(s)}]^2}{[Na_2CO_{3(s)}]} = [CO_{2(g)}]\,[H_2O_{(g)}]$$

$$K_c = [CO_{2(g)}]\,[H_2O_{(g)}]$$

Also, $K_p = p_{CO_2} \times p_{H_2O}$

where $p_{CO_2}$ and $p_{H_2O}$ are the partial pressures of $CO_2$ and $H_2O$, respectively.

To find the partial pressures of $CO_2$ and $H_2O$, we first find $K_p$. $K_p$ and $K_c$ are related by $K_p = K_c(RT)^{\Delta ng}$

where $\Delta$ ng = the number of moles of gaseous products –the number of moles of gaseous reactants.

R is the gas constant,
$R = 0.0821$ L atm mole $^{-1}K^{-1}$
T – absolute temperature, $T = 125° + 273° = 398°K$
$n = 2 - 0 = 2$

$$K_p = K_c(RT)^2$$

$$K_p = 2.34 \times 10^{-4} \frac{mole^2}{L^2} [0.0821L \text{ atm mole}^{-1}K^{-1}]^2[398° K]^2$$

$$K_p = 0.25 \text{ atom}^2$$

$K_p = p_{CO_2}\ p_{H_2O} = p^2$, since in this case $p_{CO_2} = p_{H_2O}$

$$p_{CO_2} = p_{H_2O} = K_p$$

$$p_{CO_2} = p_{H_2O} = 0.25 = 0.5 \text{ atom.}$$

Sodium bicarbonate can be used as a fire extinguisher because upon contact with fire, it decomposes, forming $CO_2$ which blankets the flame (density of $CO_2$ is higher than air). $O_2$ needed to sustain the flame is cut off and the fire is therefore extinguished.

3. (a) $H_2SO_4$, a strong acid, reacts with sodium hydroxide, NaOH, a strong base, by a double replacement reaction. The balanced reaction is:

$$H_2SO_4 + 2NaOH \rightarrow Na_2SO_4 + 2H_2O$$

By this pattern, a strong acid and base double exchange positive and negative ions to form a salt plus water.

(b) Normality of an acid solution equals equivalents of the substance per liter of solution. Equivalent of an acid equals:

$$\frac{\text{molecular weight in grams}}{\text{number of H atoms per molecule.}}$$

For $H_2SO_4$, molecular weight is:

$H (2 \times 1) + S (32) + O (4 \times 16) = 98$

Therefore, equivalents equal $\frac{98}{2} = 49$ grams

A one normal solution of $H_2SO_4$ has one equivalent, 49 g, of acid per liter of solution. 2.45 grams per two liters = 1.225 grams per solution in the stated problem. Therefore, this 1.225 grams is only a fraction of the needed amount for a one normal concentration, 49 g.

$$\frac{1.225}{49} = .025N$$

(c) Molarity of a solution equals moles of a substance per liter of solution. One mole of $H_2SO_4$, 98 grams, in a liter of solution produces a solution strength of one molar or 1M. The 1.225 grams per liter in the solution given is only a fraction of this:

$$\frac{1.225}{98} = 0.0125M$$

(d) $pH = \log \frac{1}{(H+)}$

where the concentration is in moles per liter as furnished by the acid. From sections(C) and (B), .0125 moles of fully dissociated $H_2SO_4$ produce twice that amount of hydrogen ions as reflected by the normality = .025. Note by its formula that one $H_2SO_4$ particle yields two hydrogen ions each. Thus, .0125 is doubled.

$$pH = \log \frac{1}{(H+)}$$

$$= \log \frac{1}{.025}$$

$$= \log \frac{1}{2.5 \times 10^{-2}}$$

$$= \log \frac{10^2}{2.5}$$

$$= 2 - .3979$$

$$= 1.6$$

During division of numbers, corresponding logs are subtracted.

## PART C

4. Elements are the simplest compilations of matter. As substances, they are composed of only one kind of atom and cannot be broken down into simpler substances by ordinary chemical means. A diamond consists of a tremendous number of carbon atoms. In a pure diamond, only atoms of this element are found. The same observation is made for sulfur powder. Thus an element is homogenous, uniform in composition throughout. Elements are systematically organized into metals and nonmetals in the periodic table. C, H, O and N, (using the chemical symbols) for example, are the most abundant elements in living organisms. Si and O are the most common elements in the earth's crust.

Compounds are homogeneous substances formed by the union of two or more different kinds of atoms through chemical reactions. In molecules such as $C_6H_{12}O_6$ and $H_2O$, the atoms bond covalently into molecules by outer shell electron sharing. The forestated formulas reveal the identity of atoms bonded in the molecule along with the number of atoms of each. A compound has properties unique from the separate uncombined elements in it. Thus the compound FeS behaves differently from separate iron or sulfur atoms. The elements combine in a definite, unvarying proportion by weight. FeS is always 63.63% iron.

Elements in a mixture are not chemically combined. They can combine in any proportion and retain their individual properties. Thus a mixture of iron filings and sulfur powder can contain 10% iron or 90% iron. A magnet can separate iron from the mix while a solvent can dissolve the sulfur from the iron.

5. The periodic table organizes about 106 elements, metals and nonmetals, systematically. Over one hundred years ago, the Russian chemist Mendeleev arranged the sixty-three elements then known by

order of increasing atomic mass. In this original arrangement, Mendeleev noted a pattern of increase or decrease in a characteristic among a row of elements, e.g., density, boiling point, etc. At the end of a row the pattern of increase or decrease would then reoccur and start a new row, hence acting periodically. This periodicity named the table. In the process of aligning rows, colinear vertical columns of elements were also composed.

The modern periodic table sequences the larger known group of elements, about 60% metals and 40% nonmetals, by order of increasing atomic number. The smaller group of nonmetals is organized to the right of a stepline separating them from the larger group of metals. Seven horizontal rows or periods also produce the vertical columns of elements termed families. Families IA and IIA give way to transition element families IB through VIIIB to the right. Families IIIA through VIIIA are to the right of them.

For the A families the Roman numeral also signifies the number of valence electrons. Distinct metals have three or fewer outer shell electrons with the most active further down in a family. Further down in a family, metals add energy shells one by one through successive periods. Valence electrons become further away from the nucleus, more loosely held and thus more easily oxidized. A metal's activity is related to its ease of oxidation or electron loss. Transition elements, B families, have one, two or three valence electrons, indicative of metals.

Nonmetals are more active near the top of their given family. Thus fluorine, at the top of family VIIA attracts electrons most actively, a distinct nonmetal property.

Each cell of the table centers the element's symbol with atomic number to the upper right and atomic mass to the lower left. Some tables also include the electron shell arrangement from the inside out: i.e. 2–7 for the nonmetal fluorine or 2–8–8–1 for the metal potassium.

6. (a) Combination or Synthesis

$$Fe + S \rightarrow FeS$$
$$H_2 + CO_2 \rightarrow H_2CO_3$$

The molecules or products to the right contain a larger number of atoms than each of the individual reactant units. In the first reaction, one atom plus one atom yields FeS with two atoms. In the second example, two plus three atoms yield a six-atom molecule.

(b) Analysis or Decomposition

$$2H_2O \rightarrow 2H_2 + O_2$$

$$H_2CO_3 \rightarrow H_2 + CO_2$$

This is opposite in pattern to synthesis. Larger molecules form molecules with fewer individual atoms.

(c) Single Replacement

$$Zn + H_2SO_4 \rightarrow ZnSO_4 + H_2$$

$$2K + 2HCl \rightarrow 2KCl + H_2$$

A more active metal replaces the first metal in a compound, thus liberating that formerly combined metal.

(d) Double Replacement

$$HCl + NaOH \rightarrow NaCl + H_2O$$

$$H_2SO_4 + 2KOH \rightarrow K_2SO_4 + 2H_2O$$

Both the cations (positive) and anions (negative) exchange partners. This is more characteristic of a reaction between an acid and base. They yield a salt plus water. The salt derives its cation from the base and anion from the acid. The left over hydrogens, $H^+$, and bases, $OH^-$, combine to yield the water.

7. An acid can be defined as a compound that dissociates hydrogen ions in solution: $HCl \rightarrow H^+ + Cl^-$. Acids taste sour, such as the citric acid in a lemon or lime. They turn blue litmus paper pink and register below the point of neutrality, 7, on the pH scale. They are

caustic to metals and tissues. Acids react with bases to form a salt and water in a double replacement reaction:

$$HCl + NaOH \rightarrow NaCl + H_2O$$

HCl, hydrochloric acid, is commonly used in manufacturing many products requiring acid conditions. High levels of this are secreted in the stomach with a pH of 1–3. Sulfuric acid, $H_2SO_4$, is used to clean corrosive metals, manufacture products and in battery fluid.

Bases form hydroxyl ions by dissociation in solution: $NaOH \rightarrow Na^+ + OH^-$. NaOH, sodium hydroxide, is the household agent lye. Bases are used to manufacture soaps (NaOH, KOH), lime water $Ca(OH)_2$ and fertilizers ($NH_4OH$). Bases taste bitter and are slippery to the touch. They turn pink litmus blue and register above 7 on the alkaline side of the pH scale. They are the other reactant with an acid in the previously described double replacement reaction.

Salts are derived from such acid-base reactions. Their production leads to neutralization. They have a pH of 7 in solution. Examples include sodium chloride → NaCl, table salt; sodium bicarbonate → $NaHCO_3$, baking soda; and $CuSO_4$, copper sulfate, for copper plating. In a neutralization reaction, a salt derives its positive ion from the acid and negative ion from reacting base. Salts are usually very soluble in water from their high degree of ionization:

$$NaCl \rightarrow Na^+ + Cl^-$$

8. Buffers are compounds that maintain a constant pH in a solution. Chemical additions to aquatic systems can potentially alter an optimally needed pH. Components of a buffer system react to remove that source of acidity or alkalinity and maintain the constant, optimal pH.

Buffers usually exist in compound pairs: a weak acid and salt of that acid. One of the most common in the human body is carbonic acid, $H_2CO_3$ and sodium bicarbonate, $NaHCO_3$.

Acid accumulation is met by the salt of that pair for reaction:

$$HCl + NaHCO_3 \rightarrow NaCl + H_2CO_3$$

This double replacement reaction removes a strong source of acidity, HCl. The formed carbonic acid, $H_2CO_3$, is weaker. It dissociates less for less of a pH threat. Thus tendency toward acidosis is opposed.

Base accumulation is met by the acid of the buffer pair:

$$NaOH + H_2CO_3 \rightarrow NaCO_3 + H_2O$$

This double replacement reaction removes the strong base source and fights alkalosis.

The extracellular fluid bathing body cells has an optimal pH of 7.35 –7.45. Cells immersed in this ECF cannot tolerate departures from this optimum. By-products of exercise or diet constantly threaten this optimum. Buffers mobilize to prevent permanent, radical pH changes.

In aquatic ecosystems, limestone buffers of stream beds oppose acidity. Normal rain is acidic, about 5.5. Acid rain is more concentrated, 4.5, and often overwhelms a stream's capacity to maintain its aquatic pH optimum.

# THE ADVANCED PLACEMENT EXAMINATION IN

# CHEMISTRY

# TEST III

# THE ADVANCED PLACEMENT EXAMINATION IN

# CHEMISTRY

## ANSWER SHEET

1. (A) (B) (C) (D) (E)
2. (A) (B) (C) (D) (E)
3. (A) (B) (C) (D) (E)
4. (A) (B) (C) (D) (E)
5. (A) (B) (C) (D) (E)
6. (A) (B) (C) (D) (E)
7. (A) (B) (C) (D) (E)
8. (A) (B) (C) (D) (E)
9. (A) (B) (C) (D) (E)
10. (A) (B) (C) (D) (E)
11. (A) (B) (C) (D) (E)
12. (A) (B) (C) (D) (E)
13. (A) (B) (C) (D) (E)
14. (A) (B) (C) (D) (E)
15. (A) (B) (C) (D) (E)
16. (A) (B) (C) (D) (E)
17. (A) (B) (C) (D) (E)
18. (A) (B) (C) (D) (E)
19. (A) (B) (C) (D) (E)
20. (A) (B) (C) (D) (E)
21. (A) (B) (C) (D) (E)
22. (A) (B) (C) (D) (E)
23. (A) (B) (C) (D) (E)
24. (A) (B) (C) (D) (E)
25. (A) (B) (C) (D) (E)
26. (A) (B) (C) (D) (E)
27. (A) (B) (C) (D) (E)
28. (A) (B) (C) (D) (E)
29. (A) (B) (C) (D) (E)
30. (A) (B) (C) (D) (E)
31. (A) (B) (C) (D) (E)
32. (A) (B) (C) (D) (E)
33. (A) (B) (C) (D) (E)
34. (A) (B) (C) (D) (E)
35. (A) (B) (C) (D) (E)
36. (A) (B) (C) (D) (E)
37. (A) (B) (C) (D) (E)
38. (A) (B) (C) (D) (E)
39. (A) (B) (C) (D) (E)
40. (A) (B) (C) (D) (E)
41. (A) (B) (C) (D) (E)
42. (A) (B) (C) (D) (E)
43. (A) (B) (C) (D) (E)
44. (A) (B) (C) (D) (E)
45. (A) (B) (C) (D) (E)
46. (A) (B) (C) (D) (E)
47. (A) (B) (C) (D) (E)
48. (A) (B) (C) (D) (E)
49. (A) (B) (C) (D) (E)
50. (A) (B) (C) (D) (E)
51. (A) (B) (C) (D) (E)
52. (A) (B) (C) (D) (E)
53. (A) (B) (C) (D) (E)
54. (A) (B) (C) (D) (E)
55. (A) (B) (C) (D) (E)
56. (A) (B) (C) (D) (E)
57. (A) (B) (C) (D) (E)
58. (A) (B) (C) (D) (E)
59. (A) (B) (C) (D) (E)
60. (A) (B) (C) (D) (E)
61. (A) (B) (C) (D) (E)
62. (A) (B) (C) (D) (E)
63. (A) (B) (C) (D) (E)
64. (A) (B) (C) (D) (E)
65. (A) (B) (C) (D) (E)
66. (A) (B) (C) (D) (E)
67. (A) (B) (C) (D) (E)
68. (A) (B) (C) (D) (E)
69. (A) (B) (C) (D) (E)
70. (A) (B) (C) (D) (E)
71. (A) (B) (C) (D) (E)
72. (A) (B) (C) (D) (E)
73. (A) (B) (C) (D) (E)
74. (A) (B) (C) (D) (E)
75. (A) (B) (C) (D) (E)
76. (A) (B) (C) (D) (E)
77. (A) (B) (C) (D) (E)
78. (A) (B) (C) (D) (E)
79. (A) (B) (C) (D) (E)
80. (A) (B) (C) (D) (E)
81. (A) (B) (C) (D) (E)
82. (A) (B) (C) (D) (E)
83. (A) (B) (C) (D) (E)
84. (A) (B) (C) (D) (E)
85. (A) (B) (C) (D) (E)

# ADVANCED PLACEMENT CHEMISTRY EXAM III

## SECTION I

Time: 1 Hour 45 Minutes
85 Questions

**Note**: For all questions referring to solutions, assume that the solvent is water unless otherwise stated.

**Directions:** Each set of lettered choices below refers to the numbered statements immediately following it. Select the one lettered choice that best fits each statement. A choice may be used once, more than once, or not at all in each set.

**Questions 1 - 4**

(A) grayish solid
(B) greenish-yellow gas
(C) pale-yellow gas
(D) reddish-brown gas
(E) reddish-brown liquid

1. $Br_2$

2. $Cl_2$

3. $F_2$

4. $I_2$

**Questions 5 - 7**

(A) barium
(B) beryllium
(C) calcium
(D) magnesium
(E) strontium

5. Major component of limestone

6. Least active by oxidation

7. Least dense

**Questions 8 - 11 refer to energy levels and the maximum number of electrons they can hold.**

(A) eight
(B) eighteen
(C) thirty-two
(D) twelve
(E) two

8. K

9. L

10. M

11. N

**Directions:** Each of the following questions or incomplete sentences is followed by five possible answers. Choose the best one in each case.

12. Select the indicator that changes color at a pH of 4.

(A) malachite green
(B) methyl orange
(C) methyl violet
(D) phenolphthalein
(E) thymol blue

13. A Fahrenheit temperature is converted to a corresponding Centigrade (Celsius) value by the equation: $C = \frac{5}{9}(F-32)$. A recorded Fahrenheit value is 82 degrees. Its value on the Kelvin temperature scale is approximately:

(A) 355.0 K
(B) 27.7 K
(C) 83.7 K
(D) 246 K
(E) 300.7 K

14. Consider the balanced equation:

$$Fe + S \rightarrow FeS$$

Approximately what amount of sulfur in grams must be available for 28 grams of iron to react to form the compound?

(A) 1
(B) 8
(C) 16
(D) 32
(E) 64

15. An element's atom most commonly has the following listing of

subatomic particles:

6 protons, 6 neutrons, 6 electrons.

Select the following listing of particles that reveal an isotope to this given atom.

(A) 6 protons, 6 neutrons, 4 electrons

(B) 6 protons, 6 neutrons, 8 electrons

(C) 6 protons, 8 neutrons, 6 electrons

(D) 8 protons, 6 neutrons, 6 electrons

(E) 8 protons, 8 neutrons, 6 electrons

16. Select the element with an atomic number of 15 and 5 electrons in its valence shell.

(A) chlorine
(B) nitrogen
(C) oxygen
(D) phosphorus
(E) sulfur

17. Aluminum has three electrons in its valence shell. Sulfur has six electrons at its similar site. A probable compound between them has the formula:

(A) $AlS$
(B) $AlS_2$
(C) $AlS_3$
(D) $Al_2S_3$
(E) $A1_3S_2$

18. The correct ranking of halogens, from most reactive to least reactive, is:

(A) Cl-F-Br-I

(B) F-C1-Br-I

(C) He-I-Br-C1

(D) I-Br-C1-F

(E) Ne-C1-Br-I

19. In the periodic table, metals with low melting points appear just to the left and below the nonmetals. A metal with a probable low melting point is:

(A) cadmium

(B) iron

(C) lithium

(D) selenium

(E) rubidium

20. Select the <u>incorrect</u> statement about the kinetic theory of gases.

(A) Average energy of each particle is the same regardless of mass.

(B) Distances between their molecules are large.

(C) Their molecules are imperfectly inelastic.

(D) Their motion is constant.

(E) Velocity of molecules increases with increasing temperature.

21. 96 grams of oxygen are produced by chemical reaction from ozone. The number of moles of ozone required to produce this amount of oxygen is:

(A) 1

(B) 2

(C) 3

(D) 5

(E) 10

22. Consider the reaction under standard lab conditions:

$$N_2 + 3H_2 \rightarrow 2NH_3$$

If 6 grams of hydrogen gas reacts, the volume it occupies in liters is:

(A) 2

(B) 3

(C) 22.4

(D) 67.2

(E) 100

23. Consider a mixture of oxygen and hydrogen. Assuming kinetic energy $K_e = \frac{1}{2} mv^2$, the velocity of oxygen's molecules is less than hydrogen's by a factor of:

(A) 1/2

(B) 1/4

(C) 1/8

(D) 1/16

(E) 1/32

24. Consider the following balanced equation:

$$Zn^{+} + H_2SO_4 \rightarrow ZnSO_4 + H_2$$

Zinc's oxidation number changes in this reaction from:

(A) 0 to +2

(B) 0 to +4

(C) +2 to +4

(D) +2 to –2

(E) +2 to –4

25. An example of a dipole molecule is:

(A) $CH_4$

(B) $H_2$

(C) $H_2O$

(D) NaC1

(E) $O_2$

26. Which of the acids below has the formula $HBrO_2$?

(A) bromic

(B) bromous

(C) hydrobromic

(D) hypobromous

(E) perbromic

27. A correct ranking of elements in order of increasing electronegativity is:

(A) Ca-Li-Ba-K-Ca

(B) H-I-Na-K-Ca

(C) Ca-A1-P-S-F

(D) F-Br-H-Al-Rb

(E) Na-O-C-Ca-Li

28. The formula for tin (II) hydroxide is:

(A) CuO

(B) $Ni(OH)_2$

(C) $Ni(OH)_3$

(D) $Sn(OH)_2$

(E) $Sn(OH)_4$

29. Determine the incorrect statement:

(A) In a first order gaseous reaction, if we decrease the volume of the container where the reaction is taking place, the velocity of the reaction will decrease.

(B) A catalyst creates a new path for the reaction which requires a smaller activation energy.

(C) Activation energy is constant for a certain reaction.

(D) By increasing the temperature of a reaction, we increase the amount of molecules with sufficient energy to react.

(E) All intermolecular collisions result in chemical reactions.

30. $N_2 + 3H_2 \leftrightarrow 2NH_3 \uparrow + \text{heat}$

In this reversible reaction, the equilibrium shifts to the right because of all the following factors except:

(A) adding heat

(B) adding reactant amounts

(C) formation of ammonia gas

(D) increasing pressure on reactants

(E) yielding an escaping gas

31. Consider the following reversible reaction:

$$H_2 + 3N_2 \leftrightarrow 2NH_3$$

Its equilibrium constant "K" is expressed as:

(A) $\dfrac{[NH_3]}{[N_2]\ [H_2]^3}$

(B) $\dfrac{[NH_3]^2}{[N_2]^3\ [H_2]}$

(C) $\dfrac{[NH_3]}{[N_2]\ [H_2]}$

(D) $[NH^3]$

(E) $[N_2]^2\ [H_2]^3$

32. Select the compound that is <u>not</u> a conductor in aqueous solution:

(A) $CH_3OH$

(B) $CuSO_4$

(C) HCl

(D) NaCl

(E) NaOH

33. $H_2CO_3 \leftrightarrow H^+ + HCO_3$

.5 moles/liter of carbonic acid dissociates hydrogen and carbonic ions at .1 mole per liter, each in a lab aqueous setting. Its ionization constant is:

(A) $1x10^{-2}$

(B) $2x10^{-2}$

(C) $1x10^{2}$

(D) $2x10^{1}$

(E) $2x10^{2}$

34. Consider the reaction:

$$2Al + 3S \rightarrow Al_2S_3$$

The oxidation numbers of aluminum and sulfur in the product are, respectively:

(A) 1,1
(B) –2,3
(C) 2,–3
(D) 3,2
(E) 3,–2

35. Complete ionization of a calcium hydroxide molecule yields:

(A) $Ca^{++}$, $OH^-$
(B) $Ca^{++}$, $2OH^-$
(C) $2Ca^{++}$, $OH^-$
(D) $2Ca^{++}$, $2OH^-$
(E) $3Ca^{++}$, $OH^-$

**Questions 36 – 38 refer to the valence electron dot formulas in the figure below. The letters merely identify the different atoms. They do not stand for actual known elements.**

Valence Electron Dot Formulas

```
      •          ••
      A          D


      •          ••
    • X •       :Y •
      •          •


           ••
          :Z :
           •
```

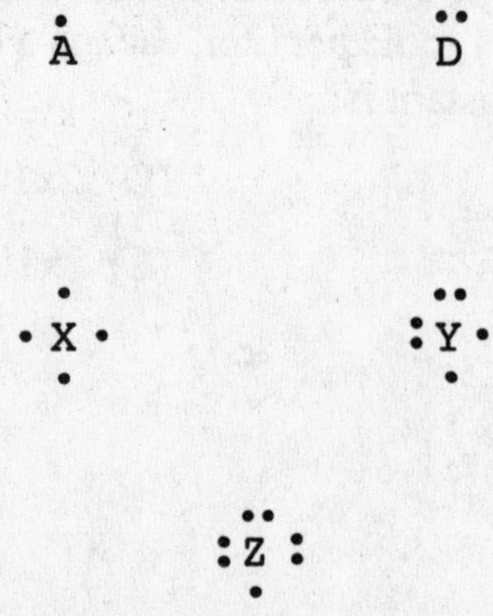

36. A possible compound by covalent bonding is:

(A) $AD_2$
(B) $AX_3$
(C) $XZ_4$
(D) $YZ$
(E) $ZD_2$

37. An unlikely compound by covalent bonding is:

(A) $AZ$
(B) $DZ_2$
(C) $DX_4$
(D) $XY_2$
(E) $XZ_4$

38. The most active metal is:

(A) A
(B) D
(C) X
(D) Y
(E) Z

39. Oxygen's valence shell electron configuration can be symbolized as:

(A) $s^2 p^2$
(B) $s^2 p^4$
(C) $s^4 p^2$
(D) $s^4 p^4$
(E) $s^4 p^6$

40. Calcium has an atomic number of 20. Its electron configuration can be summarized as:

(A) $1s^2, 2s^2, 2p^6, 3s^2, 3p^6, 4s^2$

(B) $1s^2, 2s^2, 2p^6, 3s^2, 3p^4, 4s^4$

(C) $1s^2, 2s^4, 2p^4, 3s^2, 3p^6, 4s^2$

(D) $1s^1, 2s^4, 2p^4, 3s^2, 3p^6, 4s^4$

(E) $1s^1 \cdot 2s^3 \cdot 2p^5 \cdot 3s^1 \cdot 3p^6 \cdot 4s^2$

41. Consider the following unbalanced equation. Coefficients are missing:

$$\_\_\_ NH_3 + \_\_\_ O_2 \rightarrow \_\_\_ NO + \_\_\_ H_2O$$

To balance the equation, the four consecutive coefficients from left to right are:

(A) 4, 5, 4, 6

(B) 4, 4, 5, 6

(C) 5, 4, 5, 6

(D) 5, 5, 4, 6

(E) 6, 5, 4, 4

42. Consider the following balanced equation. How many moles of hydrogen sulfide react with one mole of oxygen?

$$2H_2S + 3O_2 \rightarrow 2SO_2 + 2H_2O$$

(A) 2/3

(B) 3/2

(C) 1

(D) 2

(E) 3

43. Hydrogen, nitrogen, and oxygen combine in the following amounts to form a compound:

$H = 3.18g$ $O = 152.64g$ $N = 44.52g$

The probable formula for the compound is:

(A) $HNO_2$
(B) $HNO_3$
(C) $H_2NO_3$
(D) $H_2N_2O$
(E) $H_3NO_2$

44. Select the <u>incorrect</u> statement about radiation.

(A) Alpha rays exhibit low penetrating power.
(B) Alpha ray particles consist of 2 neutrons and 2 protons.
(C) Beta ray particles can move close to the speed of light.
(D) Beta ray particles possess a positive charge.
(E) Gamma rays lack a possession of charge.

$${}_{92}U^{238} \rightarrow {}_{90}Th^{234} \rightarrow {}_{91}Pa^{234} \rightarrow$$

45. The next disintegration product in the given radioactive decay of uranium is:

(A) ${}_{88}Ra^{226}$
(B) ${}_{86}Pa^{226}$
(C) ${}_{90}Th^{230}$
(D) ${}_{92}U^{234}$
(E) ${}_{90}U^{232}$

46. A negative subatomic particle which is about equal in mass to the proton is the:

(A) electron
(B) hyperon
(C) meson
(D) neutron
(E) positron

47. In 1919, Rutherford bombarded nitrogen gas with high speed alpha particles. Complete the following equation summarizing the results.

$$_{7}N^{14} + _{2}He^{4} \rightarrow _{8}O^{17} + \_\_\_\_$$

(A) $_{1}H^{1}$
(B) $_{1}H^{2}$
(c) $_{2}He^{2}$
(D) $_{2}He^{4}$
(E) $_{10}Ne^{18}$

48. Identify the element which is converted to the phosphorous isotope and neutron when it collides with alpha particles.

$$\_\_\_\_\_\_ + _{2}He^{4} \rightarrow _{15}P^{30} + _{0}H^{1}$$

(A) $_{13}Al^{27}$
(B) $_{7}N^{14}$
(C) $_{11}Na^{23}$
(D) $_{25}Mn^{55}$
(E) $_{15}P^{31}$

49. Light nuclei combine to yield somewhat heavier, stable nuclei with energy release. This is a definition of:

(A) atomic fission
(B) atomic fusion
(C) binding energy
(D) chain reaction
(E) radioactivity

50. Select the incorrect statement about the chemical activity at electrodes during electrolysis.

(A) Anions give up electrons.
(B) Cations take up electrons.
(C) Oxidation occurs at the anode.
(D) Proton transfer occurs in the reactions.
(E) Reduction occurs at the cathode.

**Question 51 refers to the Figure below.**

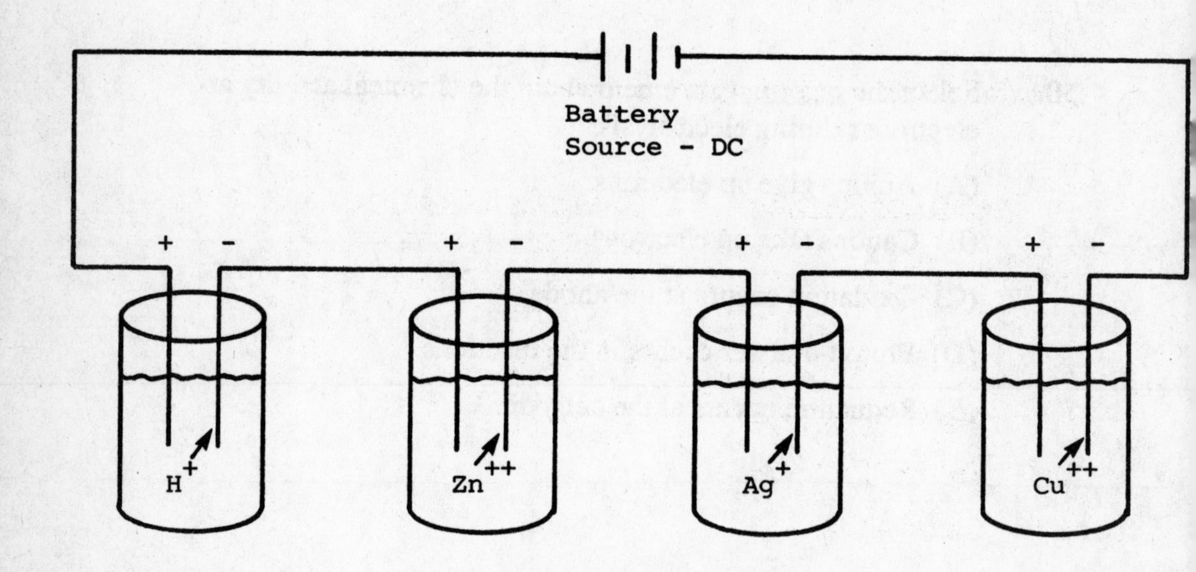


51. One faraday of electricity is passed through the series of solutions with different electrolytes. Select the correct statement about substances liberated at the electrodes.

(A) 4.002 grams of hydrogen

(B) 65.54 grams of zinc

(C) 65.37 grams of copper

(D) 107.86 grams of silver

(E) 196.96 grams of gold

52. The approximate quantity of grams of aluminum needed to furnish $6.02 \times 10^{23}$ electrons in an electrolytic cell is:

(A) 3
(B) 9
(C) 18
(D) 27
(E) 81

**Question 53 refers to the chart below.**

| Half Cell Reaction | Standard Electrode Potentials (volts) |
|---|---|
| $K \leftrightarrow K^{+} + e^{-}$ | +2.92 |
| $Ca \leftrightarrow Ca^{++} + 2e^{-}$ | +2.87 |
| $Mg \leftrightarrow Mg^{++} + 2e^{-}$ | +2.34 |
| $Zn \leftrightarrow Zn^{++} + 2e^{-}$ | .762 |
| $Cu \leftrightarrow Cu^{++} + 2e^{-}$ | –.344 |
| $Ag \leftrightarrow Ag^{+} + e^{-}$ | –.7995 |
| $Pt \leftrightarrow Pt^{++} + 2e^{-}$ | –1.2 |

53. A potential difference of 1.1068 is produced in an electrolysis set-up with the two half cell elements of:

(A) calcium and silver
(B) copper and zinc
(C) magnesium and platinum
(D) potassium and silver
(E) zinc and silver

54. Select the metal with the highest ionization energy.

(A) cesium
(B) lithium
(C) potassium
(D) rubidium
(E) sodium

55. Consider this reaction:

$$C_8H_{18} + O_2 \rightarrow CO_2 + H_2O$$

Under standard conditions, the volume of air in liters required for complete combustion of 228 grams of octane is about

(A) 22.4
(B) 2.50
(C) 560
(D) 1560
(E) 2800

56. A buffer solution was prepared by mixing 100 ml of a 1.2M $NH_3$ solution and 400 ml of a 0.5M $NH_4Cl$ solution. What is the pH of this buffer solution, assuming a final volume of 500 ml and $K_b = 1.8 \times 10^{-5}$?

(A) 1.08
(B) 4.96
(C) 5.8
(D) 9.03
(E) 8

57. Determine $\Delta G°$ (free energy at standard conditions) for the reaction below, considering $K_p = 8$ at 25°C.

$$2A(g) \leftrightarrow B(g)$$

(A) +50.82KJ

(B) 5082J

(C) –1240J

(D) –4830J

(E) –5.153KJ

58. Which of the expressions below represents the correct rate law of the reaction?

$$2A + B \rightarrow C$$

| Experiment | A | B | initial rate (mole/l·sec) |
|---|---|---|---|
| 1 | 1 | 1 | 1.2 |
| 2 | 2 | 1 | 4.8 |
| 3 | 1 | 2 | 2.4 |
| 4 | 3 | 1 | 10.8 |
| 5 | 1 | 3 | 3.6 |

(A) rate = K[A] [B]

(B) rate = $K[A]^2$ [B]

(C) rate = K[A]

(D) rate = K[B]

(E) rate = K[A] $[B]^2$

59. For the previous reaction, determine the value of K

(A) 1.2 $l^2/mole^2$ sec

(B) 1.2 $sec^{-1}$

(C) 1.2 sec

(D) 1.2 $l^{-1}$ $mole^{-1}$sec

(E) 1.2 $l^{-1}$ $mole^{-1}$ $sec^{-1}$

60. Given the reaction below:

$$H_2(g) + \frac{1}{2} O_2(g) \rightarrow H_2O$$

If all the heat generated from the reaction (57,800 cal/mole) is transformed into kinetic energy of the water vapor molecules generated, and this vapor is going to be used as a means of propulsion for a special rocket, determine the velocity of the "jet" of molecules leaving the rocket.

(A) 50 km/h

(B) 50 cm/s

(C) 185 km/h

(D) 500 m/s

(E) 18,500 km/h

61. In a 0.1M solution of acetic acid in water, 1% of the acid is dissociated. Determine the pH of the solution.

(A) 11

(B) 3

(C) 5

(D) 7

(E) 8

62. The table below gives solubility product constants for a few salts of Ag. From the table, we can conclude that the saturated solution with greatest value for $[Ag^+]$ is:

| Salt | Solubility Product Constants |
|---|---|
| AgI | $8.3 \times 10^{-17}$ |
| AgBr | $5.3 \times 10^{-13}$ |
| AgCl | $1.8 \times 10^{-10}$ |
| $AgIO_3$ | $3.0 \times 10^{-8}$ |
| $AgBrO_3$ | $5.3 \times 10^{-5}$ |

(A) AgI

(B) AgBr

(C) $AgIO_3$

(D) AgCl

(E) $AgBrO_3$

63. If 200 ml of $N_2$ at 25°C and pressure 400 mm Hg, and 200 ml of $O_2$ at 25°C and pressure 300 mm Hg are placed in a vessel with volume 700 ml, what is the total pressure of the mixture?

(A) 100 mm Hg
(B) 200 mm Hg
(C) 500 mm Hg
(D) 700 mm Hg
(E) 800 mm Hg

64. In which of the compounds below may CIS-TRANS isomerism occur?

(A) $CH_4$
(B) $C_6H_6$
(C) $C_2H_4$
(D) $C_2H_2(CH_3)_2$
(E) $C_2H_6$

65. In organic chemistry, the so-called "aromatic behavior" consists of:

(A) addition reactions
(B) substitution reactions
(C) oxidation reactions
(D) reduction reactions
(E) polymerization

66. Methane, $CH_4$, has a heat of formation of −18 in kilocalories per mole. Select the correct statement for a reaction producing 48 grams of methane:

(A) 9 kilocalories are absorbed
(B) 18 kilocalories are absorbed
(C) 18 kilocalories are released
(D) 36 kilocalories are released
(E) 54 kilocalories are released

67. A certain substance with molecular weight 62g x $mole^{-1}$ will cause an aqueous solution to have a change of freezing point of 1.86°/m, where m stands for molal. What would be the change in the freezing point of an aqueous solution containing 200 g of $H_2O$, if we added 8g of this substance?

(A) 2.4° C

(B) 1.2° C

(C) 3.6° C

(D) 0.6° C

(E) cannot be determined

68. A given atom has an atomic mass of 23 and an atomic number of 11. Select the incorrect statement about its atomic structure.

(A) Eight electrons are in its outermost energy shell.

(B) Its number of electrons is 11.

(C) Its number of protons is 11.

(D) Most of its mass is in the nucleus.

(E) The number of neutrons is 12.

69. The most common element in the earth's crust, by weight, is:

(A) aluminum

(B) calcium

(C) iron

(D) oxygen

(E) silicon

70. The Russian chemist Mendeleev first arranged 63 known elements in order of their increasing:

(A) atomic number

(B) atomic weight

(C) boiling point

(D) electron number

(E) silicon

71. The greatest bond energy in kcal per mole is found with:

(A) C–C
(B) H–Br
(C) H–Cl
(D) H–F
(E) H–I

72. An example of a metal which is ductile and malleable is:

(A) Au
(B) Cd
(C) Hg
(D) K
(E) Na

73. The same two elements may combine in different proportions to yield different compounds. The ratio of weights of the first element to the second in the two compounds reduces to small whole numbers.

This is the law of:

(A) conservation of mass
(B) definite proportions
(C) multiple proportions
(D) reversible volumes
(E) thermodynamics

74. Inspection of the periodic table allows all of the following except:

(A) determining the valence of transition elements
(B) finding outer shell electron numbers of alkali and alkali earth elements
(C) labeling an element as metal or nonmetal
(D) locating atomic number and mass of an element
(E) reading outer shell electron numbers of halogen elements

75. A 300 ml volume of gas experiences a pressure change from 1000 mm of mercury to 760 mm of mercury, with other laboratory factors held constant. Its new volume in ml is:

(A) 131.6
(B) 228.0
(C) 394.8
(D) 500.8
(E) 1000.0

76. For a certain atom, the atomic number is 19 while its atomic mass is 39. The number of neutrons in its nucleus is:

(A) 1
(B) 8
(C) 19
(D) 20
(E) 39

77. 11.2 liters of a gas are collected in a lab under standard conditions. The number of molecules in this volume is:

(A) 112,000
(B) $3.01 \times 10^{23}$
(C) $6.02 \times 10^{23}$
(D) $12.04 \times 10^{23}$
(E) 1,112,000

**Questions 78 and 79 refer to the list of elements in the activity series below.**

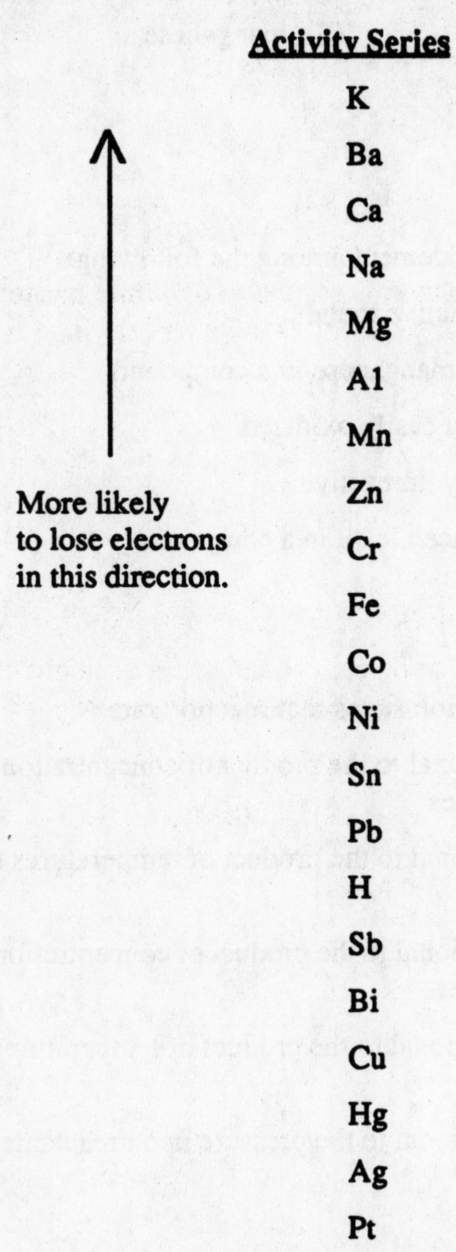
More likely
to lose electrons
in this direction.

**Activity Series**

K
Ba
Ca
Na
Mg
Al
Mn
Zn
Cr
Fe
Co
Ni
Sn
Pb
H
Sb
Bi
Cu
Hg
Ag
Pt
Au

78. Among manganese, lead, iron, chromium, and copper, the least easily oxidized metal is:

(A) chromium
(B) copper
(C) iron
(D) lead
(E) manganese

79. Select the incorrect statement among the following:

(A) gold is the least active metal
(B) iron will replace manganese in a compound
(C) potassium is most easily oxidized
(D) silver is relatively unreactive
(E) sodium will replace nickel in a compound

80. The Law of Mass Action states that reaction rate is:

(A) directly proportional to the product of concentrations of reacting molecules
(B) directly proportional to the product of temperatures of reactants
(C) inversely proportional to the product of concentrations of reacting molecules
(D) inversely proportional to the product of temperature of reactants
(E) inversely proportional to the pressure upon reactants in a mixture

81. $HA + B \rightarrow H^{+} + B + A^{-}$

By the Bronstead theory, the acid in this equation is:

(A) A

(B) HA

(C) AB

(D) B

(E) HB

82. Consider the outline of a water molecule in the drawing below. The regions most likely to be attracted towards another water dipole's negative end are:

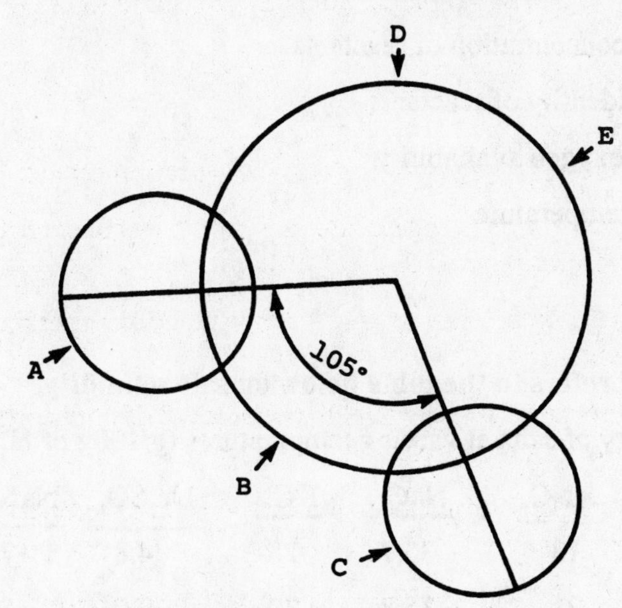


(A) A & B

(B) A & C

(C) B & D

(D) C & D

(E) D & E

83. For nitrous dioxide, $\Delta H_f$ = +8 (kcal/mole). Select the correct statement for a reaction producing 23 grams of this compound.

(A) 4 kilocalories are absorbed

(B) 8 kilocalories are absorbed

(C) 8 kilocalories are released

(D) 16 kilocalories are absorbed

(E) 16 kilocalories are released

84. In a lab, each of the following factors will vary to affect reaction rate <u>except</u>:

(A) catalyst used

(B) concentration of reactants

(C) identity of reactants

(D) oxygen availability

(E) temperature

**Question 85 refers to the table below on salt solubility.**

Solubility of salts at varying temperatures (g/100g of $H_2O$)

| T(°C) | $KNO_3$ | NaCl | KCl | $Na_2SO_4$ | $NaNO_3$ |
|---|---|---|---|---|---|
| 0 | 13 | 35.7 | 28 | 4.8 | 73 |
| 10 | 21 | 35.8 | 30 | 9.0 | 80 |
| 20 | 31 | 36.0 | 34 | 19.5 | 85 |
| 30 | 45 | 36.3 | 37 | 40.9 | 92 |

85. The salt showing the least solubility change over temperature variation is:

(a) potassium chloride

(B) potassium nitrate

(C) sodium chloride

(D) sodium nitrate

(E) sodium sulfate

# ADVANCED PLACEMENT CHEMISTRY EXAM III

## SECTION II

**Directions**: The percentages given for each individual part indicates the scoring for this section. Spend approximately 35 minutes on Parts A and B together and 40 minutes on Part C.

The methods and steps used to solve a problem must be shown clearly. Partial credit will be given if work is shown. Pay attention to significant figures.

### Part A
(30 Percent)

1. ______ $Fe$ + ______ $H_2O$ → $Fe_3O_4$ + ______ $H_2$

The following steps refer to the above stated unbalanced equation. Solve for each step in the series.

(a) Balance the equation. Print the correct coefficient in the blank next to each substance:

______ $Fe$ ______ $Fe_3O_4$

______ $H_2O$ ______ $H_2$

(b) Calculate the formula weight for each of the four substances:

$Fe$ ______ $Fe_3O_4$ ______

$H_2O$ ______ $H_2$ ______

(c) Assume that 100% of 42 available grams of iron react. Calculate the amounts in grams involved for the other three substances:

$H_2O$_____ $Fe_3O_4$_____ $H_2$_____

(d) Assuming STP, calculate the volume of hydrogen gas produced from the 42 grams of iron.

(e) How does the produced volume volume of hydrogen change at 1000 mm Hg of pressure and 20 degrees Centigrade?

## Part B
(30 Percent)

**Solve ONE of the two problems presented. Only the first problem answered will be scored.**

2. Consider the following unbalanced equation and answer each of the following steps.

$$\_\_\_ HI \rightarrow \_\_\_ H_2 + \_\_\_ I_2$$

(a) Balance the equation. Print the correct coefficient in the blank next to each substance.

_____HI _____$H_2$ _____$I_2$

(b) Print the oxidation number next to each element in hydrogen iodide.

_____H      _____I

(c) Assume that 2 moles of hydrogen iodide were originally placed in a one-liter container. At a given temperature, 25% is dissociated at equilibrium. Calculate the equilibrium concentration of each substance.

_____HI      _____$H_2$      _____$I_2$

(d) Calculate the value for the equilibrium constant at this point.

3. Consider the following unbalanced equation and answer each of the following steps.

$$H_2CO_3 \leftrightarrow H^+ + HCO_3^-$$

(a) Balance the equation. Print the correct coefficient in the blank next to each substance.

_____$H_2CO_3$      _____$H^+$      _____$HCO_3^-$

(b) 6.2 grams of carbonic acid are dissolved in one liter of solution. The solution's molarity is:

(c) The ionization constant of carbonic acid is:

$K_a = 3 \times 10^{-7}$

Calculate the concentration of hydrogen ions in moles/liter in this solution of carbonic acid.

(d) Calculate the pH of this solution.

## Part C

(40 Percent)

Choose THREE of the following five topics. Only the first three answers will be scored. Your answers will be scored based on their accuracy, relevance of the details chosen and appropriateness of the descriptive material used. Be as specific as possible and use illustrative examples and equations where helpful.

4. What is the Pauli exclusion principle, and how does it help us to construct models for electronic configurations?

5. Outline the activity series of metals. Explain how it predicts the relative ability of metals to replace one another in a salt. Include the changing oxidation states of metals involved in a simple replacement reaction.

6. Compare and contrast the three physical states of water. Discuss water's unique properties over these states. Include heat of vaporization, changing density, maximum density, and phenomena of large water bodies in the environment. Relate these properties to the water molecule's behavior as a dipole.

7. Explain, with examples, how the following factors influence reaction rate:

   (a) identity of reactants

   (b) temperature of reaction mixture

(c) pressure/absence of a catalyst

(d) reactant concentrations

If reversible, briefly state several conditions that can shift the reversible process to completion.

8. Discuss organic chemistry as an important branch of the discipline. What sets it apart from other subdivisions of chemistry? Why is the application of organic chemistry important to understanding biological phenomena?

# ADVANCED PLACEMENT CHEMISTRY EXAM III

## ANSWER KEY

| | | | | | |
|---|---|---|---|---|---|
| 1. | E | 29. | E | 57. | E |
| 2. | B | 30. | A | 58. | B |
| 3. | C | 31. | B | 59. | A |
| 4. | A | 32. | A | 60. | E |
| 5. | C | 33. | B | 61. | B |
| 6. | B | 34. | E | 62. | E |
| 7. | D | 35. | B | 63. | B |
| 8. | E | 36. | C | 64. | D |
| 9. | A | 37. | C | 65. | B |
| 10. | B | 38. | A | 66. | E |
| 11. | C | 39. | B | 67. | B |
| 12. | B | 40. | A | 68. | A |
| 13. | E | 41. | A | 69. | D |
| 14. | C | 42. | A | 70. | B |
| 15. | C | 43. | B | 71. | D |
| 16. | D | 44. | D | 72. | A |
| 17. | D | 45. | D | 73. | C |
| 18. | B | 46. | E | 74. | A |
| 19. | A | 47. | A | 75. | C |
| 20. | C | 48. | A | 76. | D |
| 21. | B | 49. | B | 77. | B |
| 22. | D | 50. | D | 78. | B |
| 23. | B | 51. | D | 79. | B |
| 24. | A | 52. | B | 80. | A |
| 25. | C | 53. | B | 81. | B |
| 26. | B | 54. | B | 82. | B |
| 27. | D | 55. | E | 83. | A |
| 28. | D | 56. | D | 84. | D |
| | | | | 85. | C |

# ADVANCED PLACEMENT CHEMISTRY EXAM III

# DETAILED EXPLANATIONS OF ANSWERS

## SECTION I

1. (E) 2. (B) 3. (C) 4. (A)
Each is a straightforward fact from common lab observation.

5. (C) 6. (B) 7. (D)
Limestone is mainly calcium carbonate, $CaCO_3$, with some iron and magnesium carbonates plus other impurities. For any periodic family of metals, metals are more easily oxidized when moving down the family. Beryllium is at the top of the alkali earth family, II A, in the periodic table. Although magnesium's atomic mass (24) is greater than beryllium's (9), it is within the sphere of an atom with one more energy shell. Thus this metal is slightly less dense than beryllium while also being tough, malleable, and ductile.

8. (E) 9. (A) 10. (B) 11. (C)
K, L, M, and N rank the energy levels from the inside out with a 2–8–18–32 capacity for electrons.

12. (B)
This is an observed fact from laboratory experience. Other color change values are: malachite green–12, methyl violet–1, phenolphthalein–9, thymol blue –3, 9.

13. (E)

By computation with the formula, 82 degrees F becomes 27.7 degrees C. Centigrade to Kelvin conversion requires adding the constant 273: K=C+273.

14. (C)

By consulting the periodic table, iron atoms have an atomic weight of about 56 while sulfur is 32. They combine in an atom to atom ratio, or a unitary gram equivalent to a unitary gram equivalent ratio. Thus 56 grams of iron combine with 32 grams of sulfur. By proportion, if iron is halved from 56 to 28, sulfur is halved from 32 to 16.

15. (C)

Isotopes of an atom have the same atomic number but a different atomic mass, due to a varying number of neutrons. Choice (C) fits this requirement.

16. (D)

The electron array is 2–8–5 among three energy levels. Checking phosphorus in the periodic table shows phosphorus in family five with the atomic number 15.

17. (D)

Aluminum, as a metal, has three electrons to offer in its outer shell. As a nonmetal, sulfur needs two electrons to fill its outer capacity of eight by the octet rule. Two aluminum atoms with three electrons each can satisfy 3 sulfur atoms each in need of 2 electrons. The subscripts stand for the number of atoms of each element.

18. (B)

Halogens are family seven of the periodic table. In a nonmetal family, the higher the element's position in the family, the greater its activity. Note the ranking of halogens from top to bottom in family seven.

19. (A)

Using this principle, notice only cadmium (Cd) fits this table location. It is below and to the left of the metal (left side) and nonmetal (right side) stepline.

20. (C)

In the kinetic theory, all molecules are perfectly elastic. Upon collision, they rebound with perfect bounce and without energy loss.

21. (B)

96 grams of $O_2$ total 3 moles by division: 96 grams divided by 32 grams (the molecular weight of oxygen). Two moles of ozone are equivalent to 96 grams. Its molecular weight is 48, or 3x16. Thus a mole is 48 grams. Two moles are therefore 96 grams. The balanced equation is: $2\ O_3 \longrightarrow 3\ O_2$

22. (D)

A gram molecular weight or mole of any gas occupies 22.4 liters. One mole of hydrogen gas weighs two grams. Six grams constitute three moles. 3 x 22.4 equals 67.2.

23. (B)

Since the kinetic energy is given by:

$K_e = \frac{1}{2}mv^2$ we can conclude:

$$\frac{1}{2}m_{H_2} v_{H_2} = \frac{1}{2}m_{O_2} v^2_{O_2}$$

$$\Rightarrow 2v^2_{H_2} = 32v^2_{O_2}$$

$$v_{O_2}^{2} = \frac{2}{32} v_{H_2}^{2} \Rightarrow v_O \quad \frac{1}{16} v_H$$

24. (A)

Zinc lacks an oxidation number initially in an uncombined state. It then loses two electrons and becomes +2 to couple with the –2 sulfate radical in $ZnSO_4$.

25. (C)

A dipole is an electrically asymmetrical molecule due to the unequal sharing of electron pairs between the spheres of bonding atoms. The two shared electron pairs of water spend more time in the command of oxygen's sphere than hydrogen's with its lower attracting power. Sodium chloride is not molecular but ionic. Methane ($CH_4$), hydrogen gas and oxygen gas share electron pairs equally and are thus nonpolar molecules.

26. (B)

By rules of nomenclature, a suffix, a prefix, or both are added to the word bromine to indicate its oxidation state, as shown in the table below:

| compound | oxidation state of Br | name |
|---|---|---|
| $HBrO_3$ | +5 | bromic acid |
| $HBrO_2$ | +3 | bromous acid |
| HBrO | +1 | hypobromous acid |
| HBr | –1 | hydrobromic acid |
| $HBrO_4$ | +7 | perbromic acid |

**Note:** All bromine oxidation states are computed algebraically, considering that each O atom is –2 and H atom is +1.

27. (D)

Electronegativity is the relative attracting force of the elements for electrons in a covalent bond. Nonmetals attract electrons, and fluorine is the most active nonmetal. Bromine is beneath it in the halogen family and is less active than fluorine, but it is still a very reactive nonmetal. Hydrogen is somewhat like a metal, but with less electron-attracting power. Aluminum and other metals in the middle of the periodic table

28. (D)

Sn, from the Latin word "stannous," is the symbol for tin. The radical "OH" has a valence of –1 and since Sn has a +2 valence the formula for tin (II) hydroxide becomes $Sn(OH)_2$.

29. (E)

Not all collisions result in chemical reactions: only the collisions between molecules with an amount of energy greater than or equal to the activation energy result in such chemical reactions.

30. (A)

Adding reactant amounts increases frequency of collision for more product formation. Removing ammonia, an escaping product gas, creates a void filled by an equilibrium shift to the right to form more $NH_3$. In the equation, 4 gas volumes form 2. By Le Chatelier's principle, an altered equilibrium reacts to a stress to relieve the stress. Pressurizing the high volume reactants forces the reaction to the right to relieve this stress. The reaction, however, is exothermic, producing heat, and therefore the <u>right</u>-<u>to</u>-<u>left</u> direction absorbs heat. Adding heat throws it in this direction.

31. (B)

"K," the proportionality for reaction rate, is derived by the multiplication of product molar amounts divided by the multiplication of reactant's molar amounts. Coefficients in the balanced equation translate into exponents outside the bracketed molar amounts of the molecules.

32. (A)

$CH_3OH$ is nonpolar methyl alcohol or wood alcohol. Strong acids (HC1), strong bases (NaOH), or salts ($CuSO_4$, NaCl) have ionizing properties that make them electrolytes, capable of conducting an electric current in water.

33. (B)

The ionization constant is computed by the multiplication of product mole amounts over the mole amount of ionizing reactant. By division:

$$\frac{.1 \times .1}{.5} = \frac{.01}{.5} = 0.02 = 2 \times 10^{-2}$$

34. (E)

Aluminum, a metal, is electropositive, +3. Three electrons from each of its <u>two</u> atoms fulfill the nonmetal sulfur's need for two electrons. So sulfur is –2.

35. (B)

Calcium hydroxide's formula is $Ca(OH)_2$. One molecule thus dissociates into one calcium ion ($Ca^{+2}$) and two hydroxyl ($OH^-$) ions.

36. (C)

Element X, with four electrons to offer, can satisfy four Z atoms, each in need of one electron for an outer, stable configuration of 8.

37. (C)

"A" can satisfy one "Z" atom. "D" has two valence electrons to satisfy two" Z" atoms. "X," with four outer electrons, can fulfill two "Y" atoms, each in need of two electrons. "X," with four valence electrons, can fill four "Z" atoms, each needing one electron.

38. (A)

Metals have four or fewer outer shell electrons, which they tend to lose. The most active have only one loosely held valence electron, which is easily lost.

39. (B)

Oxygen, in family six of the periodic table, has six valence electrons. Two of them saturate the smaller *s* orbital with 4 remaining for the larger *p* orbital.

40. (A)

Since oxygen has atomic number 20, the first shell can hold a maximum of 2 electrons. Eight plus eight fill out the next two shells with two left over for a fourth outer shell. *s* and *p* refer to the two different shaped orbital spheres within the three energy levels beyond the first.

41. (A)

Four ammonia molecules offer the 12 hydrogens needed for 6 water molecules. Six water molecules require 6 oxygens. Five $O_2$ molecules offer these 6 as a reactant, plus the four additional oxygens (5x2 total) for the four NO molecules among the products.

42. (A)

If 2 moles of $H_2S$ react with 3 moles of oxygen, 2/3 moles react with 1 mole: $\frac{2}{3} = \frac{X}{1}$; X=2/3.

43. (B)

Solution:

First calculate the number of moles of each element present.

1 mole H = 1.0g
1 mole O = 16.0g
1 mole N = 14.0g

Therefore, number of moles of H is

$$\frac{3.18\text{g}}{1\text{ g/mole}} = 3.18 \text{ moles H}$$

number of moles of O is

$$\frac{152.64\text{g}}{16\text{ g/mole}} = 9.54 \text{ moles O}$$

number of moles of N is

$$\frac{44.52\text{g}}{14\text{ g/mole}} = 3.18 \text{ moles N}$$

The smallest number is used to find the simplest ratio in which the elements combine as

$$H: \frac{3.18}{3.18} = 1 \qquad O: \frac{9.54}{3.18} = 3 \qquad N: \frac{3.18}{3.18} = 1$$

Therefore the simplest molecular formula is $HNO_3$.

44. (D)

All statements are true except this one. Beta ray particles, known as high-speed electrons, have a negative charge.

45. (D)

The first step results from emission of an alpha particle, loss of 2 protons, and 2 neutrons. Thus atomic number decreases by 2 (2 protons) and atomic mass drops by 4. The second step involves emission of a beta particle (electron), which increases atomic number by one, but does not affect atomic mass. Note that the same alteration occurs from protactinium, Pa, to ${}_{92}U^{234}$ in this well-known sequence from uranium's isotope of 238.

46. (E)

Each of the three basic subatomic particles has a counterpart particle with equal mass, but opposite electrical charge. The positron (proton-electron) has approximately the same mass as the proton but with negative charge.

47. (A)

An alpha particle has two protons and two electrons. Incorporation of one of its protons and two neutrons increases the original nitrogen's atomic number and mass to 8 (7+1) and 17 (14+3), respectively. A free proton, or hydrogen ion, $_1H^1$, is left over.

48. (A)

The helium nucleus, actually an alpha particle with two protons and two neutrons, adds two protons to aluminum to increase atomic number from 13 to 15. Adding one of its neutrons as well, mass is increased from 27 to 30. One free neutron remains. Working backwards from $_{15}P^{30}$ yields $_{13}Al^{27}$ by subtraction.

49. (B)

This is a straight-forward definition of atomic fusion as opposed to a splitting or fission.

50. (D)

During electrolysis, anions, or negative ions, move to an anode and release electrons (oxidation). The <u>electrons</u> move through a metallic conductor to the cathode where the electrons are accepted (reduction). Thus the electrons move from one electrode to another.

51. (D)

For every faraday of electricity that is passed through a series of electrolytes, one gram equivalent weight of an element is released at an electrode. The corresponding numbers for atomic weights are found from the periodic table. Silver, Ag, is the only correctly stated gram weight.

52. (B)

A monovalent element such as hydrogen, $H^+$, requires its entire gram equivalent weight to furnish 6.02 x $10^{23}$ electrons. In this case, each atom furnishes one electron. Aluminum is trivalent, furnishing three electrons per atom. Therefore, only one-third of its gram equivalent weight is needed. 1/3 x 27 (Al)=9 grams.

53. (B)

The potential difference is calculated by determining the more easily oxidized metal's potential, and subtracting the standard electrode potential from it. Only a subtraction of .762 (Zn)–(–.344) (Cu) yields a difference of 1.1068. Zinc loses electrons more easily than copper.

54. (B)

Ionization energy is the energy needed to move one or more electrons from the neutral atoms. All choices are alkali metals from group #1 of the periodic table. Although all these metals tend to lose their single valence electron and ionize, lithium has only two energy shells, with its electron close to the nucleus attracting sphere. Therefore, more energy is needed to attract its valence electron from the atom.

55. (E)

After balancing the reaction we obtain:

$$2\,C_8H_{18} + 25\,O_2 \rightarrow 16\,CO_2 + 18\,H_2O$$

Octane's molecular weight is 114: C(8x12) + H(18x1). Thus one mole weighs 114 grams. From the balanced equation, two moles (228 grams) react with twenty-five moles of oxygen gas. At STP, one mole of a gas occupies 22.4 liters. By proportion, the twenty-five reacting moles occupy 560 liters (22.4x25). Since oxygen is about one-fifth of the atmosphere, 560 is multiplied by five for the air volume required.

56. (D)

The total amount of $NH_3$ added is:

$$1.2 \frac{mole}{l} \times 0.1\ l = 0.12 \text{ mole}$$

The total amount of $NH_4^+$ added is:

$$0.5 \frac{mole}{l} \times 0.4\ l = 0.2 \text{ mole}$$

Therefore the "new" concentrations for a total volume of (0.1 + 0.4) l are:

$$[NH_3] = \frac{0.12}{0.5} = 0.24 \text{ M}$$

$$[NH_4^+] = \frac{0.2}{0.5} = 0.4 \text{ M}$$

From the reaction below:

$$NH_3 + H_2O \leftrightarrow NH_4^+ + OH^-$$

We obtain the expression for $K_b$:

$$K_b = \frac{[NH_4^+]\,[OH^-]}{[NH_3]}$$

$$[OH^-] = \frac{(1.8x10^{-5}) \times (0.24)}{0.4}$$

$$[OH^-] = 1.08x10^{-5}$$

$$pOH = -\log [OH^-]$$

$$pOH = -(-5 + 0.334)$$

pOH = 4.966 and pH = 14– pOH = 9.0334

pH $\simeq$ 9.03

57. (E)

$\Delta G^{o}$ is given by the expression below:

$$\Delta G^{o} = -2.303RT \log K_p$$

$$\Delta G^{o} = -2.303 \times (8.314) \times 298 \times \log 8$$

$$= -2.303 \times 8.314 \times 298 \times 0.9031$$

$$= -5152.9 \text{ J} = -5.153 \text{ KJ}$$

58. (B)

Solution:

Let the rate be expressed as follows:

$$\text{rate} = K[A]^n[B]^m$$

To find n, examine the data and find the ratios of rates for reactions where [B] was kept constant as:

$$\frac{\text{rate}_2}{\text{rate}_1} = \frac{K[2A]^n[B]^m}{K[A]^n[B]^m} = \frac{4.8}{1.2} = 4$$

Simplifying we have

$$2^n = 4 = 2^2$$

$\rightarrow$ $n = 2$

or $$\frac{\text{rate}_4}{\text{rate}_1} = \frac{K[3A]^n}{K[A]} \quad \frac{[B]^m}{[B]^m} = \frac{10.8}{1.2} = 9$$

$\rightarrow$ $3^n = 9 = 3^2$

$n = 2.$

Similarly for m consider reactions where [A] = constant.

$$\frac{\text{rate}_3}{\text{rate}_1} = \frac{K[A]^n[2B]^m}{K[A]^n[B]} = \frac{2.4}{1.2} = 2$$

or $2^m = 2 = 2^1$

$\rightarrow$ $m = 1$

or $$\frac{\text{rate}_5}{\text{rate}_1} = \frac{K[A]^n\,[3B]^m}{K[A]^n\,[B]^m} = \frac{3.6}{1.2} = 3$$

$3^m = 3$

$m = 1. \therefore \text{ rate} = K[A]^2[B].$

59. (A)

K is given by the expression below:

$$\text{rate} = K[A]^2[B]$$

$$1.2\,\frac{\text{mole}}{1 \times s} = K\left[\frac{1\text{ mole}}{1}\right]^2 \left[\frac{1\text{ mole}}{1}\right]$$

$$K = 1.2\text{ liter}^2/\text{mole}^2\text{sec.}$$

60. (E)

Suppose that all the heat will be transformed so $Q=E_c$, where Q is the heat and $E_c$ is the kinetic energy.

The kinetic energy ($E_c$) is given by $E_c = \frac{1}{2}mv^2$.

Also the heat is given per unit of mole and the mass corresponds to the mass of 1 mole of water. After transforming 57,800 cal into KJ, we obtain:

$$2.41 \times 10^{12} = \frac{1}{2} \times 18 \times v^2$$

$$v = 5.16 \times 10^5\text{ cm/s or } 18{,}500\text{ km/h}$$

61. (B)

The dissociation equation is given below:

$$CH_3COOH \leftrightarrow H^+ + CH_3COO^-$$

If the dissociation is 1%, it means that 0.1 x 0.01 molecules dissociated and generated 0.1 x 0.01 molecules of $H^+$ or $[H^+] = 1 \times 10^{-3}$, so pH = 3.

62. (E)

The solution with greatest $[Ag^+]$ is the solution with the greatest solubility product, which is $AgBrO_3$.

63. (B)

According to Dalton's law, "the total pressure of a mixture of gases is the summation of the pressures of the individual gases if they occupied alone the volume of the mixture."

$$\left.\frac{P_1V_1}{T_1} = \frac{PV}{T}\right\} \quad \begin{array}{l} T = T_1 \\ V = 700 \\ P = ? \end{array}$$

$$400x200 = Px700$$

$$P = \frac{400x200}{700} = \frac{80000}{700}$$

$$\left.\frac{P_2V_2}{T_2} = \frac{P'V'}{T^1}\right\} \quad \begin{array}{l} T' = T_2 \\ V = 700 \\ P = ? \end{array}$$

$$300x200 = P^1x700$$

$$\frac{300x200}{700} = P^1 = \frac{600}{7} \quad Ptotal = P+P^1 = \frac{800+600}{7} = \frac{1400}{7} = 200 \text{ mm}$$

64. (D)

CIS-TRANS isomerism occurs only in compounds which divide the molecule into two parts, so that two different configurations are possible. Two possibilities are shown below:

```
H       CH3   CH3      CH3
 \     /         \     /
  C = C           C = C
 /     \         /     \
H       CH3     H       H
```

65. (B)

The aromatic compounds are very stable, due to resonance, so substitution is the most likely to happen.

66. (E)

The negative sign to the heat per mole in kilocalories means that the heat is liberated. One mole of methane is 16 grams: 12+4(1). Therefore three moles liberates 54 kilocalories or 3 x 18.

67. (B)

The first step consists of determining the molarity of the solution, which is obtained by dividing the mass of the solute by its molecular weight:

$$\text{Molarity} = \frac{8}{62} = 0.13 \text{ mole}$$

To obtain the molality, it is necessary to divide the molarity by the mass of solvent as below:

$$\text{Molality} = \frac{\text{molarity}}{K_p \text{ of solvent}} = \frac{0.13}{0.2} = 0.645\text{m}$$

The change in freezing point $\Delta t_f$ is:

$$\frac{1.86°\text{C}}{\text{M}} \times 0.645\text{M} = 1.2°\text{C}$$

68. (A)

By definition, atomic number is the number of protons or electrons. With 11 electrons, its electron arrangement is 2-8-1 over three energy levels. If the atomic number is 23, however, 12 neutrons must add to 11 protons for this total mass by simple subtraction. Note its outer level has one electron.

69. (D)

Oxygen constitutes about 49%, followed by silicon (25%), aluminum (7%), iron (5%), and calcium (3.5%) in the earth's crust.

70. (B)

This is a historical fact. The modern periodic table ranks a larger array of elements by atomic number.

71. (D)

Fluorine's high electronegativity (electron attracting power) yields a strong force with the metallic-acting hydrogen. Fluorine is the most active nonmetal, topping chlorine, bromine, and iodine, which possess decreasing electronegativity moving down the halogen family, (#7 in the periodic table). Carbon, with a valence of four, is neither strongly metallic nor nonmetallic. Its electronegativity is less than most strong nonmetals.

72. (A)

"Au" is the symbol for gold, which is malleable (capable of being hammered into a desired shape) and ductile (capable of being drawn into a wire). Metals such as Fe, Cu, Ni, and Pt in the middle of the periodic table have these properties. Cadmium (Cd), mercury (Hg), potassium (K), and sodium (Na) lack this table location and these properties.

73. (C)

For example, by varying valences, either 63.54 grams or 31.77 grams of copper can combine with 8 grams of oxygen in two different oxide compounds. These two numbers reduce in a simple ratio of 2 to 1.

Conservation of mass states that the total mass of reactants equals the total mass of products in a chemical reaction. In spite of occasional multiple effects of combination, elements for a specific compound unite in an unvarying, definite proportion. For example, water ($H_2O$) is always 88.9% oxygen and 11.1% hydrogen. The other two choices are not legitimate laws.

74. (A)

The wide span of intermittent transition elements in the table are not under a vertical group (i.e. 1, 2, 6, or 7) to name the outer shell electron number or valence. Alkali, alkali earth, or halogen families are groups 1, 2, or 7 to convey this fact. Metals are demarcated to the left of the step line with nonmetals to the right. Atomic number and mass are statistics within the cell for each element.

75. (C)

Boyle's law states that gas volume varies by inverse proportion with a pressure change. Pressure is reduced here so the gas will expand:

$$300 \times \frac{1000}{760} = 394.8\text{ml}$$

76. (D)

Atomic number is the number of protons (19). Atomic mass is the number of protons plus neutrons (39). Thus the number of neutrons must be 20, by simple subtraction.

77. (B)

One mole of any gas fills 22.4 liters. It also contains $6.02 \times 10^{23}$ molecules, a constant called Avogadro's number. 11.2 l is one-half of a mole, and is one-half of 22.4.

78. (B)

Oxidation is the loss of electrons. Copper, Cu, is lowest on the series among the list of choices with a tendency to do this.

79. (B)

Gold is at the bottom of the series list, and silver is near the bottom. Their tendency to oxidize and react is low. Potassium, at the top, is most likely to react. A metal must be above another to replace it in a compound. Sodium (Na) is above nickel (Ni). Iron (Fe) is not above manganese (Mn).

80. (A)

With increasing reactant concentrations, collisions between reactant molecules become more and more likely. The reactant capability multiplies. Choice (B) is a somewhat correct statement but the reaction rate does not increase by multiplying reactant temperatures. Choice (E) is a correct statement, but does not directly refer to the mass action idea of reactant amounts.

81. (B)

By this theory, acids are hydrogen ion (proton) donors in solution. The HA reactant yields $H^+$ by dissociation in solution.

82. (B)

Regions A and C are the positive hydrogen end of the water dipole with its asymmetrical distribution of shared electrons. The electrons are shared more in the larger oxygen end with its increased electron attracting power. As water molecules get together, the A-C positive end orients toward the adjacent water molecule's negative pole.

83. (A)

The free energy amount is positive, meaning that energy is absorbed rather than released (negative sign). One mole of NO, nitrogen monoxide, is 46 grams [(16 x 2) + 14]. 23 grams is one-half of a mole and therefore absorbs one-half the molar energy, which is 4.

84. (D)

Oxygen availability does not vary in the lab, and is normally not manipulated in an experiment. Increased temperature or reactant concentrations increase the probability of reactant molecule collisions. A catalyst can alter reaction rate, and substances vary in reactivity.

85. (C)

NaCl is sodium chloride. The amount dissolved changes by only .6 gram over 30 degrees. This is much less than potassium nitrate ($KNO_3$)–32g, potassium chloride (KCl)–9g, sodium sulfate ($Na_2SO_4$)–36.1g or sodium nitrate ($NaNO_3$)–19g.

# SECTION II

## PART A

1. (a) The balanced equation is:

$$3Fe + 4H_2O \rightarrow Fe_3O_4 + 4H_2$$

Start with the element iron. <u>THREE</u> iron atoms are required to produce the $Fe_3O_4$ molecule with its three iron atoms. <u>FOUR</u> $H_2O$ molecules are required to produce the four oxygen atoms in $Fe_3O_4$. By multiplication, the eight hydrogen atoms in $H_2O$ (4x2) also produce <u>FOUR</u> molecules of the diatomic hydrogen gas.

(b) Use the periodic table. Write atomic weights. Multiply by the number of atoms of an element. Add these totals in a molecule.

Fe = <u>56</u>

$H_2O$ = <u>18</u>: (2x1) + 16

$Fe_3O_4$ = <u>232</u>: (3x56) + (4x16)

$H_2$ = <u>2</u>: (2x1)

(c) Mole amounts in the balanced equation are:

3Fe: 3x56 = 168 grams　　　$4H_2$: 4x2 = 8 grams

$4H_2O$: 4x18 = 72 grams　　　$1Fe_3O_4$: = 232 grams

42 grams is one-fourth of the 168 grams. Therefore, the gram amounts of the other three substances are reduced proportionately to maintain the relative yet definite amounts:

$H_2O$: .25x72 = 18 grams

$Fe_3O_4$: .25x232 = 58 grams

$H_2$: .25x8 = 2 grams

(d) One mole of gas occupies 22.4 liters at STP. One mole of $H_2$ is produced in this reaction, so the volume is 22.4 l.

(e) Using the relationship below:

$$\frac{P_1V_1}{T_1} \quad \frac{P_2V_2}{T_2}$$

and substituting:

$$\frac{760 \times 22.4}{273} = \frac{1000 \times V_2}{293}$$

$$V_2 = \frac{293 \times 760 \times 22.4}{273 \times 1000} = 18.3$$

## PART B

2. (a) The balanced equation is:

$$2HI \rightarrow H_2 + I_2$$

Both products, hydrogen and iodine, are diatomic, thus 2HI will produce one of each molecule.

(b) The oxidation numbers are given below:

$= 1.1 \times 10^{-2}$

$\underline{+1}$ H
$\underline{-1}$ I

Hydrogen, acting as a metal, loses one electron and becomes +1, while iodine, acting as a nonmetal, receives one electron and becomes electronegative –1.

(c) For the reaction below, the concentration at the beginning and at equilibrium are illustrated below:

| | $2HI$ | $\rightarrow$ | $H_2$ + | $I_2$ |
|---|---|---|---|---|
| beginning | 2 | | O | O |
| equilibrium | 2-2 x 0.25 | | 0.25 | 0.25 |

At equilibrium there are 1.5 moles of HI and 0.25 moles of $H_2$ and $I_2$.

(d) The equilibrium constant is given by the product of the concentrations of products divided by the product of concentrations of reactants. All concentrations are raised to powers given by their stoichiometric coefficients.

Since the volume is 1 liter, the concentrations are given in (c) and correspond to 1.5, 0.25, and 0.25 of HI, $I_2$, and $H_2$, respectively.

$$K = \frac{(0.25)\,(0.25)}{(0.75)^2} = 1.1 \times 10^{-2}$$

3. (a) The equation is already balanced. One molecule of carbonic acid, $H_2CO_3$, furnishes one hydrogen ion, $H^+$, and one bicarbonate ion, $HCO_3^-$, upon dissociation.

(b) The molecular weight of carbonic acid is 62: $H_2$ (2x1) + C(12) + $O_3$ (3x16). A one molar solution is one mole of a substance per liter of solution; 6.2 grams is 10% or .1 of that total. By simple proportion:

$$\frac{1M}{62} = \frac{X}{6.2} \qquad X = .1M$$

(c) The ionization constant is the product of the moles of ionic products divided by the product of the moles of reactants.

$$K_a = \frac{[H^+]\,[HCO_3^-]}{[H_2CO_3]}$$

$K_a$ is given:

$$3 \times 10^{-7} = \frac{[H^+]\,[HCO_3^-]}{[H_2CO_3]}$$

Each molecule of $H_2CO_3$ produces one $H^+$ ion and one $HCO_3^-$ ion. Therefore:

$$H^+ = HCO_3^-$$

$$H_2CO_3 = 0.1 - [H^+]$$

Since $[H^+]$ is small relative to .1, because carbonic acid is a weak acid, assume

$$0.1 - [H^+] \cong 0.1$$

Therefore, by substitution

$$K_a = \frac{[H^+][H^+]}{0.1} = 3 \times 10^{-7}$$

$$[H^+]^2 = (1 \times 10^{-1})(3 \times 10^{-8})$$

$$[H^+] = \sqrt{3 \times 10^{-8}}$$

$$[H^+] = 1.42 \times 10^{-4}$$

(d) pH, by definition, is – log of the hydrogen ion concentration. From (c), the hydrogen ion concentration (moles/liter) is:

$$[H^+] = 1.42 \times 10^{-4}$$

$$pH = -\log [H^+] \quad \text{or}$$

$$pH = \log \frac{1}{H}$$

$$= \log \frac{1}{1.42 \times 10^{-4}}$$

$$= \log \frac{10^4}{1.42}$$

$$= 4 - .1523 \text{ (subtract logs to divide)}$$

$$= 3.8477$$

## PART C

4. Pauli's exclusion principle permits us to predict the maximum number of electrons in a certain orbital and the maximum number of electrons in a subshell, given by $2n^2$. According to Pauli's exclusion principle, "no two electrons in any one atom may have all four quantum numbers the same." In other words, the same orbital can have the same value for n, l and m, but since their spin is different, the fourth number will be different. In effect, this limits the number of electrons in any given orbital to two, and it also requires that the spins of these two electrons be in opposite directions.

5. The activity series is a vertical listing of metals. This listing places the most active metals at the top, arranging metals in order of decreasing activity in descending order through the list. An example of the listing of some prominent metals in this series is as follows:

K
Ca
Na
Mg
Al
Zn
Fe
Ni
Sn
Pb
Cu
Hg
Ag
Pt
Au

Thus, in this series, potassium (K) is the most active metal, and gold (Au) is the least active metal. Activity is evaluated by the metal's ability to lose valence electrons. Loss of the outer valence electrons is termed oxidation. The greater the metal's activity or reactivity, the easier it is oxidized. Potassium, a very reactive metal, easily donates its one, loosely-held valence electron. Calcium (Ca), is also very reactive, but not quite as much as potassium. At the other end of the spectrum, silver (Ag), platinum (Pt), and gold (Au) are very unreactive. They are seldom found chemically combined in nature. On the other hand, the easily oxidized potassium or sodium can react with water vapor in the air. They must therefore be stored in inert settings.

Reactive metals lose valence electrons by oxidation, and become electronegative. Thus, in the compound $CaCl_2$, calcium loses its two outer negative electrons. Its oxidation state changes from 0, indicative of a neutral atom, to +2. Using the electrons, the nonmetal chlorine becomes –2. A nonmetal active above another in the series can replace it in a salt. For example: $Pb + CuSO_4 \rightarrow PbSO_4 + C$

The more active lead (Pb) replaces copper (Cu). Its oxidation state changes from 0, neutral atom, to +2 as it loses two valence electrons. It is oxidized. Copper receives those electrons, changing from +2 to 0. It receives those two electrons and is liberated. Rather than being oxidized, it gains electrons and is reduced.

6. The three physical states of water are: solid, liquid, and gas. The difference between the three states is in the relative movement of water molecules and also the intermolecular distances. Molecules move most rapidly and are furthest apart in the gaseous state. Thus, the highly compressible or expandable gas (or water vapor) has an indefinite volume. Its shape conforms to the shape of its container, as the molecules move quite independently. Intermolecular distance lessens in the liquid state as molecules slide over one another. This fluidity makes the liquid conform to its container's shape. The volume is definite, however. There is little room for compression, and the attracting molecules are much less likely to separate.

As liquid water molecules cool, they become progressively closer together. This increasing compactness increases water's density. Their maximum closeness is, however, at four degrees Centigrade. At this temperature the molecules set up in a fixed pattern. At lower temperatures, the molecules get further apart. Water expands and density decreases. Thus, solid ice (0 degrees C) floats on liquid water. In cold winter temperatures, ponds therefore freeze from the top to the bottom, preserving and insulating aquatic life.

$H_2O$ molecules display a positive (hydrogen) and negative (oxygen) end. Two shared electron pairs are not held equally, spending more time in oxygen's sphere. Water dipoles, with positive and negative ends, therefore orient their opposite ends to one another for weak intermolecular forces. Larger than normal amounts of heat energy are needed to separate water molecules for a liquid to gas conversion: i.e. Water has a high heat of vaporization. As a "heat sponge," large bodies of water can absorb land mass heat without changing physical state. Thus, water has a moderating effect on local climate.

7. A good example of reactant identity on reaction rate centers on metals and nonmetals. Metals, particularly toward the lower lefthand corner of the periodic table, tend to lose electrons by oxidation. A metal that is more easily oxidized will react more rapidly in the presence of a nonmetal. A more active nonmetal such as fluorine, chlorine, or oxygen will more readily accept electrons and suffer reduction. Metals and nonmetals react to form compounds at different rates depending on reactant identity. Salts are formed by an acid and base reaction with varying rates, again depending on identity: i.e.HCl (hydrochloric acid) and NaOH (sodium hydroxide) react very quickly to form the salt NaCl plus $H_2O$.

Increased temperature causes increased rates of movement of reacting particles. This increases the probability for collision and contact for reacting particles. In general, reaction rates double for every 10 degrees Celsius rise in temperature.

Catalysts alter chemical reaction rates without suffering chemical change in the process. Most are positive, thus accelerating the rate of a reaction. They are believed to lower the resistance offered by the energy of activation required to start the reaction.

The concentration factor is best explained by the Law of Mass Action: the speed of a chemical reaction is directly proportional to the product of the concentrations of the reacting molecules. Therefore, increasing concentrations favor greater probability for reactants' collision, contact, and thus reaction. If a reaction is reversible as follows, many factors would favor the formation of the product C.

$$A + 3B \leftrightarrow 2C$$

For example, it will shift to the right if:

a) The product is a gas or precipitate. Its consequent escape creates a void, leading to increased formation of C.

b) Temperature – If, for example, the left-to-right reaction is exothermic, adding heat shifts it to the left. This direction absorbs heat and alleviates the stress. The opposite condition shifts it to the right.

(c) Pressure – If the substances are gases, 3 gas volumes of B combine with a gas volume of A to yield 2 gas volumes of C. If pressure is increased, the shift is to the right (from 4 volumes to 2) to relieve the stress.

8. Inorganic chemistry is the branch dealing with noncarbon compounds, with a few exceptions. By definition, organic chemistry is the study of compounds containing carbon. Simple carbon compounds such as carbon dioxide, carbon monoxide and carbonate compounds are not included.

More specifically, compounds with elaborate carbon skeletons in their molecules are organic. Carbon, with its valence of four, readily bonds with other carbon atoms. Each time a carbon atom bonds to another carbon, three remaining valence electrons remain in each for further bonds.

Thus, the potential for configurations is tremendous, with either a chain or ring of carbon atoms as the backbone. This complexity may lead to a biological ability to store energy, encode information, etc.

Consider one prominent family of organic compounds, the alkanes. These hydrocarbons show C to C and C to H covalent bonds:

```
                  H
                  |
methane:      H - C - H
                  |
                  H

                      H   H
                      |   |
ethane:           H - C - C - H
                      |   |
                      H   H
                  |   |   |
propane:      H - C - C - C - H
                  |   |   |
                  H   H   H

                  H   H   H   H
                  |   |   |   |
butane:       H - C - C - C - C - H
                  |   |   |   |
                  H   H   H   H

                  H   H   H   H   H
                  |   |   |   |   |
pentane:      H - C - C - C - C - C - H
                  |   |   |   |   |
                  H   H   H   H   H
```

These molecules can also exist in carbon rings, yielding isomers or varying structural formulas of the same molecule. C to C double or triple bonds alter the C to H ratios in other organic families.

Unlike inorganic compounds, organic compounds are generally insoluble in water. They do dissolve in many organic solvents such as carbon tetrachloride and alcohol. They seldom ionize and react more slowly than inorganic substances. Many are readily combustible.

Four important families of organic compounds have biological significance. Carbohydrates supply energy. Small molecules, such as monosaccharide glucose, flow in the blood of the human body. This blood sugar supplies immediate energy to cells. Large carbohydrates such as polysaccharides glycogen (animals) and starch (plants) store energy. Lipids are the most efficient energy storers on a per gram basis. They also build body structure, combining with proteins in cell membranes. By trapping heat within a body, lipid or fat layers also insulate. In addition to structural contributions, all enzymes and some hormones are proteins. Enzymes as organic catalysts speed up reactions. Nucleic acids such as DNA and RNA are the essence of gene structure and function.

# THE ADVANCED PLACEMENT EXAMINATION IN

# CHEMISTRY

# TEST IV

# THE ADVANCED PLACEMENT EXAMINATION IN

# CHEMISTRY

## ANSWER SHEET

1. Ⓐ Ⓑ Ⓒ Ⓓ Ⓔ
2. Ⓐ Ⓑ Ⓒ Ⓓ Ⓔ
3. Ⓐ Ⓑ Ⓒ Ⓓ Ⓔ
4. Ⓐ Ⓑ Ⓒ Ⓓ Ⓔ
5. Ⓐ Ⓑ Ⓒ Ⓓ Ⓔ
6. Ⓐ Ⓑ Ⓒ Ⓓ Ⓔ
7. Ⓐ Ⓑ Ⓒ Ⓓ Ⓔ
8. Ⓐ Ⓑ Ⓒ Ⓓ Ⓔ
9. Ⓐ Ⓑ Ⓒ Ⓓ Ⓔ
10. Ⓐ Ⓑ Ⓒ Ⓓ Ⓔ
11. Ⓐ Ⓑ Ⓒ Ⓓ Ⓔ
12. Ⓐ Ⓑ Ⓒ Ⓓ Ⓔ
13. Ⓐ Ⓑ Ⓒ Ⓓ Ⓔ
14. Ⓐ Ⓑ Ⓒ Ⓓ Ⓔ
15. Ⓐ Ⓑ Ⓒ Ⓓ Ⓔ
16. Ⓐ Ⓑ Ⓒ Ⓓ Ⓔ
17. Ⓐ Ⓑ Ⓒ Ⓓ Ⓔ
18. Ⓐ Ⓑ Ⓒ Ⓓ Ⓔ
19. Ⓐ Ⓑ Ⓒ Ⓓ Ⓔ
20. Ⓐ Ⓑ Ⓒ Ⓓ Ⓔ
21. Ⓐ Ⓑ Ⓒ Ⓓ Ⓔ
22. Ⓐ Ⓑ Ⓒ Ⓓ Ⓔ
23. Ⓐ Ⓑ Ⓒ Ⓓ Ⓔ
24. Ⓐ Ⓑ Ⓒ Ⓓ Ⓔ
25. Ⓐ Ⓑ Ⓒ Ⓓ Ⓔ
26. Ⓐ Ⓑ Ⓒ Ⓓ Ⓔ
27. Ⓐ Ⓑ Ⓒ Ⓓ Ⓔ
28. Ⓐ Ⓑ Ⓒ Ⓓ Ⓔ
29. Ⓐ Ⓑ Ⓒ Ⓓ Ⓔ
30. Ⓐ Ⓑ Ⓒ Ⓓ Ⓔ
31. Ⓐ Ⓑ Ⓒ Ⓓ Ⓔ
32. Ⓐ Ⓑ Ⓒ Ⓓ Ⓔ
33. Ⓐ Ⓑ Ⓒ Ⓓ Ⓔ
34. Ⓐ Ⓑ Ⓒ Ⓓ Ⓔ
35. Ⓐ Ⓑ Ⓒ Ⓓ Ⓔ
36. Ⓐ Ⓑ Ⓒ Ⓓ Ⓔ
37. Ⓐ Ⓑ Ⓒ Ⓓ Ⓔ
38. Ⓐ Ⓑ Ⓒ Ⓓ Ⓔ
39. Ⓐ Ⓑ Ⓒ Ⓓ Ⓔ
40. Ⓐ Ⓑ Ⓒ Ⓓ Ⓔ
41. Ⓐ Ⓑ Ⓒ Ⓓ Ⓔ
42. Ⓐ Ⓑ Ⓒ Ⓓ Ⓔ
43. Ⓐ Ⓑ Ⓒ Ⓓ Ⓔ
44. Ⓐ Ⓑ Ⓒ Ⓓ Ⓔ
45. Ⓐ Ⓑ Ⓒ Ⓓ Ⓔ
46. Ⓐ Ⓑ Ⓒ Ⓓ Ⓔ
47. Ⓐ Ⓑ Ⓒ Ⓓ Ⓔ
48. Ⓐ Ⓑ Ⓒ Ⓓ Ⓔ
49. Ⓐ Ⓑ Ⓒ Ⓓ Ⓔ
50. Ⓐ Ⓑ Ⓒ Ⓓ Ⓔ
51. Ⓐ Ⓑ Ⓒ Ⓓ Ⓔ
52. Ⓐ Ⓑ Ⓒ Ⓓ Ⓔ
53. Ⓐ Ⓑ Ⓒ Ⓓ Ⓔ
54. Ⓐ Ⓑ Ⓒ Ⓓ Ⓔ
55. Ⓐ Ⓑ Ⓒ Ⓓ Ⓔ
56. Ⓐ Ⓑ Ⓒ Ⓓ Ⓔ
57. Ⓐ Ⓑ Ⓒ Ⓓ Ⓔ
58. Ⓐ Ⓑ Ⓒ Ⓓ Ⓔ
59. Ⓐ Ⓑ Ⓒ Ⓓ Ⓔ
60. Ⓐ Ⓑ Ⓒ Ⓓ Ⓔ
61. Ⓐ Ⓑ Ⓒ Ⓓ Ⓔ
62. Ⓐ Ⓑ Ⓒ Ⓓ Ⓔ
63. Ⓐ Ⓑ Ⓒ Ⓓ Ⓔ
64. Ⓐ Ⓑ Ⓒ Ⓓ Ⓔ
65. Ⓐ Ⓑ Ⓒ Ⓓ Ⓔ
66. Ⓐ Ⓑ Ⓒ Ⓓ Ⓔ
67. Ⓐ Ⓑ Ⓒ Ⓓ Ⓔ
68. Ⓐ Ⓑ Ⓒ Ⓓ Ⓔ
69. Ⓐ Ⓑ Ⓒ Ⓓ Ⓔ
70. Ⓐ Ⓑ Ⓒ Ⓓ Ⓔ
71. Ⓐ Ⓑ Ⓒ Ⓓ Ⓔ
72. Ⓐ Ⓑ Ⓒ Ⓓ Ⓔ
73. Ⓐ Ⓑ Ⓒ Ⓓ Ⓔ
74. Ⓐ Ⓑ Ⓒ Ⓓ Ⓔ
75. Ⓐ Ⓑ Ⓒ Ⓓ Ⓔ
76. Ⓐ Ⓑ Ⓒ Ⓓ Ⓔ
77. Ⓐ Ⓑ Ⓒ Ⓓ Ⓔ
78. Ⓐ Ⓑ Ⓒ Ⓓ Ⓔ
79. Ⓐ Ⓑ Ⓒ Ⓓ Ⓔ
80. Ⓐ Ⓑ Ⓒ Ⓓ Ⓔ
81. Ⓐ Ⓑ Ⓒ Ⓓ Ⓔ
82. Ⓐ Ⓑ Ⓒ Ⓓ Ⓔ
83. Ⓐ Ⓑ Ⓒ Ⓓ Ⓔ
84. Ⓐ Ⓑ Ⓒ Ⓓ Ⓔ
85. Ⓐ Ⓑ Ⓒ Ⓓ Ⓔ

# ADVANCED PLACEMENT CHEMISTRY EXAM IV

## SECTION I

Time: 1 Hour 45 Minutes
85 Questions

**Note**: For all questions referring to solutions, assume that the solvent is water unless otherwise stated.

**Directions:** Each set of lettered choices below refers to the numbered statements immediately following it. Select the one lettered choice that best fits each statement. A choice may be used once, more than once, or not at all in each set.

**Questions 1 – 3**

(A) I
(B) Kr
(C) Sr
(D) Cs
(E) At

1. Which element has the characteristics of a metalloid?

2. Which element is an inert gas?

3. Which of the elements is the most metallic?

**Questions 4 – 6**

(A) Calcium carbonate
(B) Stannous fluoride
(C) Acetic acid
(D) Formaldehyde
(E) Acetylsalicylic acid

4. Used as a preservative in embalming procedures

5. Used daily as a dental cleanser

6. Commonly found in aspirin

**Directions:** Each of the following questions or incomplete sentences is followed by five possible answers. Choose the best one in each case.

7. Allotropes are best described as

(A) having the same composition but occurring in different molecular structures.

(B) being without definite shape.

(C) having both acid and base properties.

(D) elements with more than one molecular or crystalline form and with different physical and chemical properties.

(E) having the same number of protons and electrons, but a different number of neutrons

8. Hydrolysis of sodium carbonate yields

(A) a strong acid and a strong base

(B) a weak acid and a strong base

(C) a weak acid and a weak base

(D) a strong acid and a weak base

(E) none of the above

9. What volume does 34 grams of ammonia gas occupy at standard temperature and pressure? (assume ideal behavior)

(A) 22.4 l

(B) 44.8 l

(C) 11.2 l

(D) 0 l

(E) none of the above

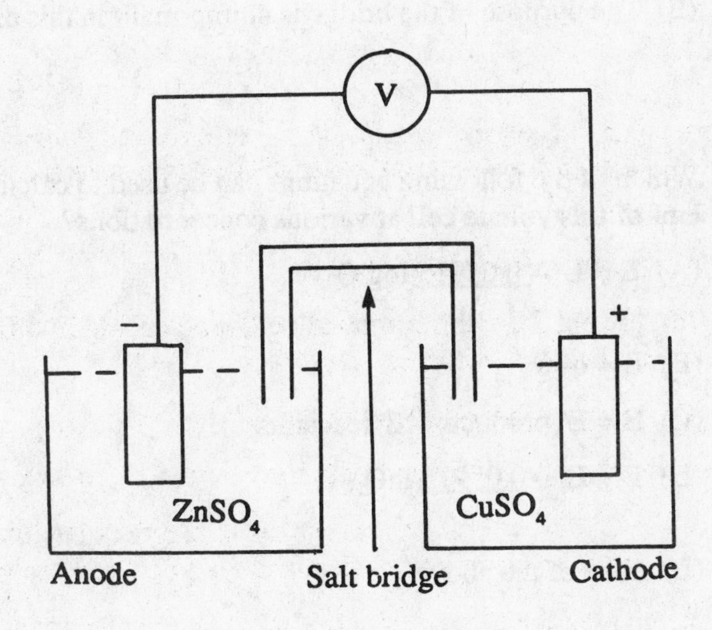


Given two solutions, $ZnSO_4$(1M) and $CuSO_4$(1M), answer **questions 10 – 12** based on this diagram.

10. What reaction, if any, takes place at the cathode?

(A) $Cu^{2+} + 2e^- \rightarrow Cu$

(B) $Zn^{2+} + 2e^- \rightarrow Zn$

(C) $Cu \rightarrow Cu^{2+} + 2e^-$

(D) $Zn \rightarrow Zn^{2+} + 2e^-$

(E) none of the above

11. What is the purpose of the salt bridge?

(A) To allow the two solutions to mix.

(B) To allow copper to migrate to the other cell and vice-versa.

(C) To allow the positive and negative ions to migrate.

(D) To allow counterclockwise flow of current.

(E) The purpose of the bridge is unimportant in this diagram.

12. Which of the following equations can be used to calculate the emf of this voltaic cell at various concentrations?

(A) $E = E^{\circ} - \frac{0.05915}{n} \log Q$

(B) $E = q - w$

(C) $E = E^{\circ}$ products $- E^{\circ}$ reactants

(D) $E = E^{\circ} - \frac{0.05915}{n} \ln Q$

(E) none of the above

13. Which of the following is not a homogeneous mixture?

(A) Sugar in water

(B) Salt in water

(C) Sand in water

(D) Gasoline

(E) Soft drinks

14. Rutherford's most significant discovery in nuclear chemistry is that

(A) the volume occupied by an atom is largely empty space

(B) the gold atom has a dense nucleus

(C) alpha particles radiate through gold foil

(D) the negative electron region is unaffected by the positive $\alpha$–particle

(E) the $\alpha$ –particle passes mostly undeflected because of the extremely small size of the atom

15. Which of the following represents the structure of a noble gas?

(A) $^{19}_{9}X$

(B) $^{21}_{10}X$

(C) $^{17}_{8}X$

(D) $^{15}_{7}X$

(E) $^{13}_{6}X$

16. What is the functional group $R-\overset{\overset{\displaystyle O}{||}}{C}-OH$ representative of?

(A) Ethers

(B) Alcohols

(C) Acids

(D) Aldehydes

(E) Esters

17. Which of the following concepts explains the filling of atomic orbitals?

(A) Heisenberg Uncertainty Principle

(B) Pauli Exclusion Principle

(C) Aufbau Process

(D) Hund's rule

(E) Schrödinger's equation

18. Which of the following contains the largest <u>number</u> of atoms or molecules?

(A) 49 g of Fe

(B) 5 liters of $H_2$ at STP

(C) 150 g of ethanol

(D) 10 liters of ozone at STP

(E) 80 g of calcium carbide

19. The nomenclature for atomic orbitals consists of a number and a letter. These depend on, respectively

(A) the principal and magnetic quantum numbers

(B) the principal and spin quantum numbers

(C) the azimuthal and magnetic quantum numbers

(D) the principal and azimuthal quantum numbers

(E) the spin and azimuthal quantum numbers

20. The abbreviated electronic configuration of an element of atomic number 42 can be:

(A) $[Kr]5s^1 4d^5$

(B) $[Kr]5s^2 4d^4$

(C) $[Kr]4d^6$

(D) $[Kr]5s^2 5p^4$

(E) $[Kr]5s5p^5$

21. Which of the following is/are paramagnetic?

I. $Fe^{2+}$
II. $Fe^{3+}$
III. $CO^{3+}$
IV. $Ni^{2+}$

(A) I and IV

(B) I only

(C) II only

(D) I and II

(E) All the ions listed.

22. Which of the following diagrams describes the electron density in the $d_{x^2-y^2}$?

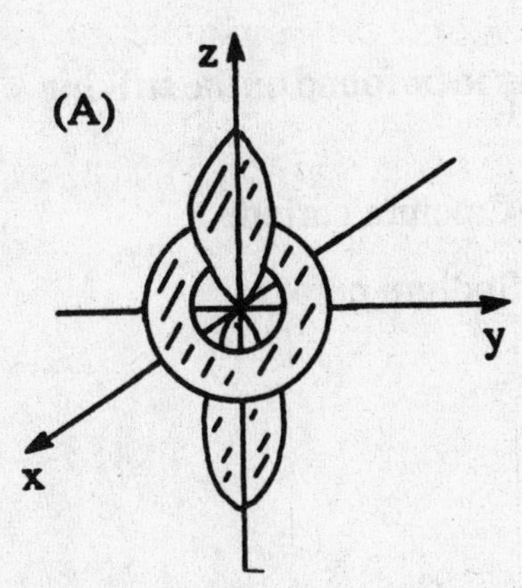


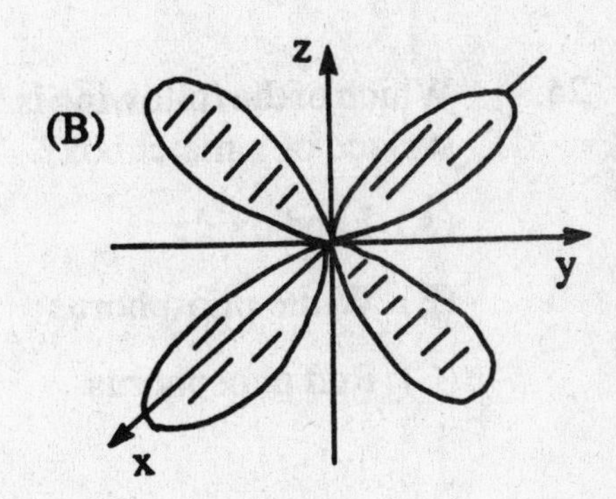


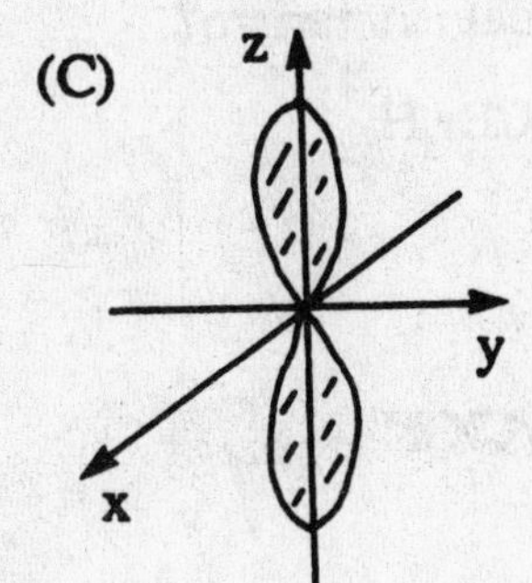


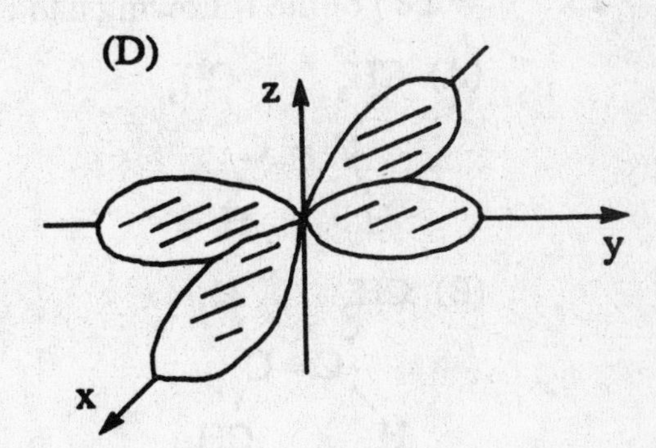


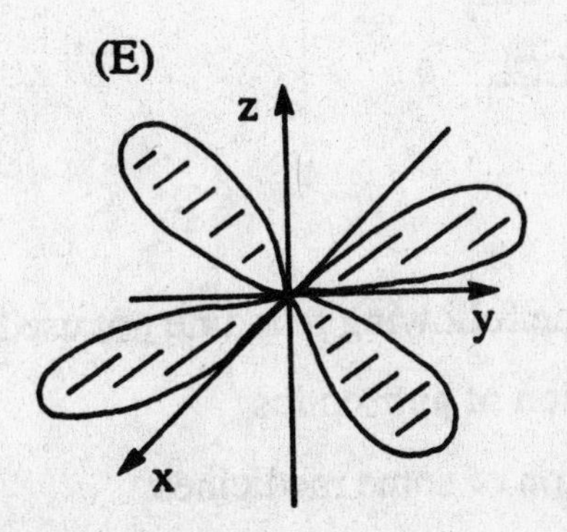

23. What will be the shape of a molecule when the central atom is surrounded by five atoms and the molecule does <u>not</u> contain lone pairs of electrons?

(A) tetrahedral
(B) triangular
(C) triangular bipyramid
(D) planar
(E) octahedral

24. Which of the following is most likely to be found on the striking surface of a matchbox?

(A) Lead oxide
(B) White phosphorus
(C) Red phosphorus
(D) Calcium carbide
(E) Sodium peroxide

25. Which of the following can exhibit optical isomerism?

(A)
```
CH3       CH3
   \     /
    C = C
   /     \
  H       H
```

(B)
```
CH3       H
   \     /
    C = C
   /     \
  H       CH3
```

(C) $N_2$ $CH_3$ $CCl_2$

(D) $CH_3$ $CBr_2H$

(E) HCBrClCN

26. In which of the following is iodine <u>not</u> used?

(A) Preparation of antiseptics

(B) Preparation of some medicines

(C) Photographic film

(D) Audio-recording tapes

(E) Dye compounds

27. VSEPR method is used to

(A) predict the geometries of an atom.

(B) estimate the energy levels of orbitals in an atom.

(C) estimate electronegativities of elements.

(D) predict the geometries of molecules and ions.

(E) There is no such thing as a VSEPR method.

28. At 25°C, the $K_{sp}$ for $CaSO_4$ and $Ag_2SO_4$ are $2.4 \times 10^{-5}$ and $1.2 \times 10^{-5}$ respectively. Which of the following is true?

(A) The solubility of $CaSO_4$ is twice that of $Ag_2SO_4$.

(B) The solubility of $Ag_2SO_4$ is twice that of $CaSO_4$.

(C) The solubility of $Ag_2SO_4$ is sensitive to the square of the sulphate ion concentration.

(D) The solubilities of $CaSO_4$ and $Ag_2SO_4$ are equal.

(E) The solubilities of $Ag_2SO_4$ and $CaSO_4$ in mol $liter^{-1}$ differ by a factor of 2.9.

H–O–H    O = C = O    O = S = O

29. For the molecules listed above, the resultant dipole moments are oriented as (from left to right)

(A) O, →, ←

(B) ↑, O, ↓

(C) ↑, O, ↑

(D) ↓, O, ↑

(E) ↑, ←, ↓

30. The balanced equation of the reaction of $MnO_4^{=}$ in an acidic medium to produce $MnO_4^-$ and $MnO_2$ is

(A) $3MnO_4^{=} + 4H^+ \rightarrow 2MnO_4^- + MnO_2 + 2H_2O$

(B) $3MnO_4^{=} \rightarrow 2MnO_4^- + MnO_2 + O_2$

(C) $2MnO_4^{=} + 2H_2O \rightarrow MnO_4^- + MnO_2 + 2O_2 + 2H_2$

(D) $2MnO_4^{=} + 2OH^- \rightarrow MnO_4^- + MnO_2 + O_2 + H_2$

(E) none of the above.

31. Methathesis reactions usually involve

(A) a transfer of protons

(B) an exchange of ions with the formation of one or more insoluble salts

(C) an electron and neutron transfer

(D) acid-base reactions

(E) none of the above

32. Which of the following salts are soluble?

I $(NH_4)_2CO_3$
II $CaSO_4$
III $PbCl_2$
IV $AgClO_4$

(A) I and III only

(B) I and IV only

(C) I, II and IV

(D) II, III and IV

(E) II and IV only

33. Which of the following combinations of names is right?

I Copper (II): cupric
II Cobalt (II): cobaltic
III Lead (II): plumbous
IV Tin (II): stannic

(A) I and III

(B) I, II and IV

(C) III only

(D) II only

(E) none of the above

34. Which of the following cycles is used to calculate the heat changes in the formation of an ionic compound from its elements?

(A) The Krebs cycle

(B) The cyclotron

(C) The Born–Haber cycle

(D) The Carnot cycle

(E) The periodic law

35. An important application of colligative properties of solutions is

(A) the determination of boiling points

(B) the determination of heat of fusion

(C) the determination of molecular weight

(D) the determination of atomic weight

(E) the evaluation of electronegativities

36. The diagram below is used to describe a reaction path. What are y, x and z respectively?

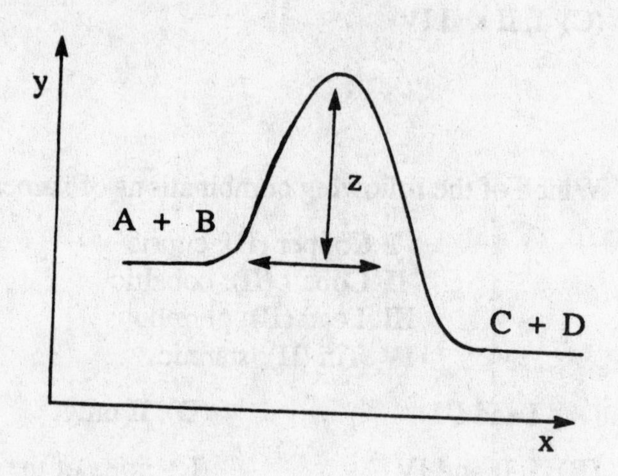


(A) temperature, volume, pressure

(B) activation energy, potential energy and temperature

(C) potential energy, reaction coordinate and activation energy

(D) distance, time, concentration

(E) time, distance, concentration.

37. The specific rate constant of the natural radioactive potassium isotope $^{40}_{19}K$ is $5.33 \times 10^{-10}$ year$^{-1}$, so the half-life of $^{40}_{19}K$ is:

(A) $1.3 \times 10^{9}$ years

(B) $1.3 \times 10^{-9}$ years

(C) $9 \times 10^{1.3}$ years

(D) $5.33 \times 10^{5}$ years

(E) $2.665 \times 10^{-10}$ years

38. Which of the following is/are both a Brønsted-Lowry base and a Lewis base?

I $NH_3$
II $BBr_3$
III $H_2O$
IV NaOH

(A) I and IV
(B) IV only
(C) I only
(D) II and III
(E) I and III

39. How many unpaired electrons does an $Ho^{3+}$ ion have in its ground state?

(A) 3
(B) 5
(C) 4
(D) 0
(E) 6

40. What is the electron configuration of $Ir^{+3}$ ion in its ground state?

(A) $[Xe]4f^{14}5d^{4}6s^{2}$
(B) $[Xe]4f^{14}5d^{6}$
(C) $[Xe]4f^{14}5d^{5}6s^{1}$
(D) $[Xe]4f^{11}5d^{7}6s^{2}$
(E) $[Xe]4f^{13}5d^{6}6s^{1}$

41. Predict the most common oxidation states of europium and scandium

(A) –1 and –3, respectively.
(B) +1 and +1, respectively.
(C) +1 and +2, respectively.

(D) +4 and +3, respectively.

(E) +2 and +3, respectively.

42. Solutions of alums are

(A) basic

(B) neutral

(C) acidic

(D) saturated with $Ca^{2+}$ ions

(E) isomorphous

**Questions 43 – 45** refer to the statement below: A voltaic cell consists of a combination of a standard silver electrode and another silver electrode in which the concentration of silver ions is $10^{-3}$M, $E^0_{Ag+/Ag}$ = 0.80V.

43. Which of the following is true?

(A) The standard electrode is the cathode.

(B) The standard electrode is the anode.

(C) There will be no electron transfer.

(D) This cell will work like a perpetual source of energy.

(E) None of the above.

44. The cell standard potential $E°_{cell}$ is

(A) 0.00 V

(B) + 0.80V

(C) – 0.80V

(D) +1.6V

(E) –1.6V

45. Which of the following is <u>false</u>?

(A) The overall cell potential depends on the difference in concentration between the two solutions.

(B) The overall cell potential is negative so no reaction takes place.

(C) The overall cell potential is positive so the cell works.

(D) Copper will be deposited at the standard electrode.

(E) The electrode immersed in the $10^{-3}$ M solution will lose weight.

46. 25.0 ml of 0.100M HCl is titrated with 0.15M NaOH. The pH of the acid solution after 10 ml of base added is

(A) 2.39

(B) 2.48

(C) 1.9

(D) 1.54

(E) 1.4

47. Which of the following is most soluble in water?

(A) Hexanol

(B) Benzene

(C) Acetic acid

(D) Acetylene

(E) Hexanoic acid

48. Which of the following has the highest boiling point?

(A) $CH_3CH_2CH_2CO_2H$

(B) $CH_3CH_2CH_2CH_2CHO$

(C) $CH_3CH_2CH_2CH_2CH_2OH$

(D) $CH_3(CH_2)_4CH_3$

(E) $CH_3CH_2 - \underset{\underset{O}{||}}{C} - CH_2CH_3$

49. Which of the following pairs of elements have almost the same atomic size?

(A) Sc and Y
(B) B and Al
(C) Al and Ga
(D) Mg and Ca
(E) Be and Mg

50. The synthesis $CO + (2n+1)H_2 \xrightarrow{\text{catalyst}} C_nH_{2n} + 2+ nH_2O$ is

(A) the Haber process
(B) the Frasch process
(C) the Fischer-Tropsch process
(D) the Born-Haber process
(E) none of the above

51. Fluorine is produced by the electrolysis of

(A) liquid hydrogen fluoride
(B) gaseous hydrogen fluoride
(C) aqueous solution of hydrogen fluoride
(D) a solution of hydrogen fluoride in molten KF•2HF
(E) calcium fluoride

52. Which of the following is used to refine silicon semiconductors?

(A) Electrolysis
(B) Zone refining
(C) Distillation
(D) Precipitation
(E) none of the above

53. Which gas has a rate of diffusion 0.25 times that of hydrogen at the same temperature and pressure?

(A) $CH_4$
(B) $PH_3$
(C) Argon
(D) $N_2$
(E) $O_2$

54. What is the total coordination number of an oxygen atom in $H_2O$?

(A) 4
(B) 3
(C) 2
(D) 8
(E) 1

55. The spin quantum number s or m is obtained as a result of

(A) the Heisenberg Uncertainty Principle
(B) solving the Schrödinger equation for the hydrogen atom
(C) Rutherford's experiment
(D) the effect of a magnetic field on an atomic spectrum
(E) spin of the nucleus

56. The ground state of $Ga^{3+}$ ion has an______ pair of electrons in its outermost shell. The ion is ______

(A) odd, paramagnetic
(B) even, paramagnetic
(C) even, diamagnetic
(D) odd, diamagnetic
(E) even, ferromagnetic

57. An alloy of iron and gallium has 63.8% iron composition. If 351.25g of the alloy is completely dissolved in $H_2SO_4$ to produce $Fe^{2+}$ and $Ga^{3+}$ ions, what is the volume of $H_2$ collected at STP (neglecting the vapor pressure of air)?

(A) 156.8 l
(B) 123.2 l
(C) 246.2 l
(D) 78.4 l
(E) 145.6 l

58. The reaction A(g) + B(g) ↔ 2C(g) + D(s) has $K_p$ of 0.65 at T = 298K. If at the beginning $P_A$ = 2 atm, $P_B$ = 5 atm, and $P_C$ = 4 atm, then initially

(A) the free energy of the reaction is zero
(B) the reaction will proceed from left to right
(C) the reaction will proceed from right to left
(D) the system will be at equilibrium
(E) the free energy of the reaction is negative

59. Oxalic acid will have ______ $K_a$'s

(A) zero
(B) 1
(C) 3
(D) 2
(E) 4

60. $C_6H_5NH_2$ is

(A) benzile ammonia
(B) benzyl ammonia
(C) hexyl ammonia
(D) phenol
(E) aniline

61. What is the volume of $H_2O$ which can be condensed after the complete combustion of 2240 l of ethane (volume of gas measured at STP?

(A) 10.8 ml
(B) 10.8 l
(C) 5.4 l
(D) 5.4 ml
(E) 3.0 l

62. Given the mechanism below for the oxidation of nitrogen (II) oxide; determine which expression gives the rate of formation of $NO_2$?

$$NO + NO \underset{k_2}{\overset{k_1}{\rightleftarrows}} N_2O_2 \quad \text{Fast equilibrium}$$

$$N_2O_2 + O_2 \xrightarrow{k_3} 2NO_2$$

(A) $k_3[N_2O_2][O_2]$

(B) $\frac{k_1}{k_2}k_3[NO]^2[O_2]$

(C) $\frac{k_2}{k_1}k_3[NO]^2[O_2]$

(D) $2\frac{k_2}{k_1}k_3[NO][O_2]$

(E) $2\frac{k_1}{k_2}k_3[NO][O_2]$

63. The pOH of a 1.0M solution of HCl is:

(A) 1
(B) 13
(C) 14
(D) 0
(E) 15

64. If the pOH of solution A is 2.5 and the pOH of B is 10.1, then which of the following is true?

(A) Solution A has a higher concentration of protons than B.
(B) Solution B is more basic than solution A.
(C) Solution A is more basic than solution B.
(D) Solution A is more acidic than solution B.
(E) Solution B has a higher concentration of hydroxyl ions than A.

65. $K_c$ for a reaction $A(g) + 3B(g) \leftrightarrow 2C(g) + 2D(s)$ has units

(A) $mole^{-2} liter^{2}$
(B) $mole^{2} liter^{-2}$
(C) $mole\ liter^{-2}$
(D) $mole^{-1}$ liter
(E) $mole^{-1} liter^{-1}$

66. The density of gaseous $SO_3$ at 15°C, 50mmHg and R = 0.0821 liter-atm/(mole–K) is:

(A) 0.223g/liter
(B) 0.223 liter $g^{-1}$
(C) 169.17g/liter
(D) 0.357 liter/g
(E) 0.357g/liter

67. Cryolite ($Na_3AlF_6$) is added to $Al_2O_3$ during electrolysis of $Al_2O_2$ to

(A) obtain higher purity Al

(B) permit the electrolysis at a lower temperature

(C) increase the solubility of $Al_2O_3$ for electrolysis

(D) increase the conductivity of $Al_2O_3$

(E) react with the oxygen in $Al_2O_2$ and hence free the Al metal

68. Which of the following is acetone?

(A) $CH_3CH_2CH_2–OH$

(B) $CH_3CH_2C$ (with H single-bonded and =O double-bonded to C)

```
         H
        /
CH3CH2C
        \\
         O
```

(C) $CH_3 – O – CH_2CH_3$

(D) $CH_3COCH_3$

(E) $CH_3CH_2CH_3$ with OH on the middle carbon

```
CH3CH2CH3
    |
    OH
```

69. A protein can be described as

(A) an addition polymer

(B) an addition copolymer

(C) a condensation polymer

(D) a polyester

(E) a polyvinyl chloride

70. What is the type of hybridization used by chlorine in the $ClF_4^-$ ion?

(A) $3sp^3d^2$

(B) $3p^3d^3$

(C) $2p^3d^3$

(D) $3sp^2d^3$

(E) $3sp^34s3d$

71. 15g of ethane reacts with chlorine to yield 15g of 1-chloropropane. The percent yield of ethyl chloride is:

(A) 46.5%

(B) 50.0%

(C) 100.0%

(D) 93.0%

(E) 23.3%

72. Which of the following compounds exhibits tautomerism?

(A) $CH_3OCH_2CH_3$

(B) $CH_3-C(OCH_3)(H)-OCH_3$ (structure: central C bonded to $CH_3$, $OCH_3$ above, $OCH_3$ to the right, and H below)

(C) 1-methylcyclopentene ring with $CH_3$ and OH on the double-bonded carbons

(D) $NH_2-\overset{\overset{\displaystyle NH_2}{|}}{C}=O$

(E) $CH_2=CH-CH_2OH$

73. What type of formula of ethane is depicted below?

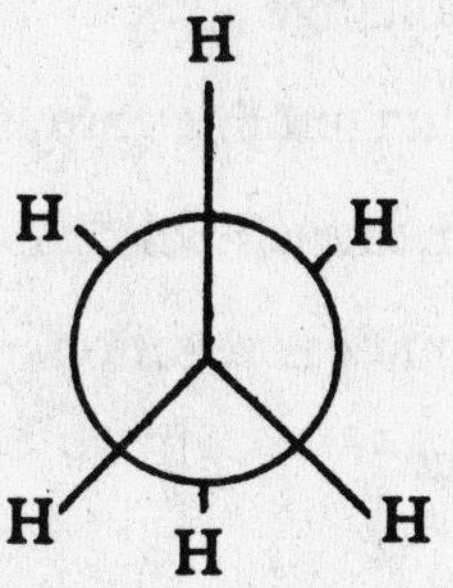


(A) Fisher projection formula

(B) Newman projection formula

(C) Lewis projection formula

(D) Kekulé projection formula

(E) Pauling projection formula

74. The reaction

$$R-\overset{\overset{\displaystyle O}{\|}}{C}H + 2Ag^+ + 3OH^- \rightarrow R\overset{\overset{\displaystyle O}{\|}}{C}-O^- + 2Ag + 2H_2O$$

is the

(A) Benedict's test
(B) Fehling's test
(C) Tollen's test
(D) protein test
(E) Fisher's test

75. Which of the following is <u>not</u> true when transmutation of an element occurs by $\beta^-$ decay?

(A) mass number does not change
(B) the initial and the product nuclides are isobars
(C) the atomic number of the new nuclide decreases
(D) the transmutation is a nuclear reaction
(E) all the above are not true

76. The version of Gibbs phase rule $f=3-p$ is valid:

(A) for a one-phase system
(B) for a three-phase system
(C) for a one-component system
(D) for a three-component system
(E) for the degree of freedom at the triple point

77. Which of the following liquid solutions show positive deviation from Raoult's law?

(A) Benzene and toluene
(B) Water and ethanol
(C) Pentane and hexane

(D) Acetone and carbon disulfide

(E) Acetone and chloroform

78. Given $VO_4^{3-}$, $CrO_4^{2-}$ and $MnO_4^-$, which of the following is true?

(A) $VO_4^{3-}$(strongest base); $HMnO_4$ (strongest acid)

(B) $H_3VO_4$ (strongest acid); $MnO_4^-$ (strongest base)

(C) $H_3VO_4$ (strongest acid); $CrO_4^{2-}$ (strongest base)

(D) $H_2CrO_4$ (strongest acid); $MnO_4^-$ (strongest base)

(E) $VO_4^{3-}$ (weakest base); $_{22}MnO_4^{2-}$ (weaker base); $CrO_4^{2-}$(weak base).

79. Brass is the name given to alloys of

(A) iron and manganese

(B) iron and copper

(C) iron and zinc

(D) copper and manganese

(E) copper and zinc

80. Stalactites and stalagmites are made of

(A) sodium carbonate

(B) sodium chloride

(C) sodium carbonate and sodium chloride respectively

(D) calcium nitrate

(E) calcium carbonate

81. Which of the following listed below is a silicate?

(A) Polyvinyl chloride
(B) Bakelite
(C) Limestone
(D) Mica
(E) Alum

82. How many resonance structures contribute to the hybrid structure of the nitrate ion?

(A) 1
(B) 2
(C) 3
(D) 4
(E) 5

83. Helium is preferred over hydrogen for filling meteorological balloons because

(A) helium is less dense than hydrogen
(B) hydrogen usually dissolves in the balloon material
(C) helium is not combustible
(D) the half-life of helium is longer then the half-life of hydrogen
(E) hydrogen condenses easily at higher altitudes

84. Which of the following is <u>false</u>?

(A) A process with $\Delta H < 0$ is more likely to be spontaneous than one with $\Delta H > 0$
(B) The rate law for a reaction is an algebraic expression relating the forward reaction rate to product concentration
(C) Ammonia is an amphoteric substance

(D) The products of a Brønsted-Lowry acid-base reaction are always a new acid and a new base

(E) One faraday is the total charge of one mole of electrons

85. At 25°C 1 liter of water contains:

(A) 55.56 moles of water

(B) 1 mole of water

(C) $6.022 \times 10^{23}$ molecules of water

(D) 0.056 moles of water

(E) 1000 moles of water

# ADVANCED PLACEMENT CHEMISTRY EXAM IV

## SECTION II

**Directions**: The percentages given for each individual part indicates the scoring for this section. Spend approximately 35 minutes on Parts A and B together and 40 minutes on Part C.

The methods and steps used to solve a problem must be shown clearly. Partial credit will be given if work is shown. Pay attention to significant figures.

### Part A
(30 Percent)

1. The temperature of a reaction is changed while initial concentrations are kept constant. The following data are obtained:

| Temperature °C | rate (mole liter $^{-1}$s$^{-1}$) |
|---|---|
| 25 | 1.75 |
| 45 | 2.6 |

(a) Calculate the activation energy.

(b) At what temperature will the rate be 3.5 times that found at 25°, with other conditions remaining constant?

(c) A catalyst is introduced into the reaction above, which lowers the activation energy for the reaction by 76%. What is the effect on the rate of the reaction at 25°C if other conditions remain constant?

### Part B
(30 Percent)

Solve **ONE** of the two problems presented. Only the first problem answered will be scored.

2. (a) Gas A decomposes by the balanced equation below:

$$A \leftrightarrow B + C$$

At 200°C the equilibrium constant is $K_c = 0.18$ mole liter$^{-1}$. What is the equilibrium composition of the system, if, initially, 0.3 mole of A, 0.2 mole of B and 0.15 moles of C were introduced into a 10 liter vessel?

(b) If, after equilibrium in the above reaction has been established, the volume of the vessel is reduced to 5 liters, explain what will happen. Also find the new equilibrium composition.

(c) What is the $K_p$ for this reaction?

3. To standardize a solution of potassium permanganate, 0.157g of sodium oxalate was dissolved in water, acidified, and titrated with the permanganate solution.

(a) Write a balanced equation for the reaction.

(b) If 15.3 ml of the potassium permanganate solution were needed, what is its molarity?

(c) What is the mass of potassium permanganate in the 15.3ml solution?

(d) Find the volume of the above potassium permanganate solution required to titrate 0.105 mole solution of ferrous sulfate.

## Part C

(40 Percent)

**Choose THREE of the following five topics. Only the first three answers will be scored. Your answers will be scored based on their accuracy, relevance of the details chosen, and appropriateness of the descriptive material used. Be as specific as possible and use illustrative examples and equations where helpful.**

4. Discuss the uses of radioactivity in society.

5. Explain the following:

   (a) Hydrogen peroxide is kept in dark brown bottles. Explain the antiseptic action of hydrogen peroxide.

   (b) Ethyl chloride acts as a fast acting local anesthetic.

   (c) A lake freezes on the surface but not at the bottom.

   (d) Aerosol cans carry a warning "Do Not Incinerate."

6. (a) Discuss the environmental impact of the uses of freons.

   (b) Explain why a moth ball packed with woolens over the summer is gone by the fall.

7. What are the differences between:

   (a) amorphous and crystalline solids?

   (b) molecular, ionic and covalent crystals?

8. Discuss the methods used in the preparation of metals.

# ADVANCED PLACEMENT CHEMISTRY EXAM IV

## ANSWER KEY

| | | | | | |
|---|---|---|---|---|---|
| 1. | E | 29. | B | 57. | B |
| 2. | B | 30. | A | 58. | C |
| 3. | D | 31. | B | 59. | D |
| 4. | D | 32. | B | 60. | E |
| 5. | B | 33. | A | 61. | C |
| 6. | E | 34. | C | 62. | B |
| 7. | D | 35. | C | 63. | C |
| 8. | B | 36. | C | 64. | C |
| 9. | B | 37. | A | 65. | A |
| 10. | A | 38. | E | 66. | A |
| 11. | C | 39. | C | 67. | B |
| 12. | A | 40. | B | 68. | D |
| 13. | C | 41. | E | 69. | C |
| 14. | A | 42. | C | 70. | A |
| 15. | B | 43. | A | 71. | A |
| 16. | C | 44. | A | 72. | C |
| 17. | D | 45. | B | 73. | B |
| 18. | C | 46. | D | 74. | C |
| 19. | D | 47. | C | 75. | C |
| 20. | A | 48. | A | 76. | C |
| 21. | E | 49. | C | 77. | D |
| 22. | D | 50. | C | 78. | B |
| 23. | C | 51. | D | 79. | E |
| 24. | C | 52. | B | 80. | E |
| 25. | E | 53. | E | 81. | D |
| 26. | D | 54. | A | 82. | C |
| 27. | D | 55. | D | 83. | C |
| 28. | E | 56. | D | 84. | B |
| | | | | 85. | A |

# ADVANCED PLACEMENT CHEMISTRY EXAM IV

# DETAILED EXPLANATIONS OF ANSWERS

## SECTION I

1. (E)

Metalloids cannot be satisfactorily identified as being either metals or non-metals for they possess some properties of each. They are usually identified as the staircase elements.

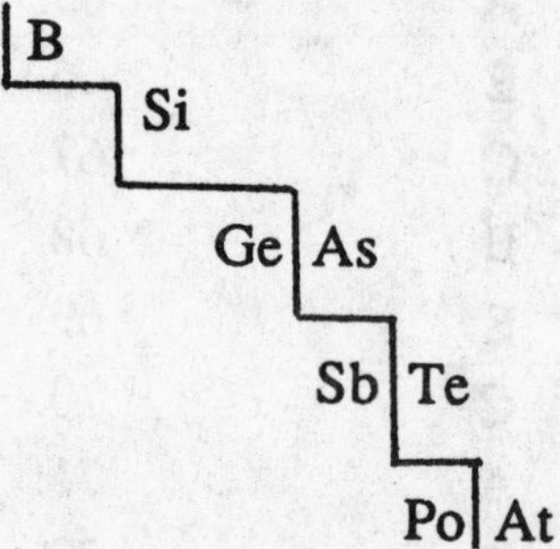


Note that At has mostly metallic properties; therefore it is not included in the staircase.

2. (B)

Krypton is a member of the noble gas family otherwise known as inert gases. These gases, due to their filled valence shells, are quite unreactive, hence the term "inert gas."

3. (D)

Metals are located on the left of the periodic table, while the non-metals are listed on the right side. Of all the elements listed, Sr is the element furthest to the left of the periodic table and, therefore, the most metallic. Sr is also located in the group IIA generally known as the alkaline-earth metals.

4. (D)

Actually the liquid form of formaldehyde, called formalin, is used to preserve tissue specimen.

5. (B)

Stannous fluoride is the active ingredient in certain brand name toothpastes which capitalize on the oxidizing effectiveness of fluoride ions.

6. (E)

$CH_3COOC_6H_4COOH$, acetylsalicylic acid, is the white crystalline compound commonly used to reduce pain and fever in the form of aspirin.

7. (D)

Allotropes are elements with more than one form due to molecular structure differences, as in $O_2$ and $O_3$ (see figure below), or as a result of differences in the arrangement of atoms or molecules, as with diamond and graphite, which are allotropic forms of carbon.

ozone oxygen

8. (B)

The reaction of a salt with water yields acid and base products.

$$Na_2CO_3 + 2H_2O \leftrightarrow \underset{\text{strong base}}{2NaOH} + \underset{\text{weak acid}}{H_2CO_3}$$

9. (B)

Since one mole of ideal gas occupies a volume of 22.4 liters at STP we have:

$$\frac{34 \text{ grams of } NH_3}{17 \text{ grams (M.W. of } NH_3)} = 2 \text{ moles of } NH_3 \times \frac{22.4 \text{ liters}}{1 \text{ mole of } NH_3}$$

$$= 44.8 \text{ liters of } NH_3$$

M.W. = molecular weight.

10. (A)

Oxidation takes place at the anode while reduction takes place at the cathode. Alternatives (C) and (D) are wrong since they both describe oxidative processes. Upon reviewing the cell solution, choice (A) is the correct answer.

11. (C)

The purpose of a salt bridge is to allow the positive and negative ions to migrate from cell to cell and thus allows a current to flow while preventing the two solutions from mixing. The current flows clockwise in this cell.

12. (A)

In order to calculate the emf value of a voltaic cell at various concentrations we use the Nernst equation

$$E = E^\circ - \frac{0.05915}{n} \log Q$$

where E = the emf for the reaction at the new concentration

$E^\circ$ = the standard electrode potential

n = the number of moles of electrons involved in the half reactions

Q = the reaction quotient

13. (C)

A homogeneous mixture is one of uniform composition. All of the choices are homogeneous except for the heterogeneous mixture of sand in water, in which case you can see the separate bits of sand dispersed in the water medium.

14. (A)

Rutherford projected a beam of ∝-particles from a radioactive source onto very thin gold foil. Most passed through without deflection, a few were diverted from their paths and very few were deflected back towards the source. Rutherford therefore concluded that: 1) the atom consists largely of empty space, 2) each atom must contain a heavy positively charged body (the nucleus) due to the repulsions.

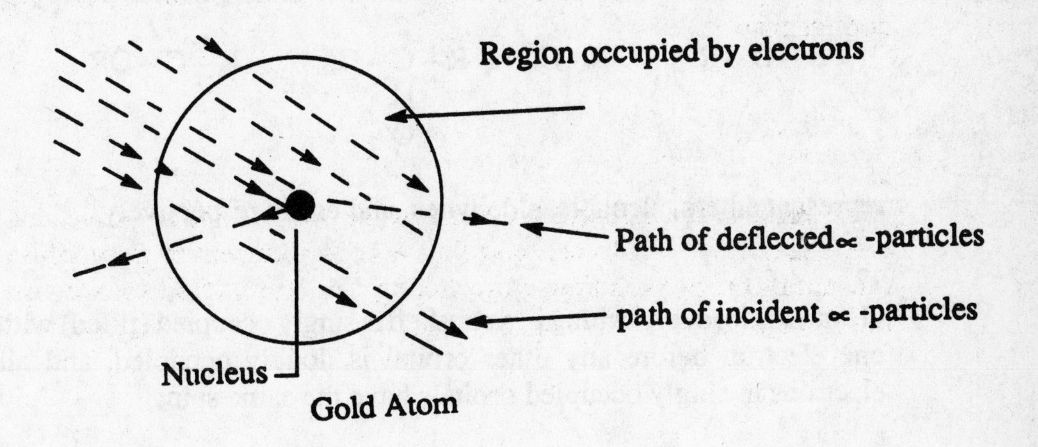


Note: ∝ -particles consist of positive charge

15. (B)

X represents an element.

$^{\text{atomic mass}}_{\text{atom number}}X$

Based solely on the atomic number we can conclude that $^{21}_{10}X$ must represent the noble gas neon.

16. (C)

The functional group is representative of an acid (e.g. acetic acid $CH_3-COOH$).

Other functional groups such as:

$$R-O-R, \quad R-COH, \quad R-\underset{\underset{O}{||}}{C}-H, \quad R-\overset{\overset{O}{//}}{C}-OR$$

represent ethers, alcohols, aldehydes, and esters respectively.

17. (D)

Hund's rule: every orbital in a subshell is singly occupied (filled) with one electron before any other orbital is doubly occupied, and all electrons in singly occupied orbitals have the same spin.

18. (C)

(A) Atomic weight of Fe = 55.8g $mole^{-1}$

$\therefore$ There are $\frac{49}{55.8} \approx 0.88$ moles of Fe

1 mole element will contain Avogadro's number of atoms which is $6.026 \times 10^{23}$

$\therefore$ No. of atoms in 55.8g of Fe $= 0.88 \times 6.026 \times 10^{23}$
$= 5.3 \times 10^{23}$ atoms.

(B) at STP 1 mole of gas occupies 22.4 l

$$\text{No. of moles of 5 l of gas is } \frac{5}{22.4} \times 1 = 0.22 \text{ moles}$$

No. of molecules present is
No. of moles x Avogadro's number =

$$0.22 \times 6.026 \times 10^{23} = 1.3 \times 10^{23} \text{ molecules}$$

(C) Ethanol has the molecular formula $C_2H_5OH$
Molecular weight of ethanol = (2x12)+6+16 = 24 + 22 = 46

$$\therefore \text{No. of moles in 150g of } C_2H_5OH = \frac{150}{46} = 3.26$$

$$\text{No. of molecules} = 3.26 \times 6.026 \times 10^{23} = 19.65 \times 10^{23}$$

(D) 10 l of $O_3$ at STP contains 10/22.4 moles of $O_3$.

$$\therefore \text{No. of moles of } O_3 = \frac{10}{22.4} \times 6.026 \times 10^{23} = 2.69 \times 10^{23}$$

(E) The No. of molecules in 80g of calcium carbide is

$$\frac{80}{\text{Mol.Wt. of } Ca_2C} \times 6.026 \times 10^{23} = \frac{80}{92.16} \times 6.026 \times 10^{23} = 5.23 \times 10^{23}$$

$$\text{Mol. Wt. of } Ca_2C = (2 \times 40.08) + 12 = 92.16$$

∴ The answer is (C) 150g of ethanol contains the largest number of molecules ($19.65 \times 10^{23}$ molecules).

19. (D)

The name of an orbital consists of a number and a letter which depend on the values of the quantum numbers n and L. The numbers correspond to the value of the main or principal quantum number n (n = 1,2,3, etc.); The letter refers to the value of L, the azimuthal quantum number: s for L = 0, p for L = 1, d for L = 2, and f for L = 3.

20. (A)

The abbreviated electronic configuration of an element of atomic number 42 is [Kr] $5s^1 4d^5$, where [Kr] stands for the electronic arrangement of Krypton, element 36, indicating the filling of all sublevels through $4p^6$. The remaining 6 electrons go into 5s and 4d orbitals. The $5s^1 4d^5$ configuration is preferred over the $5s^2 4d^4$ configuration because of the stability of the half-filled 4d sublevel.

21. (E)

An atom, ion or a molecule with one or more unpaired electrons exhibits paramagnetism (they are slightly attracted into a magnetic field). Those with all electrons paired are diamagnetic (slightly repelled by a magnetic field). All the ions listed have unpaired electrons in their valence shell. The ions have the following electron configurations:

$Fe^{2+}$: [Ar]$3d^6$ with 4 unpaired electrons in the 3d orbitals

(↑↓) (↑) (↑) (↑) (↑)

$Fe^{3+}$: [Ar]$3d^5$ with 5 unpaired electrons
$CO^{3+}$: [Ar]$3d^6$ with 4 unpaired electrons
$Ni^{2+}$: [Ar]$3d^{10}$ with two unpaired electrons in the 3d orbitals.

(↑↓) (↑↓) (↑↓) (↑) (↑)

22. (D)

The geometric shapes shown are those of (A) $d_{z^2}$(B) $d_{yz}$, (C) $p_z$, (D) $d_{x^2-y^2}$ and (E) $d_{xy}$ orbitals.

23. (C)

This problem in molecular geometry can be solved by considering an electrostatic model. This is done by arranging all electrical charges of the same sign on the surface of a given sphere so that they have maximum stability. The possible stable positions on the sphere depend on the number of charges: two charges will be arranged in diametrically opposite positions; three charges will be arranged on the circumference of a circle bounding the sphere at 120° to each other; four will occupy the vertices of a regular tetrahedron. The most stable arrangements for five and six charges will be a triangular (triagonal) bipyramid and an octahedral respectively. Therefore, for a central atom surrounded by five atoms, the geometrical shape of the molecule will be a triagonal bipyramid. $PCl_5$ is an example of such a molecule.

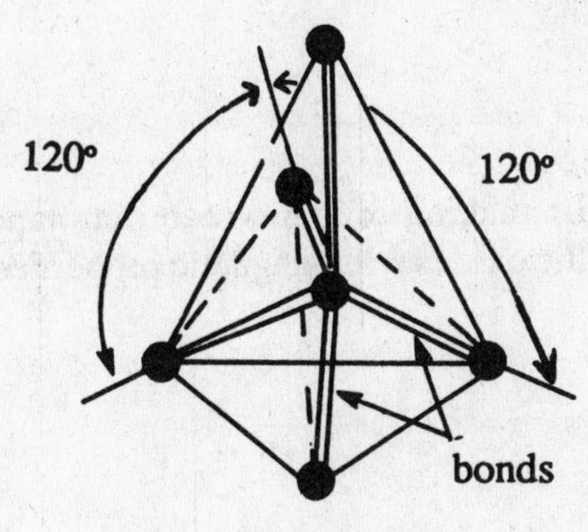

triagonal bipyramid

24. (C)

While phosphorous and red phosphorous are 2 allotropic forms of phosphorous, red phosphorous ignites by friction. Mixed with fine sand, it is used on striking surfaces of matchboxes. Although its structure is unknown, red phosphorous is less reactive than white phosphorous, which burns spontaneously in air.

25. (E)

As a rule, an organic compound with an asymmetric carbon atom will display optical activity. An asymmetric carbon atom is a carbon bonded to four different atoms or group of atoms. The only molecule given with four different atoms is (E):

```
        H
        |
  Cl –  C  – CN
        |
        Br
```

26. (D)

Iodine is <u>not</u> used in the making of audio-recording tapes. These employ a coating of metallic oxides with magnetic properties like iron oxides.

27. (D)

VSEPR is an acronym for valence-shell electron-pair repulsion. It states that bonding electron pairs and lone electron pairs of an atom will arrange themselves in space so as to minimize electron-pair repulsion around that atom. It is used to predict the geometries of various molecules and ions.

28. (E)

$K_{sp}$ is the solubility product of a slightly soluble salt.

$$\text{For } CaSO_4 \quad K_{sp} = [Ca^{2+}]\,[SO_4^{2-}]$$

If we let X represent the concentration of $SO_4^{2-}$ anions in mole liter$^{-1}$,

$CaSO_4$ dissociates in equimolar proportions so

$K_{sp} = [Ca^{2+}][SO_4^{2-}] = [X][X] = X_1^2 = 2.4 \times 10^{-5}$

$\therefore X_1 \approx 5. \times 10^{-3}$ moleliter$^{-1}$

For $Ag_2SO_4$ the $K_{sp}$ expression is different because $Ag_2SO_4 \leftrightarrow 2Ag^+ + SO_4^{2-}$ i.e. each mole of $Ag_2SO_4$ produces 2 moles of $Ag^+$ and one mole of $SO_4^{2-}$ ions

$[Ag^+] = 2X_2$ and $[SO_4^{2-}] = X_2$

$K_{sp} = [Ag^+]^2[SO_4^{2-}] = (2X_2)^2(X_2)$

$K_{sp} = 4X^3 = 1.2 \times 10^{-5}$

$X_2 = 0.0144$

$X_2 = 14.4 \times 10^{-3}$

$$\frac{X_2}{X_1} = \frac{14.4 \times 10^{-3}}{5 \times 10^{-3}} = 2.9$$

29. (B)

The dipole moment of a bond is directed from the partial positive charge to the partial negative charge or from the less electronegative to the more electronegative atom in the bond.

Example: H – Cl

$\propto +$ $\propto -$ direction of dipole moment.

$\longrightarrow$

If the molecule contains more than one bond moment, then the resultant dipole moment is the vector sum of all the bond moments.

Thus for $H_2O$ we have:

$\propto -$

Ö

$\propto +$ H / \ H $\propto +$

bond moments: ↗ ↖

i.e.: ↑ resultant dipole moment is oriented according to the resultant

For $CO_2$ we have: $\overset{\propto -}{\ddot{O}} = \overset{\propto +}{C} = \overset{\propto -}{\ddot{O}}$

bond moments ⟵+ +⟶

The vector sum is zero, therefore the resultant dipole moment is 0.

For $SO_2$

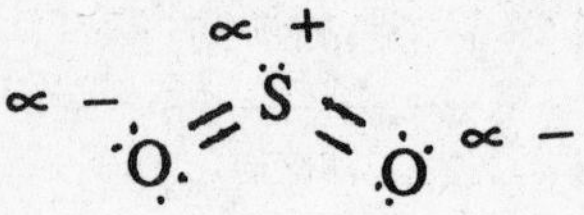

$\overset{\propto -}{O} = \overset{\propto +}{\ddot{S}} = \overset{\propto -}{O}$

bond moments ↙ ↘

resultant is vector sum  is oriented as ↓ so the combination of resultant dipole moments is ↑, O, ↓

30. (A)

The reaction is

$$MnO_4^{=} \rightarrow MnO_4^{-} + MnO_2$$

This is a type of disproportionation reaction, $MnO_4^{=}$ reacting with itself to produce $MnO_4^{-}$ and $MnO_2$ in acidic medium. "Acidic medium" informs us that the balanced equation may involve $H^+$ and/or $H_2O$ on either side of the equation. We will use the ion-electron method to balance the equation. We write two half-reactions as follows:

1st half-reaction

$$MnO_4^{=} \rightarrow MnO_4^{-}$$

The oxidation state of Mn changes from +6 to +7 in $MnO_4^-$, so the preceding is an oxidation reaction.

$$MnO_4^{2-} \rightarrow MnO_4^- + e^- \quad \text{(oxidation)}$$

2nd half reaction

$$MnO_4^{2-} \rightarrow MnO_2 \quad \text{(reduction)}$$

Here Mn is changed from +6 to +4. To account for the deficiency of oxygen on the right-hand side of the equation, we add $H_2O$, remembering that the reaction is occuring in acidic solution, which corresponds to adding $H^+$ to the left-hand side of the equation and balance it as

$$MnO_4^{2-} + 4H^+ + 2e \rightarrow MnO_2 + 2H_2O$$

For the electron gain and loss to balance, multiply the oxidation half-reaction by 2 and add the half-reactions.

$$\begin{array}{lll} 2MnO_4^{2-} & \rightarrow 2MnO_4^- + 2e^- & \text{(oxidation} \\ + \; MnO_4^{-2} + 4H + 2e^- & \rightarrow MnO_2 + 2H_2O & \text{(reduction)} \\ \hline 3MnO_4^{2-} + 4H^+ & \rightarrow 2MnO_4 + MnO_2 + 2H_2O & \end{array}$$

31. (B)

Reactions that occur in solution are usually classified into, (i) proton transfer or acid-base reactions, (ii) precipitation or metathesis, (iii) electron transfer or oxidation-reduction reactions.

Metathesis involves the exchange of ions with the formation of one or more salts. (Note: Another type of reaction which may occur not in solution is synthesis or formation of compounds from elements.)

32. (B)

From the solubility rules we know that

(1) all $(IA)^+$ and $NH_4^+$ salts are soluble, therefore I is soluble.

(2) all sulphates are soluble except sulphates of $Pb^{+2}$, $Ca^{2+}$, $Sr^{2+}$ and $Ba^{2+}$, therefore, II is not soluble.

(3) also all halides are soluble except halides of $Pb^{+2}$, $Ag^+$, $Hg^{2+}$ and $Tl^+$, therefore, $PbCl_2$ is insoluble.

(4) all salts of $NO_3^-$, $ClO_4^-$, $ClO_3^-$ and $C_2H_3O_2^-$ ions are soluble, therefore IV is soluble.

33. (A)

Cations in the common higher oxidation state are named by adding the suffix –ic to a stem.

Cations in the common lower oxidation state are named by adding the suffix –ous to a stem.

The common cations for copper, cobalt, lead and tin are

Cu: $Cu^+$ (cuprous)
$Cu^{2+}$ (cupric)

Co: $Co^{2+}$ (cobaltous)

Pb: $Pb^{2+}$ (plumbous)

Sn: $Sn^{+2}$ (stannous)
$Sn^{4+}$ (stannic)

34. (C)

The Born-Haber cycle summarizes the steps in the formation of ionic compounds from its elements. It uses Hess' law to relate the various enthalpy changes. In any closed cycle the net change in energy is zero.

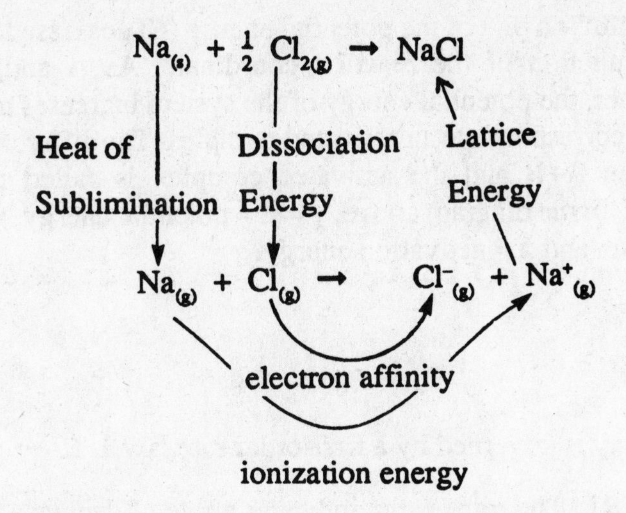


35. (C)

Properties of a solution that depend only on the concentration of the solute and not on its nature, are called colligative properties. Examples are the freezing-point depression, boiling point elevation, osmotic pressure and vapor pressure depression. Each of these can be used to find the molecular weight of an unknown solute.

36. (C)

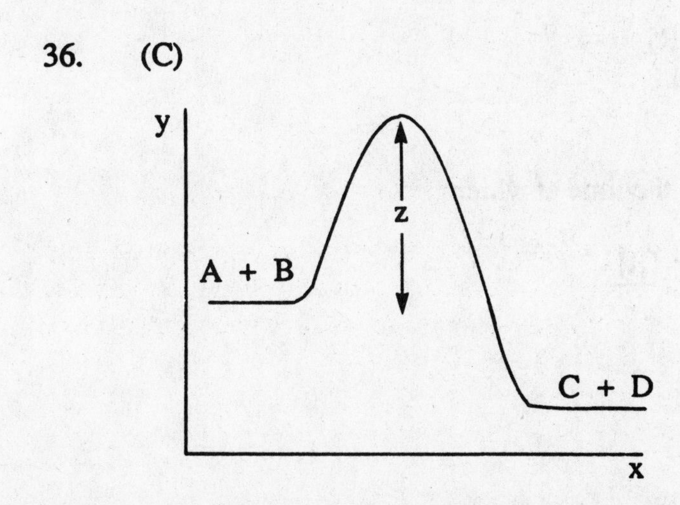

The above diagram is a plot of the potential energy (E) possessed by molecules as a function of the reaction coordinate. As A and B approach each other, the potential energy of the system increases to a maximum, which corresponds to an activated complex. The difference in energy between A+B and the activated complex is called the activation energy. In the diagram above, y = E = potential energy, x = reaction coordinate, and z = activation energy.

37. (A)

Radioactive decay is governed by a first-order rate law $R \rightarrow P$

$-\frac{d[R]}{dt} = k[R]$ The minus sign indicates a rate of decrease.

[R] - amount present at time t

k – specific rate constant

$$\frac{d[R]}{[R]} = -kdt$$

integrating gives

$$\ln[R] = -kt + c$$

at t = o, Let $[R]=[R]_o$

$$\rightarrow c = \ln[R]_o$$

$$\ln[R] = -kt + \ln[R]_o$$

or

$$\ln[R] - \ln[R]_o = -kt$$

$$\ln\frac{[R]}{[R]_o} = -kt$$

The half-life $t_{1/2}$ is the time at which

$$[R] = \frac{[R]_o}{2}$$

$$\ln\frac{[R]_o/2}{[R]_o} = -kt_{1/2}$$

$$\ln{}_{1/2} = -kt_{1/2}$$

$$t_{1/2} = -\frac{(-\ln 2)}{k} = \frac{\ln 2}{k}$$

$$t_{1/2} = \frac{0.693}{5.33\times10^{-10}} = 1.3 \times 10^9 \text{ years.}$$

38. (E)
By the Brønsted-Lowry definition, an acid is any substance that can donate protons and a base is a substance that accepts protons. By this definition, $NH_3$ and $H_2O$ are Brønsted bases because:

$$NH_3 + H^+ \rightarrow NH_4^+$$

$$H_2O + H^+ \rightarrow H_3O^+$$

By the Lewis definition an acid is an electron-pair acceptor and a base is an electron-pair donor in a chemical reaction.

By this definition we again see that $NH_3$ and $H_2O$ are Lewis bases because

```
     ..                  [    H    ]+
H – N – H + H+   →       [    |    ]
     |                   [ H–N–H   ]
     H                   [    |    ]
                         [    H    ]
```

and

```
     ..                  [  ..    ]+
H – O: + H+   →          [ H–O–H  ]
     |                   [   |    ]
     H                   [   H    ]
```

hydronium ion

$BBr_3$ is a Lewis acid because boron in $BBr_3$ has a vacant 2p orbital which can accept a pair of electrons, e.g.

$$\ddot{N}H_3 + B\,Br_3 \rightarrow H_3N \rightarrow B\,Br_3$$

NaOH is neither a Brønsted base nor a Lewis base. It is an Arrhenius base. An Arrhenius acid is a substance that produces hydrogen ions or protons in aqueous solution. An Arrhenius base produces hydroxide ions.

39. (C)

The electron configuration of holmium (Ho) with atomic number 67 is [Xe] $4f^{11}6s^2$.

Hence that of $Ho^{+3}$ is [Xe]$4f^{10}$ in the ground state. 4f has seven degenerate (i.e. equal energy) orbitals with the 10 electrons distributed by Hund's rule as: $\underline{\uparrow\downarrow}$ $\underline{\uparrow\downarrow}$ $\underline{\uparrow\downarrow}$ $\underline{\uparrow}$ $\underline{\uparrow}$ $\underline{\uparrow}$ $\underline{\uparrow}$

Thus we have 4 unpaired electrons.

40. (B)

One may assume that when removing electrons to form stable cations, the electrons would leave the orbitals in the reverse order in which they were filled. This is true to some extent; but there are many exceptions especially in the transition elements. A good rule to follow is that the electrons that are removed from an atom or ion are those with the maximum value of the principal quantum number n; and of this set of electrons the easiest to remove are those with the maximum L.

$Ir^{+3}$ is formed by removal of 3 electrons from Ir which has an electron configuration [Xe]$4f^{14}5d^7 6s^2$. So by the rule given, the first 2 electrons would leave the $6s^2$ and the third from $5d^7$ orbitals to give a ground state configuration of [Xe]$4f^{14}5d^6$.

41. (E)

The tendency to achieve completely filled and, to a lesser extent, half-filled shells, controls the chemistry of most elements of the periodic table. Therefore, Ru, with configuration $[Xe]4f^7 6s^2$, will lose the $6s^2$ electron to attain a half- filled $4f^7$ configuration, with oxidation state equal to +2. For Sc, with configuration $[Ar]3d^1 4s^2$, the most stable ion will be obtained with the loss of 3 electrons from $4s^2$ and $3d^1$ to achieve the configuration of Argon:$[Ne]3s^2 3p^6$, oxidation state equals +3.

42. (C)

Alums are double salts of the general formula $M^+M^{3+}(SO_4)_2\ 12H_2O$ where $M^+$ can be $Na^+$, $K^+$, $Ag^+$, $NH_4^+$ and other +1 ions.

$M^{3+}$ can be $Al^{3+}$, $Fe^{+3}$, $Cr^{+3}$, $Mn^{+3}$.

The most common alums are aluminum salts (therefore the name alum). Solutions of alum are acidic because of the hydrolysis of the aluminum ion.

$$Al^{3+} + H_2O \rightarrow \downarrow Al(OH)_3 + 3H^+$$

(E) is not a correct answer because although crystals of alum are isomorphous, i.e. of the same crystalline form, we cannot speak of solutions being isomorphous. Isomorphism refers to crystalline forms and not to solutions.

43. (A) 44. (A) 45. (B)

The cell in these problems is called the concentration cell.

Let the half-cell equations be:

$$\begin{array}{ll} \text{anode: } Ag \rightarrow Ag^+_{dilute} + e^- & \Sigma^\circ = -0.8V \\ \text{cathode: } Ag^+_{std} + e \rightarrow Ag & \Sigma^\circ = +0.8V \\ \hline \text{overall: } Ag^+_{std} + Ag \rightarrow Ag^+_{dilute} + Ag, & \Sigma^\circ = 0.00V \end{array}$$

$E^\circ_{cell}$ = 0.00V which answers (44). To find out if the reaction will take place as written we use a Nernst equation to evaluate the cell potential.

$$E_{cell} = E^\circ_{cell} - \frac{0.059}{n} \log \left\{ \frac{[Ag^+_{dilute}]}{[Ag^+_{std.}]} \right\}$$

n=1 in this example

$$\therefore E_{cell} = -0.059 \log \frac{10^{-3}}{1.0} = -0.059 \times (-3)$$
$$E_{cell} = 0.177$$

Since $E_{cell}$ is positive the reaction will proceed as written; if $E_{cell}$, however, had been negative then it would imply that we had written the equation in the wrong direction. From the half-cell reactions we conclude that the standard electrode is the cathode so the only correct statement in 43 is (A). 43 (D) is false since the cell will stop functioning after the concentration of $Ag^+$ ions around the electrodes evens out. The only false statement in 45 is (B) since we found from the Nernst equation that $E_{cell}$ = +0.117.

46. (D)

The number of moles of base added is

10 ml x 0.15 = 1.5 m mole

The reaction of the base and acid is:

$$HCl + NaOH \rightarrow NaCl + H_2O$$

so 1.5 m mole of HCl reacts.

The amount of acid remaining is

(25.0 ml x 0.1) –1.5 m mole = 2.5 – 1.5 = 1m mole

The new volume in which this acid exists is

$$V_T = 25 + 10 = 35 \text{ ml}$$

$$\text{concentration of protons } [H^+] = \frac{1\text{m mole}}{35 \text{ ml}}$$

$$[H^+] = 0.0286M = 2.86 \times 10^{-2}$$

$$pH = -\log(2.86 \times 10^{-2}) = 2 - \log 2.86 = 2 - 0.456 = 1.54$$

47. (C)

The most polar of the organic substances listed is acetic acid $CH_3COOH$ therefore it is the most soluble.

48. (A)

The compounds listed are approximately the same molecular weight, but the boiling point of the acid is relatively higher as a result of intermolecular hydrogen bonding between two molecules.

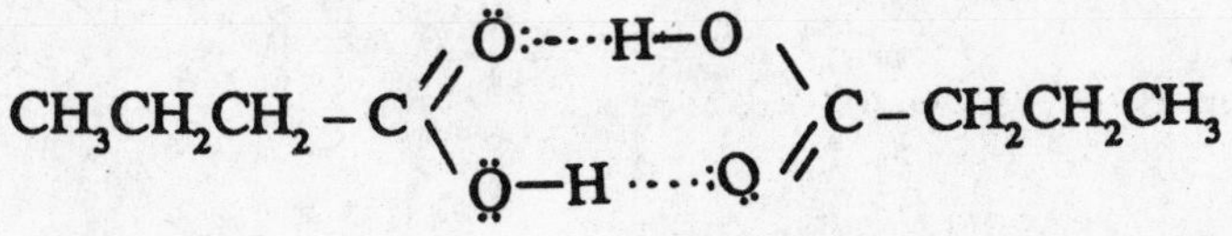

$CH_3CH_2CH_2-C(=\ddot{O}\cdots H-O)(-\ddot{O}-H\cdots O=)C-CH_2CH_2CH_3$

49. (C)

In the periodic table as one moves down a group, the atomic size increases as a result of an increase in the principal quantum number n, corresponding to a decrease in the attraction force to the nucleus. But as one moves from left to right across a period, the atomic size decreases as a result of an increase in nuclear charge while n remains constant. The decrease in atomic size as one moves across the first transition elements (n and l constant) results in gallium having the same size as aluminum, and germanium having the same size as silicon.

50. (C)

The given hydrocarbon synthesis is referred to as the Fischer-Tropsch process.

51. (D)

The source of fluorine is $CaF_2$ (calcium fluoride) also called fluorspar. But the elemental fluorine is obtained by the electrolysis of hydrogen fluoride obtained by the action of $H_2SO_4$ on $CaF_2$. Since gaseous and liquid HF are nonelectrolytes, the electrolysis takes place in molten $KF \cdot 2HF$ which forms a conducting solution.

$F^-$ ion cannot be oxidized in aqueous solution since the hydroxyl ($OH^-$) ion or the $H_2O$ molecule are more easily oxidized than the $F^-$ ion.

52. (B)

Of the listed answers, only zone refining is used to purify silicon. Zone refining involves melting a rod of an element near one end and moving the heat source slowly to the opposite end of the rod. The impurities being more soluble in the melt than in the solid will move along the rod and concentrate at one end of the rod. This end is discarded leaving a pure rod.

53. (E)

Graham's law of diffusion relates the diffusion rate and molecular weight of gases as

$$\frac{\text{rate (A)}}{\text{rate (B)}} = \sqrt{\frac{M \cdot W_B}{M \cdot W_A}}$$

$$\frac{\text{rate (A)}}{\text{rate }(H_2)} = 0.25$$

$$0.25 = \sqrt{\frac{2}{M \cdot WA}}$$

$$0.0625 = \frac{2}{M \cdot W_A}$$

$$MW_A = \frac{2}{0.0625} = 2x16 = 32$$

so the gas is $O_2$.

54. (A)

The total coordination of an atom is the number of ligands plus the number of lone pairs of electrons around that atom. Oxygen has 2 hydrogen atoms and 2 lone pairs of electrons around it.

55. (D)

There are four quantum numbers which describe an electron in an atom. Three of the quantum numbers, the principal (n), the azimuthal (l) and the magnetic quantum number $m_L$ are obtained from the solution of the Schrödinger equation. The fourth, the spin quantum number $m_s$, is needed to explain the behavior of atomic spectra in a magnetic field. The electron has spin and therefore a magnetic moment which can be directed up or down.

56. (D)

The electronic configuration of Ga is $[Ar]3d^{10}4s^24p^1$ so $Ga^+$ has configuration $[Ar]3d^{10}$. So there are 5 pairs of electrons in the outermost shell which now is 3d. 5 is an odd number.

The magnetic nature of the atom depends on the presence of an unpaired electron in the atom or ion. If the atom or ion contains one or more unpaired electrons it is paramagnetic. If it contains no unpaired electrons, it is diamagnetic.

57. (B)

351.25g of alloy contain:

Fe: 0.638 x 351.25 = 224g of Fe

or $\frac{224}{56}$

56 ≈ atomic weight of Fe

Ga: 351.25 – 224 = 127.25g of Ga

= 4 moles of Fe

or 1 mole of Ga since atomic wt. of Ga = 127.25

$H_2SO_4$ supplies $2H^+$ ions in solution per $H_2SO_4$

$\therefore Fe + 2H^+ \rightarrow Fe^{2+} + H_2\uparrow$

$2\,Ga + 6H^+ \rightarrow 2Ga^{3+} + 3H_2$

1 mole of Fe gives 1 mole of $H_2$ hence 4 moles of Fe will yield 4 moles of $H_2$.

2 moles of Ga gives 3 moles of $H_2$ hence 1 mole of Ga will yield 1.5 moles of $H_2$.

$\therefore$ 351.25g of alloy gives 5.5 moles of $H_2$

At STP one mole of a gas occupies 22.4 l

$\therefore$ The volume occupied by 5.5 moles of $H_2$ = 22.4 x 5.5 gives V = 123.2 l

58. (C)

The direction of a reaction is determined by $\Delta G$. If $\Delta G < 0$, then the reaction proceeds spontaneously as written. If $\Delta G > 0$ then it will proceed in the reverse direction of the written equation.

For the reaction A(g) + B(g)     2C(g) + Ds

$$\Delta G = \Delta G^o + RT \ln \frac{[C]^2}{[A]\,[B]}$$

$$\Delta G = \Delta G^o + RT \ln \frac{(P_c)^2}{(P_A)\,(P_B)}$$

$$\Delta G^o = RT \ln k \quad \text{and} \quad \frac{(P_c)^2}{(P_A)\,(P_B)} = Q$$

$$\Delta G = RT \ln \frac{Q}{K}$$

If at any moment $Q > K$ then $\Delta G > 0$ and the reaction from left to right is not spontaneous but spontaneous from right to left. Since $\Delta G < 0$ for this direction, the reaction will take place in this direction, i.e. right to left until equilibrium is established.

If $Q = K$, the reaction is at equilibrium.

If $Q < K$, $Q < 0$ the reaction proceeds from left to right until equilibrium is reached.

$$\text{In the example } Q = \frac{(P_C)^2}{(P_A)\,(P_B)} = \frac{4^2}{2 \times 5} = \frac{16}{10} = 1.6$$

$$Q = 1.6 > 0.65 = K_p$$

$\therefore$ $\Delta G > O$ and reaction will proceed from right to left.

59.     (D)

Oxalic acid $H_2C_2O_4$ has the structural formula

```
  O       O
   \\    //
    C - C
   /     \
 HO       OH
```

is a diprotic acid.

It will therefore have 2 dissociation constants: $K_{a_1} = 5.36x10^{-2}$ and $K_{a_2} = 5.3x10^{-5}$

60. (E)
$C_6H_5NH_2$ is aniline

$NH_2$

61. (C)
At STP 2240 L of ethane contains

$$\frac{2240}{22.4} = 100 \text{ moles of ethane}$$

The equation for the complete combustion of ethane is

$$2C_2H_6 + 14O_2 \rightarrow 4CO_2 + 6H_2O$$

1 mole of $C_2H_6$ yields 3 moles of $H_2O$
$\therefore$ 100 moles of $C_2H_6$ yields 300 moles of $H_2O$
1 mole of $H_2O$ is 18g
$\therefore$ 300 moles weighs 300 x18 = 5400g = 5.4Kg
1 Kg of $H_2O$ occupies a volume of 1l
$\therefore$ Volume of water condensed = 5.4 l

62. (B)
Given the mechanism as:

$$NO + NO \underset{k_2}{\overset{k_1}{\leftrightarrow}} N_2O_2$$

$$N_2O_2 + O_2 \xrightarrow{k_3} 2NO_2$$

Then the rate of formation of $NO_2$ is

$$\frac{d[NO_2]}{dt} = k_3[N_2O_2]\,[O_2]$$

The concentration of $N_2O_2$ is found from the equilibrium relation

$$k_1[NO]\,[NO] = k_2[N_2O_2]$$
$$[N_2O_2] = \frac{k_1}{k_2}\,[NO]^2$$

$$\text{the rate} = \frac{k_1\,k_3}{k_2}\,[NO]^2[O_2]$$

63. (C)

HCl dissociates completely

$\therefore [H^+] = 1.0$

$pOH = 14 - pH$

$pH = -\log[H^+] = -\log 1 = 0.0$

$\therefore pOH = 14 - 0 = 14$

64. (C)

Given that pOH of A is 2.5

$pH = 14 - 2.5 = 11.5$ (basic solution)

$[H+] = 10^{-11.5} = 10^{-12} \times 10^{+0.5} = 3.16 \times 10^{-12}M$

$pOH$ of $B = 10.1$

$pH = 14 - 10.1 = 3.9$ (acidic solution)

$\log[H^+] = -3.9$

$[H^+] = 10^{-3.9} = 10^{-4} \times 10^{+0.1} = 1.3 \times 10^{-4}M$

(A) and (B) state the same fact and are both wrong.

65. (A)

At equilibrium A(g) + 3B(g) 2C(g) + 2D(s) has equilibrium constant as shown below:

$$K_c = \frac{[C]^2}{[A]\,[B]^3} \qquad \frac{[mole/l]^2}{\left[\frac{mole}{l}\right]\left[\frac{mole}{l}\right]^3}$$

$$[Kc] = \quad mole^{-2}\ liter^2$$

66. (A)

For n moles of gas

$$PV = nRT$$

$$PV = \frac{m}{M}RT, \qquad M = \text{molecular weight}$$

$$m = \text{mass of gas}$$

$$\text{density } d = \frac{\text{mass}}{\text{volume}}$$

$$d = \frac{m}{V} = \frac{PM}{RT}$$

Molecular weight of $SO_3$ is

$$M_{SO} = 32+3(16) = 80 \text{ g/mole}$$

$$P = 50mmHg = 50/760 \text{ atm} = 0.066 \text{ atm}$$

$$T = 15 + 273 = 288K$$

$$R = 0.0821 \text{ liter-atm/(mole-K)}$$

$$d = \frac{0.066 \times 80}{0.0821 \times 288} = 0.223 \text{g/liter}$$

67. (B)

Electrolysis of $Al_2O_3$ takes place in a molten state. $Al_2O_3$ has a melting point of about 2000°C. Charles Hall found that a mixture of cryolite ($Na_3AlF_6$) and $Al_2O_3$ melts at about 1000°C. At present other materials are used to produce even lower melting temperatures.

68. (D)

Acetone is a ketone. Ketones are distinguished by the presence of C=O, the carbonyl group. Acetone has the structural formula

$$CH_3-\overset{\overset{\displaystyle O}{||}}{C}-CH_3$$

69. (C)

A protein is the polymer of amino acids. Amino acids have the general formula

$$H_2N-\underset{\underset{\displaystyle R_1}{|}}{\overset{\overset{\displaystyle H}{|}}{C}}-\overset{\overset{\displaystyle O}{||}}{C}-OH$$

There are two functional groups $-NH_2$ (amino group) and

$$-\overset{\overset{\displaystyle O}{||}}{C}-OH$$– carboxylic group.

In proteins, the amino acids are linked in a long chain,

```
      H  O               H  H    O
      |  ||              |  |  //
... + H2N - C - C - OH + H - N - C - C - OH + ...  -H2O
            |                    |                 ---->
            Rn                   Rn+1
```

```
          H  O       H  O
          |  ||      |  ||
... - N - C - C - N - C - C - ...
      |   |       |   |
      H   Rn      H   R2
```

Since water is eliminated in the reaction it is a condensation polymerization.

The bond formed between amino acids

```
   O
   ||
 - C - N -
       |
       H
```

is called a peptide bond. So the protein can be described as a polypeptide or a condensation polymer.

70. (A)

Chlorine is the central atom in $ClF_4^-$. The ground state electronic configuration of chlorine is

Cl: [Ne] $3s^2 3p^5$

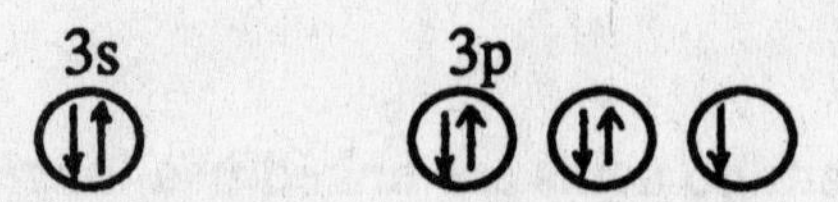

To accommodate 3F atoms and the lone pair of $F^-$ chlorine uses 2 of the 3d orbitals.

| | 3s | 3p | | | 3d | |
|---|---|---|---|---|---|---|
| Cl: [Ne] | ↑↓ | ↑↓ | ↑↓ | ↑ | ◯ | ◯ |

These six orbitals are hybridized to give 6 equivalent $3sp^3d^2$ orbitals.

| | $3s\,p^3\,d^2$ | | | | | |
|---|---|---|---|---|---|---|
| Cl: [Ne] | ↑↓ | ↑↓ | ↑ | ↑ | ↑ | ◯ |

Then the bonding of $ClF_4^-$ is

| | $3s\,p^3\,d^2$ | | | | | |
|---|---|---|---|---|---|---|
| Cl: [Ne] | ↑↓ | ↑↓ | ↑↓ | ↑↓ | ↑↓ | ↑↓ |
| | | | F | F | F | $F^-$ |

71. (A)

The reaction is

$$Cl_2 + CH_3CH_3 \rightarrow CH_3CH_2Cl + HCl$$

1 mole of ethane is (2 x 12) + 6 = 30g.

From the equation 1 mole $C_2H_6$ yields 1 mole of $C_5H_5Cl$

1 mole of $C_2H_5-Cl$ weighs (2 x 12) + 5 + 35.5 = 64.5g

30g of ethane theoretically should give 64.5g of ethyl chloride. The amount of ethyl chloride obtained from 15g of ethane is:

$$\text{amount of } C_2H_5Cl = \frac{15}{30} \times 64.5\text{g} = 32.25\text{g}$$

∴ The theoretical yield is 32.25g.

But the actual yield is 15g.

$$\therefore \text{percent yield} = \frac{15}{32.25} \times 100 = 46.5\%$$

72. (C)

Aldehydes and ketones which have $\alpha$ –hydrogens exhibit a kind of isomerism called tautomerism in which the $\alpha$ –hydrogen migrates to the carbonyl oxygen. The new isomer is both an alkene and alcohol and is called an enol. In the problem there are no aldehydes or ketones given but if we examine the compounds carefully we see that compound (C) is the enol form of a ketone.

CH₃ O H ↔ CH₃ H O H

enol form    keto form

also

CH₃ O ↔ CH₃ OH

in fact at equilibrium they all co-exist, i.e.

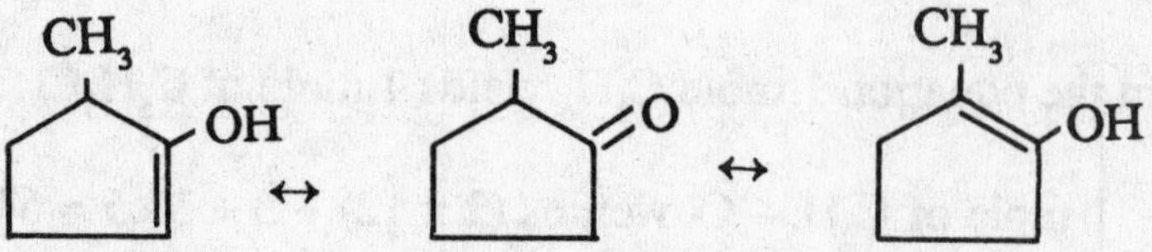

CH₃ OH ↔ CH₃ O ↔ CH₃ OH

73. (B)

Three dimensional shapes of molecules are important to the chemist and biochemist. The Newman projection is a way of showing molecular conformations of alkanes. Conformations are the arrangements of atoms depending on the angle of rotation of one carbon with respect to another adjacent carbon. A Newman projection is obtained by looking at the molecule along the bond on which rotation occurs. The carbon in front is indicated by a point and at the back by a circle, the 3 remaining valences are placed 120° to each other (but remember that the bond angles about each of the carbon atoms are 109.5° and not 120° as the Newman projection formula might suggest). e.g. for $CH_3-CH_3$

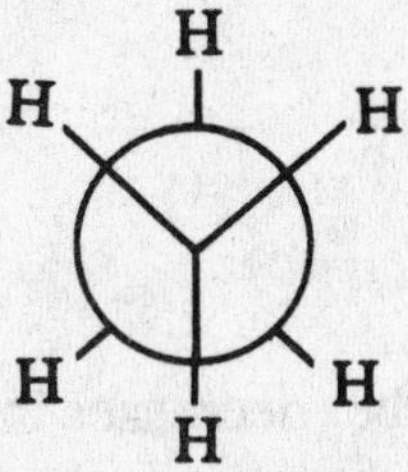

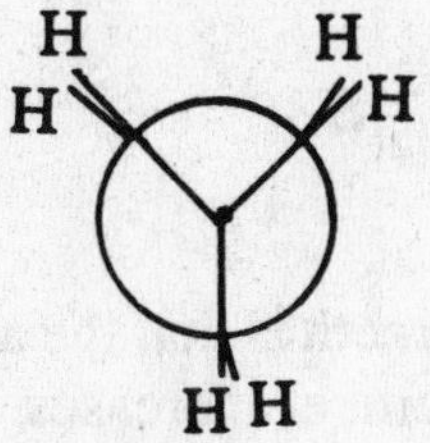

staggered conformation

eclipsed conformation

74. (C)

Aldehydes are oxidized easily to carboxylic acids by even weak oxidizing agents like the silver ion, $Ag^+$. A solution of silver nitrate in ammonium hydroxide when added to an aldehyde results in the oxidation of the aldehyde and $Ag^+$ ion is reduced to metallic silver. This test is the Tollen's test, also referred to as the silver mirror test, because the metallic silver deposits in the form of a mirror on a glass surface.

The Benedict's and Fehling's tests are also tests for aldehydes and also $\alpha$ –hydroxy ketones in sugars. In these cases the reducing agents are solutions containing copper (II) ion complexed with a tartrate (Fehling's solution) or a citrate (Benedict's solution). The copper (II) ion is reduced to copper (I) which precipitates as a brick-red copper (I) oxide.

$$R-\overset{\overset{\displaystyle O}{||}}{C}-H + \underset{\text{Fehling's or Benedict's solution}}{Cu^{2+}} \rightarrow R-\overset{\overset{\displaystyle O}{||}}{C}-OH + \underset{\text{brick-red ppt.}}{Cu_2O}$$

75. (C)

Transmutation is the conversion of a nuclide of one element into that of another element. To answer the question easily, write a balanced equation for the given transmutation.

For $\beta^-$ transmutation:

$${}^{A}_{z}A \rightarrow {}^{A}_{z+1}B + {}^{o}_{-1}e$$

The only false statement is that the atomic number decreases in a β– decay. The atomic number increases. (B) is true because isobars are defined as nuclides with the same mass numbers but different atomic numbers.

76. (C)

Gibbs phase rule in its general form is

$$f = 2 + C - p$$

where C is the number of component substances in the system and p is the number of phases present.

So if $f = 3 - p$, this implies $f = 2 + 1 - p$ or $C = 1$, therefore $f = 3 - p$ is the phase rule for a one-component system or a pure substance.

77. (D)

Raoult's law states that $P_A = X_A P_{A\,partial}$ where $P^A_A$ is the partial vapor pressure of a component in solution, $X_A$ is the mole fraction of component A and $P_A$ is the vapor pressure of pure substance A. Solutions that obey Raoult's law are called ideal. Ideal solutions form between substances in which solute-solute, solute-solvent and solvent-solvent interactions are the same.

If unlike molecules interact less strongly than like molecules do, the partial vapor pressures of the components will be greater than predicted by Raoult's law. This is known as positive deviation. Of the responses listed, only the solution of acetone and carbon disulfide would show positive deviation.

Answers (A) and (C) are examples of ideal solutions; answer (B) and (E) are examples of negative deviation.

78. (B)

The larger the positive charge on the central ion, the greater the attraction for the negative charge on the ligands. This effect results in the decreased tendency of ligands in the coordination ion to bind to other cations such as $H^+$. V in $VO_4^{3-}$ has a +5 charge; Cr in $CrO_4^{2-}$ has a +6 charge and Mn in $MnO_4^-$ has a +7 charge so the oxide ions are bound strongest to the $Mn^{+7}$, followed by $Cr^{+6}$ and then $V^{+5}$. So $MnO_4^-$ is the weakest base of the three ion complexes and hence $HMnO_4$ is the strongest acid. $VO_4^{3-}$ is the strongest base and $H_3VO_4$ the weakest acid. $CrO_4^{2-}$ and $H_2CrO_4$ have intermediate basic and acidic properties respectively.

79. (E)

Brass is an alloy of copper and zinc.

80. (E)

Stalactites and stalagmites are structures that appear in limestone ($CaCO_3$) caves. When ground water saturated with calcium bicarbonate seeps into the cave and in the presence of circulating air evaporates, calcium carbonate precipitates on the ceiling forms a stalactite. If the solution of calcium bicarbonate drips to the floor of the cave before evaporating, the calcium carbonate is precipitated on the cave floor forming a stalagmite.

$$Ca(HCO_3)_2 \xrightarrow{\text{evaporation}} CaCO_3 + H_2O + CO_2$$

81. (D)

Mica is a silicate with representative formulas $KAl_2Si_3AlO_{10}(OH)_2$ called muscovite and $K_2Li_3Al_4Si_7O_{21}(OH,F)_3$ called lepidolite. Mica has layer structure.

82. (C)

The $NO_3^-$ structure is a hybrid of the three structures below:

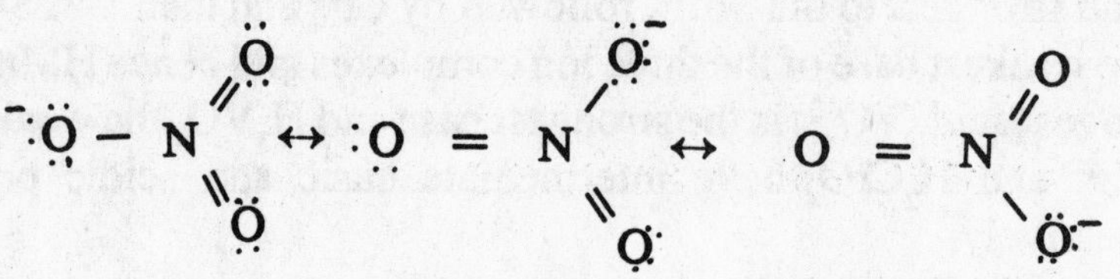

$$^{-}:\ddot{O}-N(=\ddot{O})(=\ddot{O}) \leftrightarrow :O=N(-\ddot{O}:^{-})(=\ddot{O}) \leftrightarrow O=N(=O)(-\ddot{O}:^{-})$$

83. (C)

Although helium is more dense than hydrogen, it is preferred for filling meteorological balloons and dirigibles because hydrogen is combustible while helium is not.

84. (B)

The statement in (B) is false, because the rate law is an equation which relates the rate of the reaction to the concentrations of the reactants.

Statement (A) is true because a process is spontaneous if $\Delta G < 0$. Since $\Delta G = \Delta H - T\Delta S$, if $\Delta H < 0$, $\Delta G$ is more likely to be negative than if $\Delta H < 0$, because $\Delta S$ then has to be large and positive.

Statement (C) is true because $NH_3$ can act as a base $NH_3 + H^+ \rightarrow NH_4^+$ and also as an acid $2NH_3 \rightarrow NH_4^+ + NH_2^-$ amide ion.

Statement (D) is also true. A Brønsted acid is a substance that gives protons (hence it must contain hydrogen), and the Brønsted base is a substance that accepts a proton (i.e. must have an unshared electron pair). The products of a Brønsted acid-base reaction are always a new acid and a new base, e.g.

| $CH_3COOH$ + $OH^-$ | → | $CH_3COO^-$ | + | $H_2O$ |
|---|---|---|---|---|
| acid base | | conjugate base of $CH_3COOH$ | | conjugate acid of $OH^-$. |

Statement (E) is true because the total charge of a mole of electrons (i.e. Avogadro's number of electrons) is defined as a faraday.

85. (A)
1 liter of water weighs 1000 g.

1 mole of water = 18g.

Number of moles of water in 1000g =

$$\frac{1000}{18} = 55.56 \text{ moles}$$

## SECTION II

### PART A

The relationship between the rate of a reaction, the activation energy ($E_a$), and temperature (T) is given by the Arrhenius equation:

$$k = Ae^{-(E_a/RT)} \qquad (1)$$

where k is the rate constant, A is a constant for the particular reaction, $E_a$ is the activation energy, R is the gas constant and T is the absolute temperature.
It is easier to handle equation (1) in logarithms

$$\ln k = \ln A - \frac{E_a}{RT} \qquad (2)$$

or

$$\log k = \log A - \frac{E_a}{2.303RT_2}$$

a) From the data $k_1 = 1.75$ mole liter$^{-1}$ s$^{-1}$

at $T_1 = 25+273 = 298K$

and

$k_2 = 2.6$ mole liter$^{-1}$ s$^{-1}$

at $T_2 = 45+273 = 318K$

$$\log k_1 = \log A - \frac{E_a}{2.303RT_1} \qquad (3)$$

$$\log k_2 = \log A - \frac{E_a}{2.303RT_2} \qquad (4)$$

Since A is not known, eliminate it by subtracting equation (3) from equation (4).

$$\log k_2 - \log k_1 = \frac{E_a}{2.303RT_1} - \frac{E_a}{2.303RT_2}$$

$$\log k_2 - \log k_1 = \log \frac{k_2}{k_1}$$

$$\therefore \log \frac{k_2}{k_1} = \frac{E_a}{2.303R}\left(\frac{1}{T_1} - \frac{1}{T_2}\right) \qquad (5)$$

also

$$\frac{k_2}{k_1} = \frac{rate(2)}{rate(1)}$$

$$\therefore \frac{k_2}{k_1} = \frac{rate(2)}{rate(1)} = \frac{2.63}{1.75} = 1.5$$

Substituting values in the Arrhenius equation gives:

$$\log \frac{k_2}{k_1} = \log \frac{2.63}{1.75} = \frac{E_a}{2.303R}\left(\frac{1}{298} - \frac{1}{318}\right)$$

$$\log 1.5 = \frac{E_a}{2.303R}\left[(3.356x10^{-3}) - (3.145x10^{-3})\right]$$

$$0.176 = \frac{E_a}{2.303R}(2.11x10^{-4})$$

Taking $R = 8.314 \text{ J mole}^{-1} \text{ K}^{-1}$

$$E_a = \frac{0.176x2.303x8.314}{2.11 \text{ x } 10^{-4}} = 1.5x10^4 \text{ J mole}^{-1}$$

$$\therefore E_a = 15x10^3 = 15\text{kJ mole}^{-1}$$

b) Let $T_3$(K) be the temperature at which the rate is 3.5 times that found at 25°C

Then

$$\frac{k_3}{k_1} = \frac{rate(3)}{rate(1)} = 3.5$$

since rate(3) = 3.5 rate(1)

$\therefore$ using equation (5)

$$\log 3.5 = \frac{E_a}{2.303R}\left(\frac{1}{T_1} - \frac{1}{T_3}\right)$$

$$0.544 = \frac{1.5 \times 10^4}{2.303 \times 8.314}\left(\frac{1}{298} - \frac{1}{T_3}\right)$$

$$\left(\frac{1}{298} - \frac{1}{T_3}\right) = \frac{0.544 \times 2.303 \times 8.314}{1.5 \times 10^4} = 6.944 \times 10^{-4}$$

$$\frac{1}{T_3} = \frac{1}{298} - (6.944 \times 10^{-4}) = (3.356 \times 10^{-3}) - (6.944 \times 10^{-4})$$

$$\frac{1}{T_3} = 2.66 \times 10^{-3}$$

$$T_3 = \frac{1}{2.66 \times 10^{-3}} = 375.9\text{K}$$

or

$$T_3 = 102.9^\circ\text{C}$$

c) Let $k_1$ and $k_2$ be rate constants for activation energies. $E_a(1)$ and $E_a(2)$ respectively. Then the corresponding Arrhenius equations are:

$$\log k_1 = \log A - \frac{E_a(1)}{2.303RT}$$

$$\log k_2 = \log A - \frac{E_a(2)}{2.303RT}$$

again eliminate A by subtraction

$$\log k_2 - \log k_1 = \frac{E_a(1) - E_a(2)}{2.303RT}$$

$$\log \frac{k_2}{k_1} \quad \frac{E_a(1) - E_a(2)}{2.303RT}$$

we can represent $\frac{k_2}{k_1}$ by $\frac{rate(2)}{rate(1)}$ since the rate is proportional to the rate constant.

$$\therefore \log \frac{rate(2)}{rate(1)} = \log \frac{k2}{k1} = \frac{E_a(1) - E_a(2)}{2.303RT}$$

$E_a(1) = 15kJ$

The catalyst lowers $E_a(1)$ by 76%

i.e. $E_a(2) = E_a(1) - 0.76E_a(1) = 0.24E_a(1)$

$$\log \frac{rate(2)}{rate(1)} = \frac{E_a(1) - 0.24E_a(1)}{2.303x8.314x298} = \frac{0.76E_a(1)}{5705.85}$$

$$\log \frac{rate(2)}{rate(1)} = \frac{0.76x15x103}{5705.85} = 1.998 \approx 2$$

$$\frac{rate(2)}{rate(1)} = 10^2 = 100.$$

So this reaction speeds up by a factor of 100.

## PART B

2. a) $A \leftrightarrow B + C$

The equilibrium constant $K_c$ is:

$$K_c = \frac{[B]\,[C]}{[A]}$$

$$[A] = \frac{n_A}{V}, \quad [B] = \frac{n_B}{V}, \quad [C] = \frac{n_C}{V}$$

V is volume of vessel

$n_i$ – number of moles of a component

$$K_c = \frac{\left(\frac{n_B}{V}\right)\left(\frac{n_C}{V}\right)}{\left(\frac{n_A}{V}\right)} = \frac{n_B \cdot n_C}{n_A} \cdot \frac{1}{V}$$

| | [A] | [B] | [C] |
|---|---|---|---|
| initially | $\frac{0.3}{10}$ | $\frac{0.2}{10}$ | $\frac{0.15}{10}$ |
| equilibrium | $\frac{0.3-x}{10}$ | $\frac{0.2+x}{10}$ | $\frac{0.15+x}{10}$ |

where x is number of moles of A which reacted.

$$K_c = \frac{(0.2 + x)(0.15 + x)}{(0.3-x)} \frac{1}{10} = 0.18$$

$$(0.2 + x)(0.15 + x) = 1.8(0.3-x)$$
$$0.03 + 0.35x + x^2 = 0.54 - 1.8x$$
$$x^2 + 2.15x - 0.51 = 0$$

The solution of the quadratic equation (1) is given below:

$$ax^2 + bx + c = 0$$

is

$$x_{1,2} = \frac{-b \pm \sqrt{b^2 - 4ac}}{2a}$$

$\therefore$ solution of equation (1) is

$$x_{1,2} = \frac{-2.15 \pm \sqrt{(2.15)^2 - (4x - 0.51)}}{2}$$

$$x_1 = \frac{-2.15 + 2.58}{2} = 0.216$$

$$x_2 = \frac{-2.15 - 2.58}{2} = -2.37$$

$x_2$ has no physical meaning i.e. the number of moles of A reacting cannot be negative so we discard $x_2$.

The number of moles of A reacting is x = 0.106 moles.

∴ The equilibrium composition is:

[A] = 0.3 – 0.216 = 0.084 mole
[B] = 0.2 + 0.216 = 0.416 mole
[C] = 0.15 + 0.216 = 0.366 mole

These results may be checked as follows:

$$K_c = \frac{(0.416)\,(0.366)}{(0.084) \times 10} = 0.18$$

b) When the volume is reduced to 5 liters, the equilibrium will be shifted to the left, since this will relieve the stress. B and C react to reduce the number of moles and therefore the total volume of the reacting system.

$K_c$ is still equal to 0.18

| | [A] | [B] | [C] |
|---|---|---|---|
| initially | $\frac{0.084}{10}$ | $\frac{0.416}{10}$ | $\frac{0.366}{10}$ |
| at new equilibrium | $\frac{0.084+x}{5}$ | $\frac{0.416-x}{5}$ | $\frac{0.366-x}{5}$ |

$$K_c = \frac{(0.416-x)\,(0.366-x)}{(0.084+x)\;\;5} = 0.18$$

$$(0.416-x)\,(0.366-x) = 0.9\,(0.084+x)$$
$$0.152 - 0.782\,x + x^2 = 0.0756 + 0.9x$$
$$x^2 - 1.682X + 0.076 = 0$$

$$x_{1,2} = \frac{-(-1.682) \pm \sqrt{(-1.682)^2 - (4 \times 0.076)}}{2}$$

$$x_1 = \frac{1.682 + 1.589}{2} = 1.64$$

$$x_2 = \frac{1.682 - 1.589}{2} = 0.046$$

The root $x_1 = 1.64$ has no chemical significance, because it would require more B and C to be consumed than were originally present. The correct root is $x_2 = 0.05$.

The new equilibrium concentrations are:

$$[A] = 0.084 + 0.046 = 0.1305$$
$$[B] = 0.416 - 0.046 = 0.37$$
$$[C] = 0.366 - 0.046 = 0.32$$

check:

$$K_c = \frac{(0.37)(0.32)}{(0.13)5} = 0.18$$

c) $K_p = K_c(RT)^{\Delta n}$ (1)

$$A \leftrightarrow B + C$$
$$\Delta n = 2 - 1 = 1$$

$$\therefore K_p = K_c(RT)$$

$$K_p = 0.18 \ \frac{\text{mole}}{\text{liter}} \times 0.0821 \ \frac{\text{liter–atm}}{\text{mole K}} \times (200+273)$$

$$K_p = 0.18 \times 0.0821 \times 473 \text{ atm}$$

$$K_p = 6.99 \text{ atm.}$$

If one forgets the equation (1), it can be derived as follows:

$$K_p = \frac{[P_B][P_C]}{[P_A]}$$

where $P_i$ is the partial pressure of the gas components of the system.

$$P_i = \frac{n_i RT}{V}$$

$$K_P = \frac{\left(\frac{n_B RT}{V}\right)\left(\frac{n_C RT}{V}\right)}{\left(\frac{n_A RT}{V}\right)} = \frac{n_B \cdot n_C}{n_A} \cdot (RT) = K_C(RT)$$

3. a) The reaction occurring between the permanganate and oxalate ions is an oxidation-reduction reaction, as shown below:

The ionic equation is

$$MnO_4^- + C_2O_4^{2-} \rightarrow Mn^{2+} + CO_2 \text{ (acid solution)}$$

The half-reactions are

reduction: $MnO_4^- + 8H^+ + 5e^- \rightarrow Mn^{2+} + 4H_2O$
oxidation: $C_2O_4^- \rightarrow 2CO_2 + 2e^-$

To balance the overall reaction we need $10e^-$ in each half reaction. So we multiply the oxidation half-reaction by 5 and the reduction half-reaction by 2 and add the two half-reactions

$$2MnO_4^- + 5C_2O_4^- + 16H^+ \rightarrow 2Mn^{2+} + 10CO_2 + 8H_2O$$

b) The number of moles of $Na_2C_2O_4$ dissolved is:

$$n_{Na_2C_2O_4} = \frac{0.157}{M.W._{Na_2C_2O_4}} =$$

$$M.W._{Na_2C_2O_4} = (2 \times 23) + (2 \times 12) + (4 \times 16) = 134$$

$$n_{Na_2C_2O_4} = \frac{0.157}{134} = 1.16 \times 10^{-3} \text{mole}$$

From the balanced equation we know that 5 moles of oxalate react with 2 moles of $MnO_4^-$ 1 mole oxalate reacts with 2/5 mole $MnO_4^-$ $1.16 \times 10^{-3}$ oxalate reacts with $1.16 \times 10^{-3} \times 2/5$ mole of $MnO_4^-$

$\therefore$ number of moles $MnO_4^-$ is

$$1.16 \times 10^{-3} \times \frac{2}{5} = 0.465 \times 10^{-3} \text{ moles}$$

$0.465 \times 10^{-3}$ moles of $MnO_4^-$ are contained in 15.3 ml

∴ The molarity of $KMnO_4$ solution is

$$M_{KMnO^4} = \frac{0.465 \times 10^{-3} mole}{15.3 \times 10^{-3} l}$$

$$M_{KMnO^4} = 3.039 \cdot 10^{-2} mole/l$$

Molarity of $KMnO_4$ = 0.03M.

c) The mass of $KMnO_4$ in 15.3 ml solution is:

$$M_{KMnO_4} = M.W_{KMnO_4} \times n_{KMnO_4}$$

$$M.W_{KMnO_4} = 39 + 55 + (16 \times 4) = 94 + 64 = 158$$

$$M_{KMnO_4} = 158 \times 0.465 \times 10^{-3} = 0.073g.$$

d) We need a balanced equation to solve the problem.

Reduction half-reaction

$$MnO_4^- + 8H^+ + 5e^- \rightarrow Mn^{2+} + 4H_2O$$

Oxidation half-reaction

$$Fe^{2+} \rightarrow Fe^{3+} + e^-$$

Multiply oxidation half-reaction by 5 and add to reduction half-reaction to get the overall balanced equation.

$$MnO_4^- + 8H^+ + 5Fe^{2+} \rightarrow Mn^{2+} + 5Fe^{3+} + 4H_2O$$

25 ml of $FeSO_4$ contains:

$25 \times 10^{-3} L \times 0.105 = 2.625 \ 10^{-3}$ mole of $Fe^{2+}$ ions

From the balanced equation:

5 moles $Fe^{2+}$ react with 1 mole $MnO_4^-$

1 mole $Fe^{2+}$ reacts with $\frac{1}{5}$ mole $MnO_4^-$

$2.625 \times 10^{-3}$ mole $Fe^{2+}$ reacts with $2.625 \times 10^{-3} \times \frac{1}{5}$ =

$0.523 \times 10^{-3}$ mole $MnO_4^-$

Number of moles = Molarity x Volume

$$\text{Volume} = \frac{\text{Number of Moles}}{\text{Molarity}}$$

$$\text{Volume of } KMnO_4 = \frac{0.523 \cdot 10^{-3}}{0.03} = 17.43 \times 10^{-3} L$$

Volume of $KMnO_4$ = 17.43 ml

## PART C

4. The use of radioisotopes is widespread in chemistry, biology, medicine and other areas of science and industry.

All of these uses depend on one or more of the following characteristics of radioisotopes and nuclear decompositions.

a) The chemical properties and reactions of a radioisotope are exactly the same as those of a nonradioactive atom of the same element.

b) Radiation can be detected some distance from its source.

c) Each radioisotope has a characteristic half-life.

d) Radioactive emissions interfere with normal cell growth and division.

The age of rocks containing uranium can be detected by finding the ratio of $^{238}U$ to $^{206}Pb$. $^{206}Pb$ is the stable isotope product of $^{238}U$ radioactive decay. In rocks where uranium is absent a potassium-argon method is used.

$$^{40}_{19}K + ^{0}_{-1}e \rightarrow ^{40}_{18}Ar \qquad t_{1/2} = 1.3 \times 10^9 \text{ years}$$

Again the ratio of $^{40}K$ to argon is measured.

The age of materials of organic origin such as bones, wood, etc. can be estimated by measuring the ratio of $^{14}C$ to $^{12}C$. $^{14}C$ is radioactive and is constantly produced in the upper atmosphere by bombardment of cosmic neutrons on $^{14}_{7}N$,

$$^{14}_{7}N + ^{1}_{0}n \rightarrow ^{14}_{6}C + ^{1}_{1}P$$

The $^{14}_{6}C$ begins to decay immediately with $t_{1/2}$= 5770 years, so there is an equilibrium concentration of $^{14}_{6}C$ maintained in the atmosphere. $^{14}CO_2$ in the air is taken by plants through photosynthesis and is transferred from plants to the animals who consume them. Once the plant or animal dies, the $^{14}C$ is not replaced and its concentration reduces as a result of radioactive decay.

Radioisotopes can be incorporated into a molecule or compound. Such compounds are said to be tagged or labelled. These compounds are used in medicine for diagnosis. Iron$^{-59}$ in the form of iron-citrate ($FeC_6H_6O_7$) is routinely used in the determination of red blood-cell volume, and rates of red blood-cell production and destruction.

Radiation therapy is also used in the treatment of certain tumors.

In industry, nuclear reactions are used to generate energy in fission reactions. In a fission reaction a large nucleus is split into two medium-sized nuclei. Only a few nuclei are known to undergo fission, mainly uranium$^{-235}$ and plutonium$^{-239}$.

$$^{235}_{92}U + ^{1}_{0}n \rightarrow \left[^{236}_{92}U\right] \rightarrow ^{139}_{56}Ba + ^{94}_{36}Kr + 3\,^{1}_{0}n + \text{Energy}$$

5. In the presence of light, hydrogen peroxide decomposes to form water and oxygen:

$$2H_2O_2 \xrightarrow{\text{light}} 2H_2O + O_2\uparrow$$

To prevent $H_2O_2$ from decomposing, it is stored in dark-brown bottles which prevent light from reaching the liquid.

Hydrogen peroxide acts as an antiseptic because as the hydrogen peroxides decomposes, it produces a very reactive atomic oxygen. Bacteria, which are organic in composition, are easily oxidized (and therefore killed) by the atomic oxygen.

$$2H_2O_2 \rightarrow 2H_2O + 2[O] \rightarrow O_2$$

[O] → oxidation organic material

b) Ethyl chloride, $CH_3CH_2Cl$ has a very low boiling point (about 13°C). It exists as a gas under normal conditions. It liquifies under pressure. When it is sprayed on the skin, it evaporates very quickly and in doing so, it absorbs energy from the skin. The skin and the nerve endings in the skin are cooled. The skin therefore feels numb under the action of ethyl chloride.

c) The volume of water increases as it changes from the liquid state to the solid. This is because of hydrogen bonding. In the liquid state the water molecules can be forced together, because the hydrogen bonding is randomly oriented. In the solid state, however, the molecules are arranged in a regular crystal lattice and the molecules cannot be squashed together as in the liquid state. The density of ice is therefore lower than water (at 4°C $d_{H_2O}$ = 1.000g/cm³, at 0°C $d_{ice}$ = 0.998g/cm³). So, when the lake freezes, the ice which is less dense rises to the surface of the lake. The layer of ice at the surface also acts as an insulator and so lakes rarely freeze all the way to the bottom. Fish and other aquatic life can therefore survive in the lake during the winter.

d) Aerosol cans always contain vapors although it may seem to us to be empty. They become "empty" when the pressure inside is equal to the pressure outside. If an empty can is incinerated, i.e. if the temperature is raised to very high temperatures, the volume and pressure of the vapor increases according to the gas laws. Since the volume of the can is fixed, the volume of the gas will be constant; as a result the pressure of the vapor is dramatically increased. A stage is reached where the can cannot contain the pressure built in it. It will rupture with an explosion of the expanding gases. The explosion can cause material damage or human injury.

6. a) Freons are fluorinated hydrocarbons. Freons are used as propellants of aerosol insecticides and as refrigerants. Due to their inertness they are not decomposed in the lower atmosphere. They slowly diffuse into the stratosphere where they are decomposed by ultraviolet radiation from the sun. As the freons decompose they react with the ozone. The ozone in the stratosphere acts as a barrier for the earth against ultraviolet radiation. So as the ozone layer is depleted as a result of the action on it by freons, the amount of ultraviolet radiation reaching the earth is increased. The increased levels of the ultraviolet radiation destroy crops, animals and raise the incidence of skin cancer.

b) A mothball is crystalline naphthalene. The crystal structure is the molecular type. The molecules are held in place by very weak London forces. It requires very little thermal energy for the molecules to escape from the crystal. So in summer when the temperature is relatively high the naphthalene easily evaporates, the vapors of which offers protection for woolens against moths. So by fall almost all the mothball will evaporate.

7. a) There are two types of solids: amorphous and crystalline.

Crystals are solids bounded by planar surfaces. Amorphous substances have no definite form.

Particles in a crystal are well ordered whereas those in an amorphous substance are chaotic, i.e. arranged in a random manner.

Crystals have sharp and definite melting points, while amorphous solids soften as they are heated and melt over a range of temperatures. Examples of crystals are NaCl, quartz, and of amorphous are tar and glass solids.

The differences between molecular, ionic and covalent crystals lie in the types of particles present at the lattice points, and the nature of the attractive forces between them.

The lattice of molecular crystals are occupied by molecules or atoms, in ionic crystals by ions and in covalent crystals by atoms.

The binding forces in molecular crystals consist only of atoms and nonpolar molecules for example: Ar, $O_2$ and naphthalene (the binding forces are London forces). In crystals of polar molecules like $SO_2$ the binding forces are dipole-dipole attractions. In some molecular crystals where the polar molecules contain hydrogen, eg. $H_2O$, $NH_3$ and HF, the molecules are attracted in the crystal by hydrogen bonding.

In ionic crystals such as NaCl the binding forces are electrostatic.

In covalent crystals a network of covalent bonds links the atoms throughout the solid. Examples of covalent crystals are carborundum (SiC), quartz ($SiO_2$) and diamond (C).

Molecular crystals are soft and have low melting points since their lattice energies are small.

Ionic crystals have large lattice energies therefore such crystals are hard and have high melting points. They are also brittle, because external forces can cause planes of ions to slip by one another and these planes pass from a condition of mutual attraction to one of mutual repulsion.

All the crystals are poor conductors of electricity because a) the electrons are bounded to individual molecules in molecular crystals and localized in covalent bonds in covalent crystals. b) ions in ionic crystals are fixed rigidly in place. (When ionic crystals melt they conduct electricity, but then they are no longer crystals).

## 8. Preparation of metal.

In a few cases where metals are found in the elemental states, most metals occur as oxides, sulfides, halides, silicates, carbonates and sulfates. (These are called ores).

The preparation of metals generally consists of transforming a positively charged atom into an element. i.e. $M^{n+} + ne^- \rightarrow M$.

The reaction is a reduction. The industrial reduction process is called smelting.

There are chemical and electrochemical means of reducing the metal from its oxidation state to its elemental state. The following are examples of the chemical methods:

1) Reduction by heating in air.

Oxides of less active metals decompose at high temperatures. So when sulfide ores of mercury and copper are roasted in air, the metal (instead of the metallic oxide) is formed

$$HgS + O_2 \rightarrow Hg + 8O_2$$

$$Cu_2S + O_2 \rightarrow 2Cu + 80_2$$

2) Reduction with carbon.

Moderately active metals are reduced by carbon.

$$Fe_2O_3 + 3C \xrightarrow{\Delta} 2Fe + 3CO$$

$$Fe_2O_3 + 3CO \xrightarrow{\Delta} 2Fe + 3CO_2$$

$$2SnO + C \rightarrow 2Sn + CO_2$$

3) Reduction with hydrogen.

This is used when carbon is not suitable, where formation of carbides mixed with the metal is undesireable e.g. in the production of tungsten.

$$WO_3 + 3H_2 \rightarrow W + 3H_2O$$

$$PbO + H_2 \rightarrow Pb + H_2O$$

4) Reduction with an active metal.

This is also used when carbon is not an effective reducing agent. Aluminum, sodium and calcium are used as reducing agents.

$$TiCl_4 + 2Mg \rightarrow 2MgCl_2 + Ti$$

5) Reduction by electrolysis.

Very active metals, the alkali and alkaline earth elements are produced by the electrolysis of anhydrous fused salts. The halides of these elements are used because of the lower melting points. The production of aluminum is an exception. Here molten cryolite $Na_3AlF_6$ is used as a solvent for $Al_2O_3$. Carbon serves both as anode and cathode. The reaction at the electrodes are

Anode: $6O^{2-}(l) \quad 3O_2(g) + 12e^-$

Cathode: $4Al^{3+}(l) + 12e^- \quad 4Al(l)$

---

$4Al^{3+}(l) + 6O^{2-}(l) \quad 3O_2 + 4Al$

# THE ADVANCED PLACEMENT EXAMINATION IN

# CHEMISTRY

# TEST V

# THE ADVANCED PLACEMENT EXAMINATION IN CHEMISTRY

# ANSWER SHEET

1. (A) (B) (C) (D) (E)
2. (A) (B) (C) (D) (E)
3. (A) (B) (C) (D) (E)
4. (A) (B) (C) (D) (E)
5. (A) (B) (C) (D) (E)
6. (A) (B) (C) (D) (E)
7. (A) (B) (C) (D) (E)
8. (A) (B) (C) (D) (E)
9. (A) (B) (C) (D) (E)
0. (A) (B) (C) (D) (E)
1. (A) (B) (C) (D) (E)
2. (A) (B) (C) (D) (E)
3. (A) (B) (C) (D) (E)
4. (A) (B) (C) (D) (E)
5. (A) (B) (C) (D) (E)
6. (A) (B) (C) (D) (E)
7. (A) (B) (C) (D) (E)
8. (A) (B) (C) (D) (E)
9. (A) (B) (C) (D) (E)
0. (A) (B) (C) (D) (E)
1. (A) (B) (C) (D) (E)
2. (A) (B) (C) (D) (E)
3. (A) (B) (C) (D) (E)
4. (A) (B) (C) (D) (E)
5. (A) (B) (C) (D) (E)
6. (A) (B) (C) (D) (E)
7. (A) (B) (C) (D) (E)
8. (A) (B) (C) (D) (E)
9. (A) (B) (C) (D) (E)

30. (A) (B) (C) (D) (E)
31. (A) (B) (C) (D) (E)
32. (A) (B) (C) (D) (E)
33. (A) (B) (C) (D) (E)
34. (A) (B) (C) (D) (E)
35. (A) (B) (C) (D) (E)
36. (A) (B) (C) (D) (E)
37. (A) (B) (C) (D) (E)
38. (A) (B) (C) (D) (E)
39. (A) (B) (C) (D) (E)
40. (A) (B) (C) (D) (E)
41. (A) (B) (C) (D) (E)
42. (A) (B) (C) (D) (E)
43. (A) (B) (C) (D) (E)
44. (A) (B) (C) (D) (E)
45. (A) (B) (C) (D) (E)
46. (A) (B) (C) (D) (E)
47. (A) (B) (C) (D) (E)
48. (A) (B) (C) (D) (E)
49. (A) (B) (C) (D) (E)
50. (A) (B) (C) (D) (E)
51. (A) (B) (C) (D) (E)
52. (A) (B) (C) (D) (E)
53. (A) (B) (C) (D) (E)
54. (A) (B) (C) (D) (E)
55. (A) (B) (C) (D) (E)
56. (A) (B) (C) (D) (E)
57. (A) (B) (C) (D) (E)

58. (A) (B) (C) (D) (E)
59. (A) (B) (C) (D) (E)
60. (A) (B) (C) (D) (E)
61. (A) (B) (C) (D) (E)
62. (A) (B) (C) (D) (E)
63. (A) (B) (C) (D) (E)
64. (A) (B) (C) (D) (E)
65. (A) (B) (C) (D) (E)
66. (A) (B) (C) (D) (E)
67. (A) (B) (C) (D) (E)
68. (A) (B) (C) (D) (E)
69. (A) (B) (C) (D) (E)
70. (A) (B) (C) (D) (E)
71. (A) (B) (C) (D) (E)
72. (A) (B) (C) (D) (E)
73. (A) (B) (C) (D) (E)
74. (A) (B) (C) (D) (E)
75. (A) (B) (C) (D) (E)
76. (A) (B) (C) (D) (E)
77. (A) (B) (C) (D) (E)
78. (A) (B) (C) (D) (E)
79. (A) (B) (C) (D) (E)
80. (A) (B) (C) (D) (E)
81. (A) (B) (C) (D) (E)
82. (A) (B) (C) (D) (E)
83. (A) (B) (C) (D) (E)
84. (A) (B) (C) (D) (E)
85. (A) (B) (C) (D) (E)

# ADVANCED PLACEMENT CHEMISTRY EXAM V

## SECTION I

Time: 1 Hour 45 Minutes
85 Questions

**Note:** For all questions referring to solutions, assume that the solvent is water unless otherwise stated.

**Directions**: Each set of lettered choices below refers to the numbered statements immediately following it. Select the one lettered choice that best fits each statement. A choice may be used once, more than once, or not at all in each set.

**Questions 1 – 5**

(A) carbon, oxygen
(B) chlorine, bromine
(C) aluminum, galium
(D) zinc, mercury
(E) silver, gold

1. Are members of the halogen family
2. Are transition metals
3. Exist in well-known allotropic forms
4. In their standard states at 25°C one element is a gas and the other is a liquid
5. These elements, when combined, form an amalgam

**Questions 6 – 10**

(A) $K(s) + H_2O(aq)$
(B) $O_2(g) + 2H_2(g)$
(C) $AgNO_3(aq) + NaCl(aq)$

(D) He(g) + Ar(g)

(E) NaOH(aq) + HC (aq)

6. An oxidation-reduction process that yields water as a product

7. The product of this reaction is an insoluble precipitate

8. A very exothermic process that is kinetically slow at room temperature

9. No product is expected from this reaction

10. This reaction produces a solution pH»8

**Questions 11 – 15**

(A) tetrahedral

(B) square-planar

(C) trigonal-bipyramidal

(D) trigonal-pyramidal

(E) octahedral

11. Is the geometry of $CCl_4$

12. Is the geometry of $XeF_4$

13. Is the geometry of $NH_3^-$

14. Is the geometry of $PCl_5$

15. Is the geometry of $CCl_3$

**Directions:** Each of the questions or incomplete statements below is followed by five suggested answers or completions. Select the one that is best in each case.

16. Which one of the following species has not been correctly matched with its number of neutrons and electrons?

(A) $^{14}_{7}N$ 7 neutrons, 7 electrons

(B) $^{16}_{8}O^{2-}$ 8 neutrons, 10 electrons

(C) $^{53}_{24}Cr^{2+}$ 29 neutrons, 26 electrons

(D) He 2 neutrons, 2 electrons

(E) C 7 neutrons, 6 electrons

17. Determine the oxidation state of Mn in $KMnO_4$

(A) +1

(B) +3

(C) +5

(D) +7

(E) –1

18. Which of the following graphs is a valid representation of Boyle's Law?

(A)

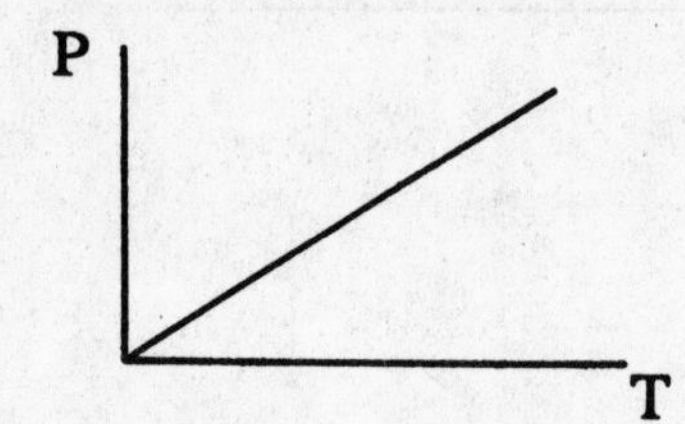


(B)

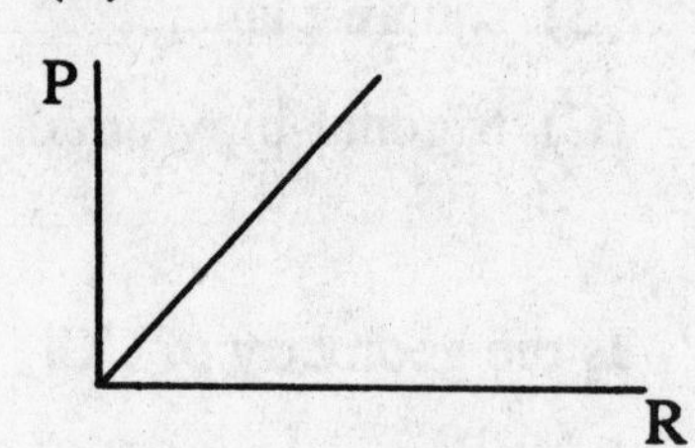


(C)

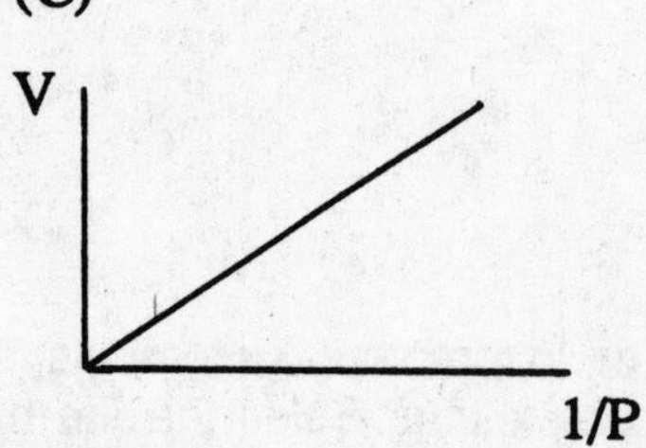


(D)

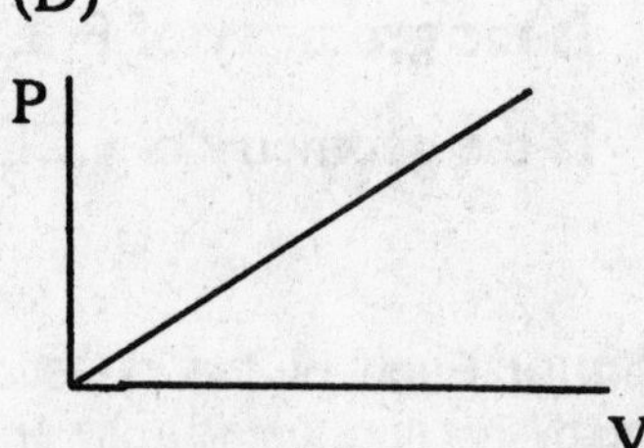


(E)

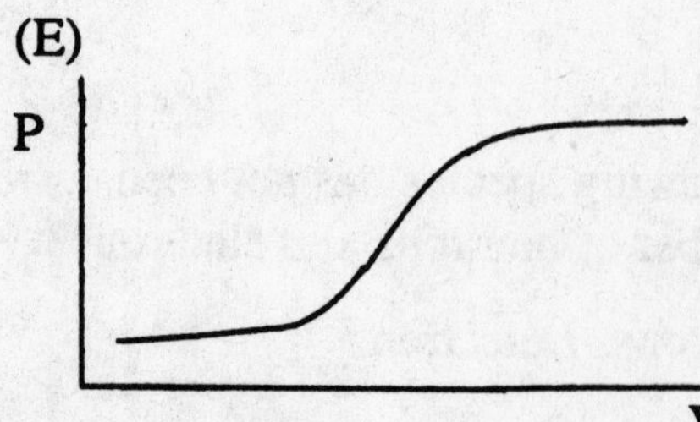

19.

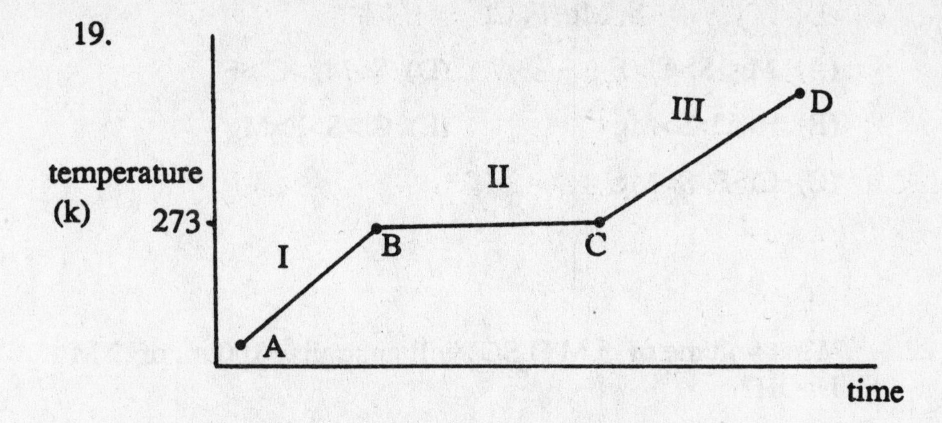


The above sketch represents a warming curve for water at 1 atm pressure. Select the one incorrect statement. (Note: $C_I$ = heat capacity of water in region I, $\Delta T$ = change in temperature between points specified, $\Delta H_f$ = heat of fusion)

(A) Water is a solid in region I of the curve

(B) Water exists in a solid/liquid equilibrium in region II

(C) The energy absorbed by a sample of mass N in the temperature range A to C is $(C_I)(\Delta T_{A,B})(N) + (\Delta H_f)(\Delta T_{B,C})(N)$

(D) The energy absorbed by a sample in the range B to C is $\Delta H_f N$

(E) In region II water doesn't gain any kinetic energy but does gain in entropy

20. According to Hund's rule, how many unpaired electrons does the ground state of iron have?

(A) 6

(B) 5

(C) 4

(D) 3

(E) 2

21. Arrange the following neutral gaseous atoms in order of decreasing atomic radius.

S, Mg, F, Cl

(A) Mg>S>Cl>F

(B) F>Cl>S>Mg

(C) Cl>F>S>Mg

(D) S>Mg>Cl>F

(E) Cl>S>F>Mg

22. What volume of .5 M $H_2SO_4$ will neutralize 100 ml of .2 M NaOH?

(A) 400 ml

(B) 200 ml

(C) 100 ml

(D) 20 ml

(E) 2 ml

23. Copper metal will replace silver ions in solution, resulting in the production of silver metal and copper ions. This indicates that:

(A) silver has a higher oxidation potential than copper

(B) a combustion reaction is occurring

(C) copper has a higher oxidation potential than silver

(D) silver is much less soluble than copper

(E) copper metal is readily reduced

24. How many moles of chlorine gas are produced when one mole of $Cr_2O_7^{2-}$ reacts in the following unbalanced raction?

$Cr_2O_7^{2-}(aq) + Cl^-(aq) \rightarrow Cr^{3+}(aq) + Cl_2(g)$ (acidic solution)

(A) 1

(B) 2

(C) 3

(D) 4

(E) 8

25. What is the pH of a $10^{-13}$M HCl solution?

(A) 1
(B) 5
(C) 7
(D) 10
(E) 13

26. Which of the following compounds have only nonpolar bonds?

(A) $KO_2$
(B) NaF
(C) HF
(D) KBr
(E) $I_2$

27. Determine the formal charge of N in acetonitrile ($CH_3CN$).

(A) –2
(B) –1
(C) O
(D) +1
(E) +2

28. Arrange the following ion pairs in order of increasing lattice energy

LiCl, $BaCl_2$, LiBr, LiI

(A) LiI, LiBr, LiCl, $BaCl_2$
(B) $BaCl_2$, LiI, LiBr, LiCl
(C) $BaCl_2$, LiCI, LiBr, LiI
(D) LiCl, LiBr, LiI, $BaCl_2$
(E) LiCl, $BaCl_2$, LiI, LiBr

29. Lithium (AW=6.941) exists as two naturally occurring isotopes, $^{6}_{3}$Li and $^{7}_{3}$Li, with relative atomic masses of 6.015 and 7.016. Find the percent abundances of the two isotopes.

(A) 23.1, 46.2
(B) 74.30, 25.70
(C) 90, 10
(D) 92.51, 7.49
(E) 94.63, 5.37

30.

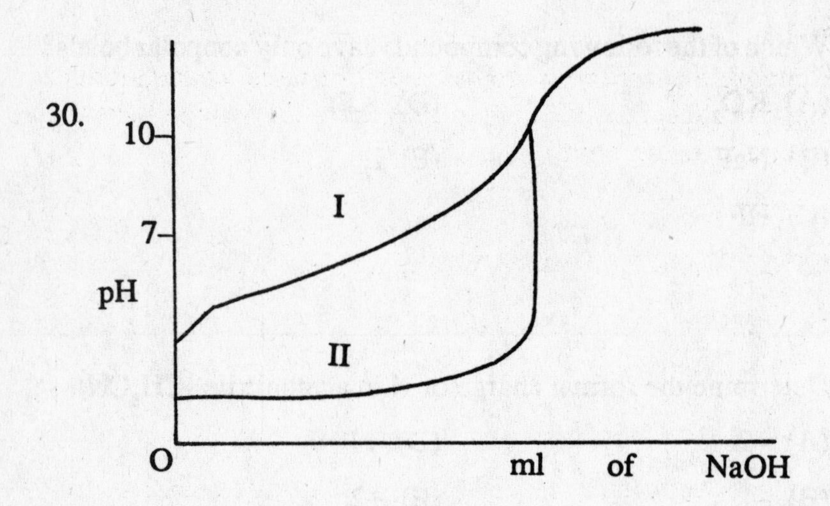


Above is a sketch of two acid-base titration curves. The curves represent the different monoprotic acids of identical concentration titrated with .100 M NaOH. Identify the correct association of acid with its titration curve.

(A) curve I HCl, curve II HBr
(B) curve I $CH_3COOH$, curve II HCl
(C) curve I HBr, curve II $CH_3COOH$
(D) curve I $HNO_3$, curve II $HNO_2$
(E) curve I $CH_3COOH$, curve II $H_2SO_4$

31. What is the oxidation state of iron in the coordination complex $Na_2K_2$ $[Fe(CN)_6]$?

(A) –2
(B) –1
(C) +1
(D) +2
(E) +3

32. Suppose that solutions are made up in 1 molal concentrations for five substances. Which of these solutions would have the lowest freezing point?

(A) NaBr
(B) $C_6H_{12}O_6$
(C) NaCl
(D) $Na_2SO_4$
(E) $CH_3COOH$

33. A professor needs to make a buffer solution for a class demonstration. He mixes together equal volumes of various solutions. Identify the combination of solutions that would <u>not</u> produce a buffer solution.

(A) 1 M $CH_3COOH$, 1 M HCl, .2 M $NaCH_3COO$
(B) 1 M $CH_3COOH$, 1 M $NaCH_3COO$
(C) 1 M $CH_3COOH$, .5 M NaOH
(D) 1 M $CH_3COOH$, 1 M HCl, 1 M NaOH, 1 M $NaCH_3COO$
(E) 2 M $NH_3$, 1 M HCl

**Questions 24 – 28** are based on the following information: A student is given a soluble sodium salt containing one of 8 possible anions: acetate, chloride, bromide, iodide, sulfide, sulfate, and phosphate.

34. What cations other than sodium will likely form soluble salts with all the anions?

(A) $Zn^{2+}$, $Pb^{2+}$, $K^+$
(B) $Pb^{2+}$, $Hg^{2+}$, $Fe^{2+}$
(C) $Ca^{2+}$, $Mg^{2+}$
(D) $K^+$, $NH_4^+$
(E) $Fe^{2+}$

35. 3-4 drops of $AgNO_3$ are added to a solution of the salt. No precipitate results. Which anions are absent?

(A) $Cl^-$, $Br^-$, $I^-$
(B) $Cl^-$, $Br^-$, $CH_3COO^-$, $I^-$
(C) $Cl^-$, $Br^-$, $I^-$, $S^{2-}$, $PO_4^{3-}$
(D) $Cl^-$ only
(E) none of the above

36. Suppose that a precipitate did form in the reaction in Question 35 and it is then treated with a few drops of nitric acid. No change is observed. Which anions can be present?

(A) $Cl^-$, $Br^-$, $I^-$, $SO_4^{2-}$
(B) $CH_3COO^-$, $NO_3^-$
(C) $Cl^-$, $Br^-$, $I^-$, $SO_4^{2-}$
(D) $Cl^-$, $Br^-$, $I^-$ only
(E) none of the above

37. The student treats some fresh solution of the salt with $BaCl_2$. A precipitate forms. Which anions could have been present?

(A) $SO_4^{2-}$, $PO_4^{3-}$
(B) $Br^-$, $Cl^-$, $I^-$
(C) $CH_3COO^-$ only
(D) $Br^-$ only
(E) $NO_3^-$ only

38. After various tests, the student has identified the anion and 4 drops of ethyl alcohol are added to a small amount of the salt. The student completes a specific test as follows: 2 drops of 18M $H_2SO_4$ and then 4 drops of ethyl alcohol. After the solution is heated a fruity odor is produced. What is the anion?

(A) $Cl^-$
(B) $S^{2-}$
(C) $CH_3COO^-$
(D) $SO_4^{2-}$
(E) $PO_4^{3-}$

39. Suppose that a new set of quantum numbers is developed in which the first three numbers are analogous to the "old set" of numbers (n, l, $m_l$), but the fourth member ($m_s$) differs in that it can accommodate only 1 electron ($m_s = 1$) instead of 2 in the "old" system ($m_s = \pm\frac{1}{2}$). What are the quantum numbers n and l that describe the outermost electron of sulfur (AN=16)?

(A) 4,1
(B) 3,1
(C) 4,0
(D) 3,0
(E) 2,–1

40. The species $O_2^-$

(A) has 2 unpaired electrons
(B) has 1 unpaired electron
(C) has 0 unpaired electrons
(D) has a bond order of $2\frac{1}{2}$
(E) has a bond order of 2

41. Choose the correct statement pertaining to the following energy diagram.

$$A \underset{k_2}{\overset{k_1}{\leftrightarrow}} \qquad C \underset{k_4}{\overset{k_3}{\leftrightarrow}} \qquad E \underset{k_6}{\overset{k_5}{\leftrightarrow}}$$

(A) The reaction A → G is endothermic

(B) $k_3$ is larger than $k_1$

(C) Species B is called an intermediate product

(D) Species E is kinetically and thermodynamically more stable than species C

(E) The rate determining step involves reaction C → E

42. The compound

$$\begin{array}{ccccc} & H & & H & \\ & | & & | & \\ H- & C & -O- & C & -H \\ & | & & | & \\ & H & & H & \end{array}$$

is best described as being a(n)

(A) alcohol

(B) alkene

(C) ether

(D) ester

(E) carboxylic acid

43. Which of the following is true of the reaction 2A(g) + B(g) C(g)?

(A) In a reaction starting with equal moles of A and B, reagent B is the limiting reagent

(B) The rate equation can be expressed as Rate = $k[A]^2 [B]$

(C) There is an increase in the entropy of the system

(D) If the reaction is endothermic, a decrease in temperature will favor formation of reactant

(E) The percent yield of the reaction is determined as (moles of C produced/moles of A consumed) x 100%

44. The reaction 2A(g) + 2B(g) → 2C(l) is a spontaneous and exothermic reaction. What are the signs of ΔG, ΔH, ΔS, and ΔE?

| | ΔG | ΔH | ΔS | ΔE |
|---|---|---|---|---|
| (A) | + | + | + | – |
| (B) | + | + | – | + |
| (C) | – | – | + | – |
| (D) | – | – | – | + |
| (E) | – | – | – | – |

45. An oxide of phosphorus contains 56.36% oxygen by mass. What is the empirical formula of the oxide?

(MW P = 30.97, MW O = 16.00)

(A) $PO_2$

(B) $PO_3$

(C) $P_2O_4$

(D) $P_2O_5$

(E) $P_3O_7$

46. Determine the density of methane gas (MW = 16.0) at 25°C and 6.00 atm

(A) 3.77 g/l

(B) 10.4 g/l

(C) 46.8 g/l

(D) 3.92 g/l

(E) 1.21 g/l

47. Calculate the number of moles of carbon dioxide produced during the combustion of 2 moles of ethane ($C_2H_6$)

(A) 2

(B) 4

(C) 6

(D) 8

(E) 10

48. A gas at 25°C occupies a 10 liter volume at P atm pressure. The gas is allowed to expand to a volume of 15 liters at 377°C. What is the new pressure?

(A) 1.45

(B) 1.45P

(C) 1.45/P

(D) 1.07

(E) 1.07/P

**Questions 49–50**

$$\begin{array}{ccccccccccc} & & H & & H & & & & H & & H \\ & & | & & | & & & & | & & | \\ H & - & C_1 & - & C_2 & = & C_3 & = & C_4 & - & N & - & H \\ & & | & & & & & & & & \\ & & H & & & & & & & & \end{array}$$

49. In the above compound what is the hybridization state for $C_3$?

(A) sp
(B) $sp^2$
(C) $sp^3$
(D) $sp^4$
(E) $sp^3d$

50. Which atom has $sp^3$ hybridization <u>and</u> one pi bond?

(A) $C_1$
(B) $C_2$
(C) $C_4$
(D) N
(E) none of the atoms

51. Which of the following statements is false?

(A) The energy of the universe is constant
(B) $\Delta E$=w for adiabatic processes
(C) The entropy of the universe tends toward a maximum
(D) Entropy is a measure of increasing disorder
(E) The entropy of a perfect crystal of a pure substance at 0°C is zero

52. A radiation source of varying wavelength is directed onto a piece of metal connected to an electron detector. The following results are noted:

| experiment | frequency of radiation | detector reading |
|---|---|---|
| 1 | $6.2 \times 10^{14}$ s$^{-1}$ | no electrons detected |
| 2 | $6.3 \times 10^{14}$ s$^{-1}$ | no electrons detected |
| 3 | $6.4 \times 10^{14}$ s$^{-1}$ | electrons detected |
| 4 | $6.5 \times 10^{14}$ s$^{-1}$ | electrons detected |

Which of the following is a true statement?

(A) The threshold frequency of the metal is $6.4 \times 10^{14}$ s$^{-1}$

(B) The experimental results illustrate the Heisenberg Uncertainty Principle

(C) The electrons in experiment 4 have greater velocity than the electrons in experiment 3

(D) More electrons are detected in experiment 4 than in experiment 3

(E) If the radiation source of experiment 2 is brought closer to the metal, at some point electrons would be detected

53. Determine ΔH (n kcal) for the reaction $2C(s) + O_2(g)$ $2CO(g)$. Additional information:

(i) $CO_2(g) \rightarrow O_2(g) + C(s)$ ΔH = 94.1 kcal

(ii) $CO_2(g) \rightarrow \frac{1}{2} O_2(g) + CO(g)$ ΔH = 67.7 kcal

(A) –52.8

(B) –26.4

(C) +26.4

(D) –161.8

(E) cannot be determined from the given information

54. The Ksp for $BaF_2$ is $1.7 \times 10^{-6}$. What is the solubility of fluoride in moles per liter?

(A) $3.4 \times 10^{-6}$

(B) $1.3 \times 10^{-3}$

(C) $1.5 \times 10^{-2}$

(D) $7.5 \times 10^{-3}$

(E) $1.7 \times 10^{-6}$

55. A galvanic cell can be represented as $Pt(s)|Sn^{2+}$ (aq,1M), $Sn^{4+}$ (aq, 1M) || $Fe^{2+}$ (aq, 1M), $Fe^{3+}$ (aq, 1M) | Pt(s). What reaction is occurring at the anode?

(A) $Pt \rightarrow Pt^{2+} + 2e^-$

(B) $Sn^{2+} \rightarrow Sn^{4+} + 2e^-$

(C) $Pt \rightarrow Sn^{2+} + 2e^-$

(D) $Fe^{2+} \rightarrow Fe^{3+} + 1e^-$

(E) $Fe^{3+} + 1e^- \rightarrow Fe^{2+}$

56. Identify the incorrect Lewis structure(s).

(A) $CH_4$

```
    H
    |
H - C - H
    |
    H
```

(B) $NH_3O$

```
    H                      H
    |      ..              |      ..
H - N - O:       ↔     H - N = O
    |      ..              |      ..
    H                      H
```

(C) $C_2H_3N$

```
    H
    |
H - C - C ≡ N:
    |
    H
```

(D) $NH_4^+$

$$\left[\begin{array}{c} H \\ | \\ H - N - H \\ | \\ H \end{array}\right]^+$$

(E) $NO_2^-$ $[\ddot{O} = N - \ddot{O}:]^- \leftrightarrow [:\ddot{O} - \ddot{N} = \ddot{O}]^-$

57. The reaction $2H_2 + NO \rightarrow H_2O + \frac{1}{2} N_2$ has a rate law of the form: rate = $k[H_2]^x [NO]^y$

Find the sum of x and y from the given data.

| experiment | $[H_2]$, M | [NO], M | rate, M $s^{-1}$ |
|---|---|---|---|
| a | $1.0 \times 10^{-3}$ | $4.6 \times 10^{-2}$ | $3.1 \times 10^{-4}$ |
| b | $2.0 \times 10^{-3}$ | $4.6 \times 10^{-2}$ | $6.2 \times 10^{-4}$ |
| c | $1.0 \times 10^{-3}$ | $1.84 \times 10^{-1}$ | $5.0 \times 10^{-3}$ |

(A) 1/2

(B) 2

(C) 3

(D) 4

(E) 5

58. $COCl_2(g) \rightarrow CO(g) + Cl_2(g)$ Kp = $6.7 \times 10^{-9}$ atm (100°C)

For the above reaction, equal pressures of the reagent and products ($P_{CO} = P_{COCl_2} = P_{Cl_2} = 1$ atm) are placed in a flask. At equilibrium (100°C) it is true that

(A) the total pressure of the system must remain at 3 atm

(B) $P_{CO}$ and $P_{Cl_2}$ must decrease from their initial values

(C) $P_{CO}$ and $P_{Cl_2}$ must increase from their initial values

(D) $P_{COCl}$ remains constant

(E) final pressures cannot be determined because a volume is not specified

59. At 298 K, $\Delta G° = 90.3$ kJ for the reaction $HgO(s) \longleftrightarrow Hg(g) + \frac{1}{2}O_2(g)$. Calculate Kp at 298 K

(A) $1.48 \times 10^{-16}$ atm $^{3/2}$

(B) $2.51 \times 10^{-14}$ atm $^{3/2}$

(C) $-2.51 \times 10^{-14}$ atm $^{3/2}$

(D) $-5.42 \times 10^{-12}$ atm $^{3/2}$

(E) $6.31 \times 10^{5}$ atm $^{3/2}$

60. Calculate the molecular weight of an unknown gas X if the ratio of its effusion rate to that of He is .378 (AW of He = 4.00)

(A) 9.47

(B) 42.3

(C) 10.6

(D) 28.0

(E) 32.0

61. What is the value of "x" in the nuclear reaction $^{31}_{14}Si \rightarrow {}^{x}_{15}P + {}^{0}_{-1}\beta$?

(A) 28

(B) 30

(C) 31

(D) 32

(E) 33

62. Aqua regia is a strong acid that can dissolve even gold and platinum. It consists of a mixture of two strong monoprotic acids. These acids are

(A) nitric and sulfuric

(C) hydrochloric and hydrofluoric

(C) hydrochloric and nitric

(D) nitric and acetic

(E) perchloric and ammonia

63. A solution is made by combining 1 mole of ethanol and 2 moles of water. What is the total vapor pressure above the solution? (At the same temperature, the vapor pressure of pure ethanol is .53 atm and the vapor pressure of water is .24 atm.)

(A) .34 atm
(B) .41 atm
(C) .43 atm
(D) .56 atm
(E) .77 atm

64. Which of the following particles cannot be accelerated in a particle accelerator?

(A) electron
(B) proton
(C) neutron
(D) alpha particle
(E) $He^+$

65. An amphoteric substance

(A) is both easily oxidized and reduced
(B) is inert and unreactive
(C) can act as both an acid and a base
(D) can dissolve many solids
(E) is hydrophobic

66. Compounds that can form hydrogen bonds with water include

(A) $CH_4$
(B) $CH_4$, HF
(C) $CH_3OH$, HF, $CCl_4$
(D) $NH_3$, HF, $CH_3OH$
(E) none of the above

67. The net ionic equation that represents the reaction of aqueous hydrochloric acid with aqueous ammonia is

(A) $HCl(aq) + NH_3(aq) \rightarrow NH_4^+(aq)$

(B) $HCl(aq) + NH_3(aq) \rightarrow NH_4^+(aq) + Cl^-(aq)$

(C) $H^+(aq) + NH_3(aq) \rightarrow NH_4^+(aq)$

(D) $H^+(aq) + Cl^-(aq) + NH_4^+(aq) + OH^-(aq) \rightarrow H_2O(aq)$
$NH_4^+(aq) + Cl^-(aq)$

(E) $H^+(aq) + NH_4^+(aq) + OH^-(aq) \rightarrow H_2O(aq) + NH_4^+(aq)$

68. All of the following are colligative properties except

(A) osmotic pressure

(B) pH of buffer solutions

(C) vapor pressure

(D) boiling point elevation

(E) freezing point depression

69. How many lattice points are found in the face-centered unit cell illustrated below?

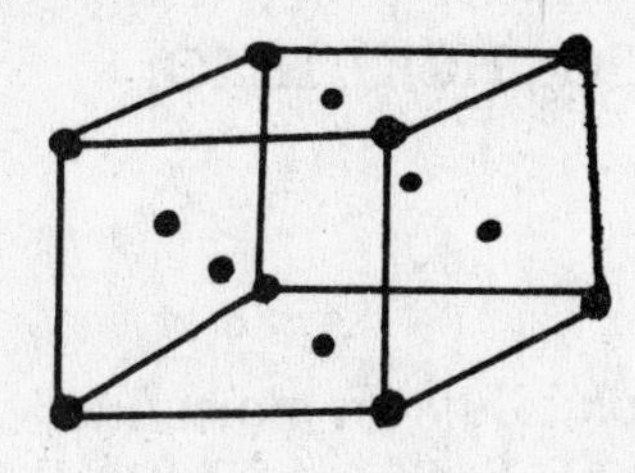

(A) 2

(B) 3

(C) 4

(D) 5

(E) 14

70. The least symmetrical unit cell, where there are unequal sides and angles, is called

(A) monoclinic
(B) triclinic
(C) orthorhombic
(D) cubic
(E) tetragonal

71. $H_2O + (CH_3)_3N \quad (CH_3)_3NH^+ + OH^-$

For the above reaction, which conjugate acid-base pair is correctly listed?

(A) $H_2O$, $(CH_3)_3N$
(B) $(CH_3)_3NH^+$, $OH^-$
(C) $H_2O$, $(CH_3)_3NH^+$
(D) $H_2O$, $H^+$
(E) none of the above

72. Arrange the acids in order of increasing strength

(A) $HCl,O_4$, $H_2SO_4$, $H_3PO_4$, $HClO$
(B) $HClO$, $HClO_4$, $H_2SO_4$, $H_3PO_4$
(C) $H_3PO_4$, $H_2SO_4$, $HClO_4$, $HClO$
(D) $HClO$, $H_3PO_4$, $H_2SO_4$, $HClO_4$
(E) $H_3PO_4$, $H_2SO_4$, $HClO$, $HClO_4$

73. Calculate the solubility (in moles/liter)of $Zn(OH)_2$ in a solu - tion buffered at pH 8. (Ksp of $Zn(OH)_2$= $1.8 \times 10^{-14}$)

(A) $1.0 \times 10^{-8}$
(B) $1.7 \times 10^{-5}$
(C) $1.8 \times 10^{+2}$
(D) $1.7 \times 10^{+5}$
(E) $1.8 \times 10^{-2}$

X-ray, I = infrared, R = radiowave)

(A) M,I,R,X

(B) R,M,I,X

(C) X,M,I,R

(D) I,X,M,R

(E) R,I,X,M

75. What would be the maximum number of emission lines for atomic hydrogen, determined from Balmer's generalized equation, when n = 1,2,3,4 or 5? $\left(\frac{1}{\lambda} = R_4\left(\frac{1}{n_f^2} - \frac{1}{n_i^2}\right)\right)$

(A) 4

(B) 6

(C) 7

(D) 8

(E) 10

76. Which of these electron configurations violates the Pauli exclusion principle?

(A) $1s^2 2s^1$

(B) $1s^2 2s^2 2p^6 3s^2$

(C) $1s^2 2s^2 2p^6 3s^1 3p^1$

(D) $1s^2 2s^2 2p^7 3s^2$

(E) $1s^2 2s^2 2p^5$

77. Indicate the incorrect electron configuration. (Note: atomic number of Ar is 18 and that of Ne is 10.)

| | |
|---|---|
| (A) Cu (AN=29) | [Ar] $4s^1 3d^{10}$ |
| (B) Cr (AN=24) | [Ar] $4s^1 3d^5$ |
| (C) C (An=6) | $1s^2 2s^2 2p^2$ |
| (D) $P^+$ (AN=15) | [Ne] $3s^2 3p^2$ |
| (E) $O^{2-}$ (AN=8) | $1s^2 2s^2 2p^4$ |

78. Calculate the de Broglie wavelength (in meters) for an electron

78. Calculate the de Broglie wavelength (in meters) for an electron moving at the velocity $2.2 \times 10^6$ $ms^{-1}$ (Note: Planck's constant (h) = $6.6 \times 10^{-34}$ Js; $m_e = 9.1 \times 10^{-31}$ kg).

(A) $3.3 \times 10^{-10}$

(B) $1.3 \times 10^{-57}$

(C) $1.6 \times 10^3$

(D) $2.7 \times 10^{-72}$

(E) $2.7 \times 10^{-8}$

79. Identify the incorrect formula

(A) $Ca_2SO_4$

(B) $H_2SO_4$

(C) $KMnO_4$

(D) LiCl

(E) $H_3O^+$

80. The incomplete combustion of hexane ($C_6H_{14}$) produces CO and $H_2O$ gases. Use a coefficient of 1 for hexane to balance the incomplete combustion reaction. What is the resulting coefficient for CO?

(A) 2

(B) 4

(C) 6

(D) 8

(E) 12

**Questions 81-82**

$$HClO + Sn^{2+} + H^+ \rightarrow Cl^- + Sn^{4+} + H_2O$$

81. In the above unbalanced reaction which reactant acts as an oxidizing agent?

(A) $H^+$

(B) $Sn^{4+}$

(C) $Sn^{2+}$

(D) HClO

(D) $Cl^-$

82. What is the oxidation state of O in C10?

(A) –2

(B) –1

(C) 0

(D) 1

(E) 2

83. The van der Waals equation for 1 mole of a real gas is $(P+a/V^2)(V-b) = RT$. Identify the false statement about the equation.

(A) The terms a and b are constants for a particular gas

(B) The a term corrects for attractive forces between molecules

(C) To determine P, one measures V, T, and R and obtains a and b from a table

(D) The b term corrects for molecular volume

(E) The values for a can vary widely since the various gases have widely varying attractive forces

84. Esterification involves the formation of an ester from a(n)

(A) acid and a ketone

(B) acid and an alcohol

(C) alcohol and an amine

(D) aldehyde and an alcohol

(E) ether and an aldehyde

85. Identify the formula of tetraaquodichlorochromium (III) chloride.

(A) $[Cr(H_2O)_4C1_2]C1$

(B) $[Cr(H_2O)_4C1_2]$

(C) $[Cr(H_2O)_4C1_3]$

(D) $[Cr(OH)_4C1_2]C1$

(E) $[Cr(OH)_4C1_3]$

# ADVANCED PLACEMENT CHEMISTRY EXAM V

## SECTION II

**Directions:** The percentages given for each individual part indicates the scoring for this section. Spend approximately 35 minutes on Parts A and B together and 40 minutes on Part C.

The methods and steps used to solve a problem must be shown clearly. Partial credit will be given if work is shown. Pay attention to significant figures.

### Part A

(30 Percent)

Solve the following problem.

1. The energy needed to remove electrons from Cs metal is 376 kJ $mol^{-1}$.

(a) Calculate the longest possible wavelength (in meters) that will remove electrons from Cs.

(b) What is the kinetic energy of the electrons when light of $2.5 \times 10^{-7}$ m strikes the cesium surface?

(c) What is the velocity of the electrons in (b)?

(d) What will happen when the light source of (b) is brought closer to the source?

### Part B

(30 Percent)

Solve **ONE** of the two problems presented. Only the first problem answered will be scored.

2. (a) Determine the rate law for the general reaction $X_2 + Y_2 \rightarrow 2XY$. Base your answer on the proposed mechanism:

(i) $X_2 \rightarrow 2X$ (fast)

(ii) $Y_2 \rightarrow 2Y$ (fast)

(iii) $X + Y \rightarrow XY$ (slow)

(b) The following information was found for a reaction, where all concentrations were kept constant:

| temperature of reaction | rate of reaction |
|---|---|
| 25°C | 2.0 mol $L^{-1}$ $s^{-1}$ |
| 35°C | 4.0 mol $L^{-1}$ $s^{-1}$ |

Calculate the activation energy for the reaction. Predict the rate of reaction at 45°C.

3. An electrochemical cell at 298K is made to generate the aqueous reaction $Cr^{2+} + Mn \rightarrow Cr + Mn^{2+}$.

(a) Calculate $E°_{cell}$.

(b) Calculate $E_{cell}$ when $[Cr^{2+}]$ = .0021 M and $[Mn^{2+}]$ = .31 M. Is the reaction spontaneous as written?

(c) Determine the value of ΔG for the reaction in (b).

(d) Suppose the concentrations in (b) were reversed. Determine the new $E_{cell}$. Is the reaction spontaneous as written?

## Part C

(40 Percent)

Choose **THREE** of the following five topics. Only the first three answers will be scored. Your answers will be scored based on their accuracy, relevance of the details chosen, and appropriateness of the descriptive material used. Be as specific as possible and use illustrative examples and equations where helpful.

4. Discuss the leveling effect phenomenon in terms of the two strong acids $HNO_3$ (nitric) and $HClO_4$ (perchloric). Which acid is expected to be stronger? (justify your answer). Suggest how this can be demonstrated experimentally by considering a solvent other than water such as methanol where there is not a leveling effect.

5. Limestone (calcium carbonate) can be converted to calcium bicarbonate when it comes in contact with water containing some dissolved carbon dioxide:

$$CaCO_3(s) + CO_2(g) + H_2O(l) \leftrightarrow Ca^{2+}(aq) + 2HCO_3^-(aq)$$

Briefly explain Le Chatelier's principle, and then use it to predict how each of the following affects the amount of bicarbonate ion produced.

(a) The pressure of $CO_2$ is increased.

(b) $CaCl_2$, a source of $Ca^{2+}$ ion, is added.

(c) Half of the calcium carbonate is removed.

(d) An inert gas, Ar, is added, which increases the total pressure.

(e) A catalyst is added to the system.

6. (a) Sketch a plot of energy (E) versus internuclear distance (r) for a covalent bond formed between two atoms. Use the sketch to explain what happens (in terms of energy, attractions, repulsions, etc.) as the distance, r, approaches zero.

(b) Explain, qualitatively, the difference in energy for the reactions below. Be sure to consider ion pairs and the ionic lattice.

(i) $Na^+(g) + Cl^-(g) \rightarrow NaCl(g)$ $-585$ kJ mol$^{-1}$

(ii) $Na^+(g) + Cl^-(g) \rightarrow NaCl(s)$ $-770$ kJ mol$^{-1}$

7. Explain / discuss the following observations:

(a) $H_2O$ has a boiling point at 373 K, and $H_2S$ has a boiling point at 212 K.

(b) A salt dissolved in water at room temperature causes the solution to become much cooler.

(c) An insect, with a density greater than that of water, can walk on water

(d) Compounds such as glucose ($C_6H_{12}O_6$) and NaCl dissolve in water yet iodine ($I_2$) does not.

8. Give explanations of the following, on the basis of your knowledge of coordination chemistry.

(a) A sample of the compound $K_3$ [$Fe(CN)_6$] is weighed in the presence and absence of a magnetic field. The weight of the sample was greater in the former case.

(b) Many transition metal complexes are colored.

(c) The complex [$Pt(NH_3)_2Cl_2$], which has a square-planar geometry, has two isomers.

(d) An octahedral complex of composition $CO(NH_3)_5$ BrCl is dissolved in water and found to consist of two ions, one of which is a bromide ion.

# ADVANCED PLACEMENT CHEMISTRY EXAM V

## ANSWER KEY

| | | | | | |
|---|---|---|---|---|---|
| 1. | B | 29. | D | 57. | C |
| 2. | E | 30. | B | 58. | B |
| 3. | A | 31. | D | 59. | A |
| 4. | B | 32 | D | 60. | D |
| 5. | D | 33. | A | 61. | C |
| 6. | B | 34. | D | 62. | C |
| 7. | C | 35. | C | 63. | A |
| 8. | B | 36. | E | 64. | C |
| 9. | D | 37. | A | 65. | C |
| 10. | A | 38. | C | 66. | D |
| 11. | A | 39. | A | 67. | C |
| 12. | B | 40. | B | 68. | B |
| 13. | D | 41. | D | 69. | C |
| 14. | C | 42. | C | 70. | B |
| 15. | D | 43. | D | 71. | E |
| 16. | C | 44. | D | 72. | D |
| 17. | D | 45. | D | 73. | E |
| 18. | C | 46. | D | 74. | B |
| 19. | C | 47. | B | 75. | E |
| 20. | C | 48. | B | 76. | D |
| 21. | A | 49. | A | 77. | E |
| 22. | D | 50. | E | 78. | A |
| 23. | C | 51. | E | 79. | A |
| 24. | C | 52. | C | 80. | C |
| 25. | C | 53. | A | 81. | A |
| 26. | E | 54. | C | 82. | A |
| 27. | C | 55. | B | 83. | C |
| 28. | A | 56. | B | 84. | B |
| | | | | 85. | A |

# ADVANCED PLACEMENT CHEMISTRY EXAM V

# DETAILED EXPLANATIONS OF ANSWERS

## SECTION I

1. (B)

Members of the group containing F, Cl, Br, I, and At are called halogens.

2. (E)

Silver and gold are transition metals since they have incompleted subshells (or are readily ionized to form incomplete d subshells).

3. (A)

Allotropes are different forms of the same element. Diamond and graphite are allotropes of carbon; dioxygen ($O_2$) and ozone ($O_3$) are allotropes of oxygen.

4. (B)

At 25° C, chlorine is a gas and bromine is a liquid. Though mercury (choice D) is a liquid, zinc is a solid.

5. (D)

An amalgam is a solution of any metal in mercury.

6. (B)

$O_2$ is reduced and $H_2$ is oxidized to form $H_2O$.

7. (C)

When $Ag^+$ and $Cl^-$ combine, a precipitate of AgCl forms immediately. This is a useful reaction for qualitative analysis.

8. (B)

This reaction is very violent and exothermic but at room temperature there is not enough activation energy to initiate the reaction.

9. (D)

The noble gases are generally inert and unreactive, especially with each other, under ordinary conditions. However, the heavier elements (Xe, Kr, Rn) will react, especially with oxygen and fluorine.

10. (A)

$K(s) + H_2O\ (aq) \rightarrow KOH\ (aq) + \frac{1}{2}\ H_2(g)$. The potassium hydroxide generates a very basic solution.

11. (A)

$CCl_4$ has 4 bonding atoms; hence, from the Valence Shell Electron-Pair Repulsion theory (VSEPR), $CCl_4$ has a tetrahedral geometry.

12. (B)

$XeF_4$ has 4 bonding atoms and 2 orbitals with lone pairs of electrons; hence, it has a square-planar geometry.

13. (D)

$NH_3$ has 3 bonding atoms and 1 orbital with a lone pair of electrons; hence, it has a trigonal-pyramidal geometry.

14. (C)

$PCl_5$ has 5 bonding atoms; hence, it has a trigonal-bipyramidal geometry.

15. (D)

$CCl_3^-$ has 3 bonding atoms and 1 orbital with a lone pair of electrons; hence, it has a trigonal-pyramidal geometry.

16. (C)

$^{53}_{24}Cr^{2+}$ has 29 neutrons and only 22 electrons. Neutral chromium possesses as many electrons as protons (AN=24), but the dipositive species loses 2 electrons.

17. (D)

The charge on the permanganate ion is –1. Thus, with oxygen possessing the normal –2 state, manganese has a +7 oxidation state.

18. (C)

Boyle's law states that at constant temperature, the volume of a gas is inversely proportional to its pressure. Mathematically, this is represented as $PV = k$, where k is a constant. Rearranging, $P = k/V$. Thus, a plot of P versus $1/V$ should yield a straight line.

19. (C)

The second part of the equation is incorrect. The energy gained in region II equals $\Delta H_f N$. There is no temperature change during the melting process. All the energy gained is involved in the change from solid to liquid structure – an entropy gain.

20. (C)

The ground state configuration of iron is [Ar] $4s^2 3d^6$. There are five d orbitals and with minimum pairing Fe has 4 unpaired electrons.

21. (A)

Atomic radius increases as the principal quantum number, n, increases, thus Cl>F. Within a row (or same principal quantum number) size decreases as the atomic number increases, thus Mg>S>Cl.

22. (D)

Use $V_A x M_A = V_B x M_B$ and solve for $V_A$: $V_A = (V_B x M_B)/M_A$ where $V_B$=100 ml, $M_B$=.2 M and $M_A$ = 1 M. Note that $M_A$= 1 M because $H_2SO_4$ is diprotic.

23. (C)

The unbalanced reaction occurring is $Cu+Ag^+ \rightarrow Cu^{2+} + Ag$. For the reaction to proceed, copper must be easier to oxidize than silver. Thus, copper must have a higher oxidation potential than silver. Choice (B) is a redox reaction. (D) is incorrect because both silver and copper metal are insoluble. (E) is incorrect because metals are not readily reduced.

24. (C)

This is a redox reaction and must be balanced accordingly:

$$1Cr_2O_7^{2-} + 6e^- \rightarrow 2Cr^{3+}$$

$$6Cl^- \rightarrow 3Cl_2 + 6e^-$$

$$1Cr_2O_7^{2-} + 6Cl^- \rightarrow 2Cr^{3+} + 3Cl_2$$

25. (C)

Be careful! The H+ concentration of pure water is $10^{-7}$ M. This concentration is much greater than the added H+ ($10^{-13}$M). Since $10^{-7} + 10^{-13} \sim 10^{-7}$, the pH is 7.

26. (E)

The polarity of a bond can be estimated by the difference in electronegativity (ΔEN) between the 2 atoms in the bond. The compounds in A to D are ionic (large ΔEN). ΔEN for $I_2$ is, of course, zero.

27. (C)

Formal charge = (valence electrons) – (nonbonding electrons) – $\frac{1}{2}$ (bonding electrons). For N in $CH_3CN$, formal charge = $5 - 2 - \frac{1}{2}(6) = 0$.

28. (A)

Lattice energy depends on (i) ionic charge (a greater charge results in a greater lattice energy)m, (ii) ionic radius ( a smaller sum of the ionic radii results in a greater lattice energy), and (iii) lattice geometry. Here, ionic charge ($Ba^{2+}$ versus $Li^+$) and ionic radius ($Cl^- < Br^- < I^-$) are the important factors.

29. (D)

Let x = fraction of $^{6}_{3}Li$ and 1–x = fraction of $^{7}_{3}Li$. 6.015x + 7.016 (1–x) = 6.941. x = .0749 or 7.49%; 1–x = .9251 or 92.51%.

30. (B)

Curve II breaks sharply to an equivalence point at pH7. This behavior is consistent with that of a strong acid (monoprotic HCl). Curve I breaks less sharply to an equivalence point pH>7. This behavior is consistent with that of a weak acid such as acetic acid. At the equivalence point the pH is basic due to water hydrolysis by acetate.

31. (D)

The oxidation states are known for the other atoms in the compound $Na^+$, $K^+$, and $CN^{-1}$. Thus, by difference, the oxidation state of iron is +2 as the compound has an overall zero charge.

32. (D)

Freezing point depression is a colligative property – it depends on the number of moles of particles (ions, molecules, etc.) in solution. The 1 molal $Na_2SO_4$ ionizes into a 3 molal concentration of ions (1 $Na_2SO_4$ $\rightarrow$ 2 $Na^+$ + $SO_4^{2-}$). This is greater than any of the other choices and will result in the solution with the lowest freezing point.

33. (A)

A buffer solution contains a weak acid (or base) and the salt of its conjugate base (or acid). Solutions B-D, when mixed, yield $CH_3COOH$ and $NaCH_3COO$. Solution E, when mixed, yields $NH_3$ and $NH_4Cl$. These are buffer solutions. Solution A yields $CH_3COOH$, HCl, and NaCl. There is no $NaCH_3COO$ remaining, hence we do not have a buffer.

34. (D)

Salts with the cations $Na^+$, $K^+$, or $NH_4^+$ are usually very soluble.

35. (C)

These ions form insoluble precipitates with silver.

36. (E)

The silver salts of $Cl^-$, Br -, $I^-$, and $S^{2-}$ remain insoluble in strongly acidic solution.

37. (A)

$Ba^{2+}$ will form insoluble salts with sulfate and phosphate.

38. (C)

In acidic solution, the acetate ion reacts with ethyl alcohol to form ethyl acetate, an ester. Esters often have fruity odors.

39. (A)

The "new" electron configuration of S is $1s^1 2s^1 2p^3 3s^1 3p^3 4s^1 3d^5 4p^1$. Thus, n=4, l=1 for the outermost S electron.

40. (B)

The MO (molecular orbital) bonding schemes for simple diatomic species must be used to answer this question. For $O_2^-$ the MO scheme is $(\sigma 1s)^2(\sigma^*1s)^2(\sigma 2s)^2(\sigma^*2s)^2(\sigma 2p)(\pi 2p)^4(\pi^*2p)^3$.
$O_2^-$ has 3 net bonding electrons and a BO (bond order) of $1\frac{1}{2}$ . It also has 1 unpaired electron (2 electrons in one of the $\pi^*$ orbitals and the 1 unpaired electron in the other $\pi^*$ orbital).

41. (D)

Species E is lower in energy than species C and is thermodynamically more stable. It is also kinetically more stable because of the larger activation energies (proportional to the height of the "hills") for the forward and reverse reactions.

42. (C)

The –C–O–C– linkage is characteristic of an ether.

43. (D)

The application of Le Chatelier's principle provides a solution: heat $+ 2A + B \rightarrow C$. The removal of heat (decrease in temperature) shifts the reaction equation toward the formation of reactants. Note that (B) would only be a true statement if this reaction was identified as the rate-limiting (or slowest) step.

44. (D)

A spontaneous, exothermic reaction must have, by definition, a negative $\Delta H$ and negative $\Delta G$. If $\Delta G$ is negative then, also by definition, $\Delta E$ is positive. Since 3 mol of gas forms 2 mol of liquid, the entropy has decreased, thus, $\Delta S$ is negative.

45. (D)

Per 100 g, the mass of P is 43.64 g and that of O is 56.36 g.

43.64 g P = 1.409 mol; 56.34 g O = 3.523 mol

P:O (mol) is 1:2.5 or 2:5 Thus, the formula is $P_2O_5$.

46. (D)

Density = mass/volume = (MW) (P)/RT = 3.92 g/L.

47. (B)

The answer is determined from the correctly balanced equation $2C_2H_6 + 7O_2 \quad 4CO_2 + 6H_2O$.

48. (B)

$P_1V_1/T_1 = P_2V_2/T_2$ where $P_1 = P$, $V_1 = 10L$, $T_1 = 298$, $T_2 = 650$ K, $V_2 = 15L$, $P_2 = ?$ $P_2 = 1.45P$

49. (A)

$C_3$ has two bonds and two $\pi$ bonds. We need two hybrid orbitals for the two $\sigma$ bonds; hence, $C_3$ is sp hybridized.

50. (E)

$sp^3$ hybridization with a $\pi$ bond implies some combination of four $\sigma$ bonds or lone electron pairs, and one $\pi$ bond. For the atoms in the compound:

$C_1$ four $\sigma$ bonds ($sp^3$)
$C_2$ three $\sigma$ bonds, one $\pi$ bond ($sp^2$)

$C_3$ two $\sigma$ bonds, two $\pi$ bonds (sp)
$C_4$ three $\sigma$ bonds, one $\pi$ bond ($sp^2$)
N three $\sigma$ bonds, one lone pair ($sp^3$)

51. (E)

Choice E would correctly state the third law of thermodynamics if the temperature was listed as 0 K.

52. (C)

The principles of the photoelectric effect are illustrated in this problem. Mathematically, $h\nu = h\nu_o + (1/2)\,mv^2$. Once the $\nu_o$ threshold frequency has been achieved (a frequency between those of exp 2 and exp 3; not necessarily at $6.4 \times 10^{14}\ s^{-1}$), any excess energy is converted into kinetic energy ($1/2mv^2$). In (E), no matter how close (or intense) the radiation source, no electrons will be detected as long as the threshold frequency has not been achieved.

53. (A)

The reaction equation is obtained as –2(i) + 2 (ii):

$$2C + 2O_2 \rightarrow 2CO_2 \quad -188.2 \text{ kcal}$$
$$2CO_2 \rightarrow O_2 + 2CO \quad + 135.4 \text{ kcal}$$

---

$$2C + O_2 \rightarrow 2CO \quad -52.8 \text{ kcal}$$

54. (C)

$BaF_2 \rightarrow Ba^{2+} + 2F^-$; $Ksp = 1.7 \times 10^{-6} = [Ba^{2+}][F^-]^2$
Let $x = [Ba^{2+}]$ and $2x = [F^-]$; $(x)(2x)^2 = 4x^3 = 1.7 \times 10^{-6}$; $x = 7.5 \times 10^{-3}$ and $2x = [F^-] = 1.5 \times 10^{-2}$.

55. (B)

In cell notation, the anodic (oxidation) reaction is indicated to the left of the double vertical bars. $Sn^{2+}$ is oxidized to $Sn^{4+}$. Note that the Pt is inert, and it serves as the electrode. Choice E represents the cathodic (reduction) reaction.

56. (B)

One of the resonance forms of (B) places 10 electrons around the N–a violation of the octet rule for a second row element. A far better structure is

```
      H
      |      ..
  H – N – O – H
      ..     ..
```

57. (C)

To determine x, use the data from exp a and b, where the $H_2$ concentration varies but the NO concentration is the same. Thus, rate (b)/rate (a) = $2 = k[H_2]_b^x[NO]_b^y / k[H_2]_a^x[NO]_a^y = (2.0 \times 10^{-3} / 1.0 \times 10^{-3})^x = 2^x$; $x = 1$. In a similar way, use exp a and c to determine y; $y = 2$.

58. (B)

The use of a reaction quotient (Q) will help determine the direction of the reaction as it achieves equilibrium.

$Q = [CO] [Cl_2] / [COCl_2] = (1) (1) / (1) = 1$ Comparing Q with Kp, we see that to achieve equilibrium there must be a decrease in product concentration and an increase in reactant concentration.

59. (A)

Use $\Delta G = -RT \ln K$ where $\Delta G = 90{,}300$ J mol$^{-1}$, T = 298 K, and R = 8.314 J mol$^{-1}$ K$^{-1}$.

60. (D)

Effusion rates of gases are related as $r_A / r_B = (MW_B / MW_A)^{1/2}$ Thus, $r_x / r_{He} = .378 = (4/x)1/2$ x = 28.0.

61. (C)

An understanding of the several radioactive decay processes is important. Here, the process is beta decay. The mass number on the left must be balanced by an equal mass on the right. Since the beta particle has a zero mass, the mass of P must be 31.

62. (C)

Hydrochloric and nitric acids are the only listed pair of strong, monoprotic acids.

63. (A)

Applying Raoult's law, we can solve the problem:

$P_A = P_A°X_A$ ($P_A$ = vapor pressure of compound A above the solution, $P_Ao$ = vapor pressure of pure compound A, $X_A$ = mole fraction of compound A). Let A = ethanol, B = water. $P_{total} = P_A + P_B = P_A°X_A + P_B°X_B$ = (.53) (.33) + (.24) (.67) = .34 atm.

64. (C)

Particle accelerators generate electric and magnetic fields that accelerate charged particles. A neutron is not charged.

65. (C)

Amphoteric or amphiprotic substances can function as either bases or acids, depending on conditions. Water, pure acetic acid, and liquid ammonia are examples of amphoteric substances.

66. (D)

A hydrogen bond is a bond formed by the coulombic attraction between a hydrogen atom that is bonded to a strongly electronegative element (usually O, F, N) and another strongly electronegative element (O, F, N). The presence of hydrogen does not by itself ensure hydrogen bonding capability (e.g. $CH_4$).

67. (C)

Since HCl is a strong acid, it dissociates completely. Thus, a chloride ion is not chemically involved in the reaction and is not in the net ionic equation.

68. (B)

pH is not a colligative property as it depends on specific properties of acids and bases (Ka, $K_b$) and not just their concentration.

69. (C)

The lattice points (LP) "belonging" to a unit cell are calculated as follows:

1/8 (no. of corner LP) + 1/4 (no. of edge LP) + 1/2 (no. of face LP) + 1 (no. of body LP). Applied to the face-centered cell we have 1/8 (8) + 1/2 (6) = 4 LP.

70. (B)

The following are the properties for each of the unit cells in the question:

| Unit cell | Edge Length | Angles |
| --- | --- | --- |
| monoclinic | $a \neq b \neq c$ | $\propto = \beta = 90^\circ \neq \propto$ |
| triclinic | $a \neq b \neq c$ | $\propto \neq \beta \neq \propto$ |
| orthorhombic | $a \neq b \neq c$ | $\propto = \beta = \propto = 90^\circ$ |
| cubic | $a = b = c$ | $\propto = \beta = \propto = 90^\circ$ |
| tetragonal | $a = b = c$ | $\propto = \beta = \propto = 90^\circ$ |

71. (E)

Conjugate acid-base pairs consist of an acid and base related by the exchange of a proton. These are $H_2O$, $OH^-$ and $(CH_3)_3N$, $(CH3)_3NH^+$.

72. (D)

The strength of oxyacids depends on the electronegativity and oxidation state of the central atom of the acid. The greater that both of these are, the greater the strength of the acid. If $H_aXO_b$ represents a general oxyacid, then b – a is proportionate to acid strength. Thus, $HC1O_4$ is a strong acid and HClO is weak.

73. (E)

$Zn(OH)_2 \leftrightarrow Zn^{2+} + 2OH^-$ $Ksp = 1.8 \times 10^{-14} = [Zn^{2+}][OH^-]^2$ A pH 8 solution means that $OH^-$ concentration is $10^{-6}$. Let $x = [Zn^{2+}]$, $10^{-6} = [OH^-]$ $1.8 \times 10^{-14} = (x)(10^{-6})^2$; $x = 1.8 \times 10^{-2}$

74. (B)

75. (E)

The 10 possible emission states are (ni $\rightarrow$ nf) 5 $\rightarrow$ 4, 5 $\rightarrow$ 3, 5 $\rightarrow$ 2, 5 $\rightarrow$ 1, 4 $\rightarrow$ 3, 4 $\rightarrow$ 2, 4 $\rightarrow$ 1, 3 $\rightarrow$ 2, 3 $\rightarrow$ 1, and 2 $\rightarrow$ 1.

76. (D)

The Pauli exclusion principle states that no two electrons in the same atom may have the same set of quantum numbers. Furthermore, only two electrons may occupy an orbital. Thus, in (D) a p configuration implies that one orbital is occupied by three electrons. Note that (C) represents an excited-state species but doesnot violate the Pauli principle.

77. (E)

The configuration of $O^{2-}$ is $1s^2 2s^2 2p^6$. Note that (A) and (B) are correct. You should be aware that certain elements (Cu, Cr, Ag, etc.)are exceptions to the normal electron filling order.

78. (A)

We use the de Broglie equation $\lambda = h/mv = 3.3 \times 10^{-10}$ m.

79. (A)

The correct formula is $CaSO_4$.

80. (C)

The balanced equation is $1C_6H_{14} + 6\frac{1}{2}O_2 \rightarrow 6CO + 7H_2O$.

81. (A)

By definition, an oxidizing agent is a species that oxidizes another species. Oxidizing agents are reduced. HClO is the oxidizing agent. $Sn^{2+}$ is the reducing agent as it is oxidized.

82. (A)

The normal oxidation state for oxygen is –2 when it is combined with other atoms. However, oxygen does have a –1 state in peroxide compounds such as hydrogen peroxide ($H_2O_2$).

83. (C)

Choices A, B, D, and E are true statements and provide a good review of the van der Waals equation. (C) is incorrect because one does not measure R since it is constant.

84. (B)

This is one of several classic organic reactions.

85. (A)

One can dissect the name: tetraaquo means four water molecules, dichloro means two chloride ions. If the chromium is +3, then there must be one outer-sphere (noncovalent) chloride remaining.

# SECTION II

## PART A

1. (a) E(atom) = $(376 \times 10^3$ J mol$^{-1}) \times$ (mol/6.02 $\times 10^{23}$ atom) = g.25 $\times 10^{-19}$ J atom$^{-1}$.

E = h o = hc/ ; = hc/E = $(6.63 \times 10^{-34}$ J s) $(3.00 \times 10^8$ m x$^{-1}$) / $6.25 \times 10^{-19}$ J = $\underline{3.18 \times 10^{-7}\ m}$

(b) E = h $_o$ + KE; KE = E − h $_o$

E = hc/ = $\frac{(6.63 \times 10^{-34} \text{ Js})(3.00 \times 10^8 \text{ m s}^{-1})}{2.50 \times 10^{-7} \text{ m}}$

= $7.96 \times 10^{-19}$J; h v$_o$ = $6.25 \times 10^{-19}$J (from (a)). Thus, KE = $7.96 \times 10^{-19}$J − $6.25 \times 10^{-9}$J = $1.71 \times 10^{-19}$ J

(c) KE = $\frac{1}{2}$mv$^2$; v = (2KE/m$_e$) = $\frac{(2)(1.71 \times 10^{-19} J)}{9.11 \times 10^{-31} \text{ kg}}$ =

$\underline{6.13 \times 10^5 \text{ m s}^{-1}}$

(d) When the light source is brought closer to the cesium metal, the light intensity increases. This causes an increase in the number of electrons that are ionized.

## PART B

2. (a) The rate is based on the slowest step (the rate-determining step), and then all intermediates in the rate equation are replaced by the equivalent expressions containing the starting reactants.

Rate = k[X] [Y]; from (i) $[X]^2 / [X_2] = K_{i;}$ $[X] = (K_i[X_2])^{\frac{1}{2}}$ and from (ii) $[Y]^2 / [Y_2] = K_{ii}$; $[Y] = K_{ii}[Y_2])$

Substituting into the rate equation:
Rate = $kK_iK_{ii}\ [X_2]^{\frac{1}{2}}\ [Y_2]^{\frac{1}{2}}$ or k' $[X\ ]^{\frac{1}{2}}\ [Y\ ]^{\frac{1}{2}}$

(b) We use $\ln(k_2/k_1) = E_a/R(1/T_1 - 1/T_2)$ where T1 = 298 K, $T_2$ = 308 K, and the rates are proportional to the rate constants

$\ln(4/2) = E_a/8.31\ J\ mol^{-1}\ K^{-1}\ (1/298\ K - 1/308\ K)$
$E_a = 8.31\ (\ln 2) / ((1/298) - (1/308)) = E_a = 5.29 \times 104\ J\ mol^{-1}$

Rate at 318 K

We can calculate the rate at 318 K by using the same equation and solving for $k_2$. We have just solved for $E_a$ and can use the data at 298 K ($rate_1$ = 2). Thus:

$\ln(k_2/k_1) = E_a / R\ (1/T_1 - 1/T_{2)}$
$\ln(Rate_2 / Rate_1) = 5.29 \times 10^4\ J\ mol^{-1} / 8.31\ J\ mol^{-1}\ s^{-1}\ (1/298\ K - 1/318\ K) = 1.34$
$\ln(Rate_2 /2) = 1.34$; $Rate_2$ (at 318 K) = $7.7\ mol\ L^{-1}\ s^{-1}$

3. From the useful information we may determine $E^\circ_{cell}$ :

| | |
|---|---|
| $Cr^{2+} + 2e^-$ | Cr −0.91 V |
| Mn | $Mn^{2+} + 2e^-$ + 1.03 V |
| $Cr^2 + Mn$ | $Cr^2 + Mn^2$ $E^\circ$ = +0.12V |

(b) We use the Nernst equation:

$E_{cell} = E^\circ - .059/n \ln([Mn^{2+}] / [Cr^{2+}])$ where n (no. of electrons transferred) = 2

$E_{cell} = .12 - .059/2\ (\ln .31 / .0021) = \underline{-0.027\ V}$

The reaction is not spontaneous as written ( a negative $E_{cell}$).

(c) $\Delta G = -nFE$

$\Delta G = -(2)\ (96{,}500\ coulomb)\ (-0.027\ V) = 5.2 \times 10^3\ J$

(d) $E_{cell} = .12 - .059/2\ (\ln .0021 / .31) = \underline{0.27\ V}$

The reaction is spontaneous (a positive $E_{cell}$).

## PART C

4. The leveling effect phenomenon involves the idea that acids stronger than the conjugate acid of the solvent have the same strength. Thus, for water:

$HNO_3 (aq) \rightarrow H_3O^+ + NO_3^-$

$HClO_4 (aq) \rightarrow H_3O^+ + ClO_4^-$

Perchloric acid is expected to be the stronger acid because of the higher oxidation state of its central atom (compared to $HNO_3$). While both acids are leveled in water, this is not true in other solvents such as methanol. Thus, in methanol, perchloric acid would be expected to ionize to a greater degree than nitric acid.

5. Le Chatelier's principle states that if a stress is applied to a chemical system, the system will react in such a way that it relieves the applied stress. For the reaction:

$K = [HCO_3^-] [Ca^{2+}] / P_{CO}$

(a) The stress of added $CO_2$ pressure is relieved by an increase in the amount of bicarbonate.

(b) The stress of added $Ca^{2+}$ is relieved by its reaction with bicarbonate to form the reactants; hence, there is a decrease in the amount of the bicarbonate.

(c) Note that $CaCO_3$ (s), as a solid, is not part of the equilibrium expression. thus, there is no change to the amount of bicarbonate formed.

(d) Since Ar is not part of the equilibrium, there should be no change in the amount of bicarbonate formed.

(e) A catalyst speeds up a reaction but does not change the final equilibrium point. There is no change in the bicarbonate formed.

6.

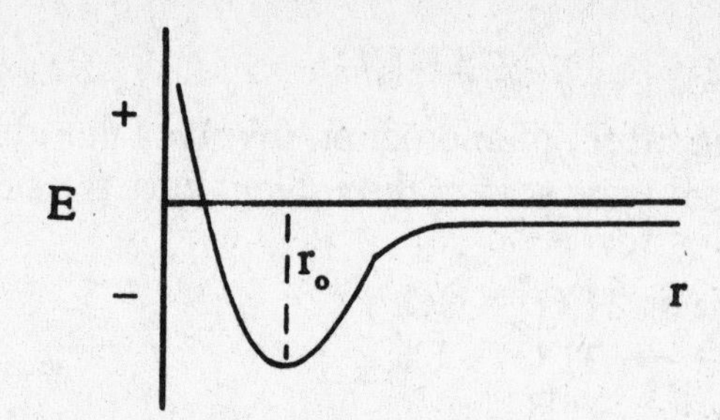


As r approaches $r_o$ (the bond length value) there is a favorable interaction between electrons and nuclei, and the energy (E) of the system reaches a minimum value; a bond is formed. As r becomes even smaller, E increases dramatically because of unfavorable nucleus-nucleus and electron-electron interactions.

(b) In the gas phase ion pair there is one electrostatic attraction, whereas in the solid lattice each ion has several nearest neighbor attractions. Thus, the ionic lattice is more stable by about 200 kJ $mol^{-1}$.

7. (a) $H_2O$ can take part in hydrogen bonding while $H_2S$ cannot. Thus, with water having stronger intermolecular attractions, it will have a higher boiling point.

(b) The $\Delta H_s$ (heat of solution) of the salt must have a positive value, i.e., the process of dissolving the salt absorbs heat from the solution causing it to become colder.

(c) The walking insect illustrates the property of surface tension – the force that pulls the water surface inward and resists an increase in the surface area of the water. The insect still does not weigh enough (even though it weighs more per unit volume than water) to overcome the surface tension and pierce the water surface.

(d) Here, one may apply the principle that "like dissolves like." Water, a polar solvent, dissolves many polar and ionic substances such as glucose and NaCl. However, iodine ($I_2$) is nonpolar and is not soluble in water.

8. (a) The compound contains unpaired d electrons and is paramagnetic. As a result it weighs more in a magnetic field.

(b) The color of a compound is caused by the compound absorbing selected wavelengths of visible light. In transition metal complexes absorption can occur by exciting electrons from one d orbital to another d orbital that is higher in energy.

(c) The complex has two geometric isomers-cis (the same type of ligands on the same side) and trans (the same type of ligands on the opposite side:

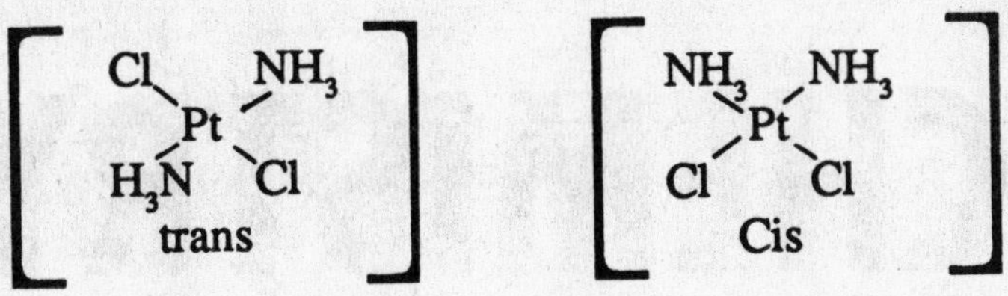

$$\left[\begin{array}{ccc} Cl & & NH_3 \\ & Pt & \\ H_3N & & Cl \end{array}\right] \qquad \left[\begin{array}{ccc} NH_3 & & NH_3 \\ & Pt & \\ Cl & & Cl \end{array}\right]$$

trans Cis

(d) The complex must be of the structure $[CO(NH_3)_5\ Cl]Br$. The chloride must be covalently bonded to the metal, while the bromide is ionic. Thus, the two ions produced in solution are of the form $[CO(NH_3)_5\ C1]^{-1}$ and $Br^{-1}$.

# THE ADVANCED PLACEMENT EXAMINATION IN

# CHEMISTRY

# TEST VI

# THE ADVANCED PLACEMENT EXAMINATION IN

# CHEMISTRY

# ANSWER SHEET

1. (A) (B) (C) (D) (E)
2. (A) (B) (C) (D) (E)
3. (A) (B) (C) (D) (E)
4. (A) (B) (C) (D) (E)
5. (A) (B) (C) (D) (E)
6. (A) (B) (C) (D) (E)
7. (A) (B) (C) (D) (E)
8. (A) (B) (C) (D) (E)
9. (A) (B) (C) (D) (E)
10. (A) (B) (C) (D) (E)
11. (A) (B) (C) (D) (E)
12. (A) (B) (C) (D) (E)
13. (A) (B) (C) (D) (E)
14. (A) (B) (C) (D) (E)
15. (A) (B) (C) (D) (E)
16. (A) (B) (C) (D) (E)
17. (A) (B) (C) (D) (E)
18. (A) (B) (C) (D) (E)
19. (A) (B) (C) (D) (E)
20. (A) (B) (C) (D) (E)
21. (A) (B) (C) (D) (E)
22. (A) (B) (C) (D) (E)
23. (A) (B) (C) (D) (E)
24. (A) (B) (C) (D) (E)
25. (A) (B) (C) (D) (E)
26. (A) (B) (C) (D) (E)
27. (A) (B) (C) (D) (E)
28. (A) (B) (C) (D) (E)
29. (A) (B) (C) (D) (E)
30. (A) (B) (C) (D) (E)
31. (A) (B) (C) (D) (E)
32. (A) (B) (C) (D) (E)
33. (A) (B) (C) (D) (E)
34. (A) (B) (C) (D) (E)
35. (A) (B) (C) (D) (E)
36. (A) (B) (C) (D) (E)
37. (A) (B) (C) (D) (E)
38. (A) (B) (C) (D) (E)
39. (A) (B) (C) (D) (E)
40. (A) (B) (C) (D) (E)
41. (A) (B) (C) (D) (E)
42. (A) (B) (C) (D) (E)
43. (A) (B) (C) (D) (E)
44. (A) (B) (C) (D) (E)
45. (A) (B) (C) (D) (E)
46. (A) (B) (C) (D) (E)
47. (A) (B) (C) (D) (E)
48. (A) (B) (C) (D) (E)
49. (A) (B) (C) (D) (E)
50. (A) (B) (C) (D) (E)
51. (A) (B) (C) (D) (E)
52. (A) (B) (C) (D) (E)
53. (A) (B) (C) (D) (E)
54. (A) (B) (C) (D) (E)
55. (A) (B) (C) (D) (E)
56. (A) (B) (C) (D) (E)
57. (A) (B) (C) (D) (E)
58. (A) (B) (C) (D) (E)
59. (A) (B) (C) (D) (E)
60. (A) (B) (C) (D) (E)
61. (A) (B) (C) (D) (E)
62. (A) (B) (C) (D) (E)
63. (A) (B) (C) (D) (E)
64. (A) (B) (C) (D) (E)
65. (A) (B) (C) (D) (E)
66. (A) (B) (C) (D) (E)
67. (A) (B) (C) (D) (E)
68. (A) (B) (C) (D) (E)
69. (A) (B) (C) (D) (E)
70. (A) (B) (C) (D) (E)
71. (A) (B) (C) (D) (E)
72. (A) (B) (C) (D) (E)
73. (A) (B) (C) (D) (E)
74. (A) (B) (C) (D) (E)
75. (A) (B) (C) (D) (E)
76. (A) (B) (C) (D) (E)
77. (A) (B) (C) (D) (E)
78. (A) (B) (C) (D) (E)
79. (A) (B) (C) (D) (E)
80. (A) (B) (C) (D) (E)
81. (A) (B) (C) (D) (E)
82. (A) (B) (C) (D) (E)
83. (A) (B) (C) (D) (E)
84. (A) (B) (C) (D) (E)
85. (A) (B) (C) (D) (E)

# ADVANCED PLACEMENT CHEMISTRY EXAM VI

## SECTION I

Time: 1 Hour 45 Minutes
85 Questions

**Note:** For all questions referring to solutions, assume that the solvent is water unless otherwise stated.

**Directions:** For each of the following questions or incomplete sentences, there are five choices. Choose the answer which is most correct.

1. How many grams of oxygen are needed for the complete combustion of 39.0 g of $C_6H_6$? The molecular weight of $C_6H_6$ is 78.0.

   $$2\,C_6H_6 + 15\,O_2 \rightarrow 12\,CO_2 + 6H_2O$$

   (A) 3.75 g
   (B) 120.0 g
   (C) 32.0 g
   (D) 60.0 g
   (E) 292.5 g

2. How many molecules are there in 22 g of $CO_2$? The molecular weight of $CO_2$ is 44.

   (A) 3
   (B) $6.02 \times 10^{23}$
   (C) 44
   (D) $9.03 \times 10^{23}$
   (E) $3.01 \times 10^{23}$

3. What is the percent carbon in sucrose, $C_{12}H_{22}O_{11}$?

   (A) 42.1
   (B) 3.5
   (C) 12.0
   (D) 6.0
   (E) 26.6

4. An atom containing which of the following number of protons, neutrons and electrons would be an isotope of hydrogen?

| | protons | neutrons | electrons |
|---|---|---|---|
| (A) | 0 | 1 | 1 |
| (B) | 1 | 2 | 1 |
| (C) | 0 | 0 | 1 |
| (D) | 2 | 1 | 2 |
| (E) | 2 | 0 | 2 |

5. A compound was found to contain only carbon, hydrogen and oxygen. The percent composition was determined as 40.0% C, 6.7% H and 53.3% O. The emperical formula of this compound is:

(A) $C_2H_4O$

(B) $C_6HO_8$

(C) CHO

(D) $CH_2O$

(E) $C_3H_6O$

6. Which one of the following species has a noble gas configuration?

(A) $Be^+$

(B) $N^-$

(C) $Mg^{2+}$

(D) $H^+$

(E) $O^{2+}$

7. Which one of the following compounds is an Iron (III) oxide?

(A) FeO

(B) $Fe_3O_4$

(C) $FeO_2$

(D) $Fe_2O_3$

(E) $Fe_2O$

8. A sample of a pure gas occupied a volume of 500 ml at a temperature of 27°C and a pressure of .4 atm. The number of moles present in this sample is:

(A) .045

(B) .008

(C) .091

(D) 8.13

(E) .182

9. The molecular weight of 0.25 g of a gas that occupies a volume of 100 ml at a pressure of 2.5 atm and a temperature of 25°C is:

(A) 28.0

(B) 12.2

(C) 2.05

(D) 20.5

(E) 24.4

10. The pair of atoms that are most likely to form a covalent compound are:

(A) H and He

(B) Na and F

(C) H and Cl

(D) Li and F

(E) Na and Cl

11. What is the hydronium ion concentration, $H_3O^+$ of a solution that has a hydroxide concentration of $1.4 \times 10^{-4}$ M?

(A) $7.2 \times 10^{-11}$

(B) $1.4 \times 10^{-10}$

(C) $1.0 \times 10^{-14}$

(D) $1.8 \times 10^{-5}$

(E) $7.0 \times 10^{-7}$

12. What is the pH of a solution that has a hydronium ion concentration, $H_3O^+$ of $1.2 \times 10^{-4}$?

(A) 3.92

(B) 4.00

(C) 3.80

(D) 5.20

(E) 7.90

13. The $H^+$ concentration of a .01 molar solution of HCN is: (The $K_i$ for HCN is $4 \times 10^{-10}$ )

(A) $2.5 \times 10^{-11}$ M

(B) .01 M

(C) $4 \times 10^{-10}$ M

(D) $2 \times 10^{-6}$ M

(E) $2 \times 10^{-5}$ M

14. What is the frequency of electromagnetic radiation that has a wavelength of 600 nm? (The velocity of light is $3.0 \times 10^8$ m/s)

(A) $2.0 \times 10^{-15}$ $s^{-1}$

(B) $2.0 \times 10^{14}$ $s^{-1}$

(C) $6.64 \times 10^{18}$ $s^{-1}$

(D) $5.0 \times 10^{14}$ $s^{-1}$

(E) $3.0 \times 10^8$ $s^{-1}$

15. The energy of electromagnetic radiation whose frequency is $8.00 \times 10^{11}$ $s^{-1}$ is: (Planck's constant is $6.63 \times 10^{-34}$ j/s)

(A) $1.2 \times 10^{45}$ j

(B) $6.02 \times 10^{23}$ j

(C) $5.30 \times 10^{-22}$ j

(D) $8 \times 10^{-46}$ j

(E) $3.0 \times 10^8$ j

16–17 Choose the letter answer which best describes the gas law that follows the graphs shown.

(A) Graham's Law

(B) Boyle's Law

(C) Charles' Law

(D) Gay – Lussac's Law

(E) Avogadro's Law

16.

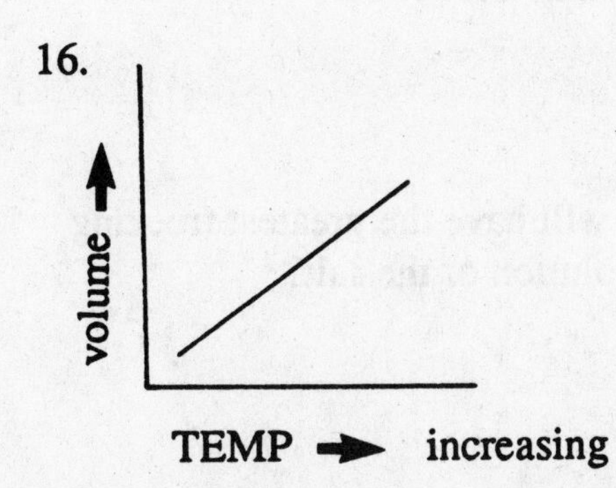


17.

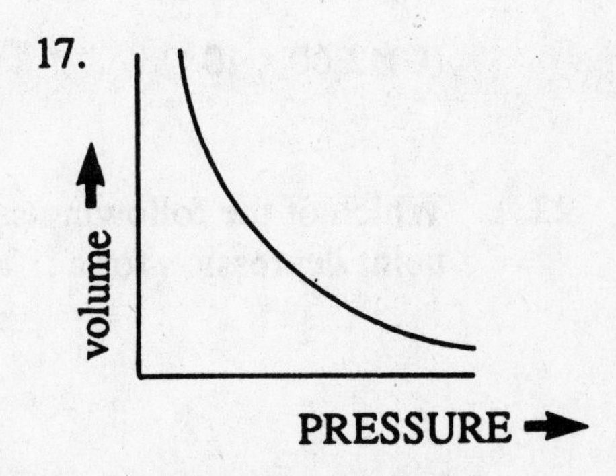

18. What is the molarity of a solution that is prepared by dissolving 32.0 g of KCl in enough water to make 425 ml of solution?

(A) 1.0
(B) 2.3
(C) $1.0 \times 10^{-3}$
(D) .425
(E) .0075

19. What would the freezing point of a solution containing 8.0 g of ethylene glycol ($C_2H_6O_2$) in 100 g of $H_2O$? (The $K^f$ for $H_2O$ is 1.86° C/M and the freezing point of pure water is 0° C.)

(A) 0.0°C
(B) –1.8°C
(C) +2.4°C
(D) + 0.2°C
(E) –2.4°C

20. Which of the following pairs are allotropes?

(A) $H_2O$ $D_2O$
(B) $O_2$ $O_3$
(C) HCl HBr
(D) $Cl_2$ $F_2$
(E) $Fe_2O_3$ $Fe_3O_4$

21. A saturated solution of AgCl was found to contain $1.3 \times 10^{-5}$ mol/l of $Ag^+$. The solubility product constant, $K_{sp}$ of AgCl is:

(A) $1.30 \times 10^{-5}$
(B) $1.14 \times 10^{-5}$
(C) $2.60 \times 10^{-5}$
(D) $1.80 \times 10^{-10}$
(E) $2.60 \times 10^{-10}$

22. Which of the following salts will have the greatest freezing point depression for a .1 M solution of the salt?

(A) $Na_2SO_4$

(B) NaCl

(C) KCl

(D) $KNO_3$

(E) $NH_4Cl$

23. The molecular orbital diagram shown below:

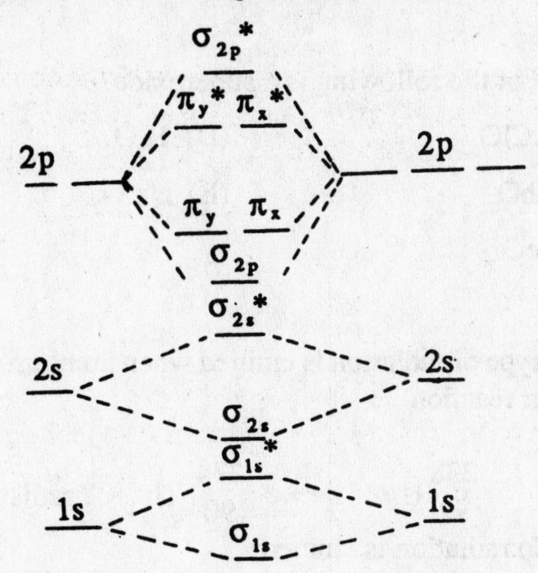


is correct for which species listed:

(A) NO

(B) CO+

(C) NO+

(D) CO

(E) $F_2$

24. If the specific heat of aluminum is .89 j/°C g, how many joules are required to heat 23.2 g of aluminum from 30°C to 80°C?

(A) 45

(B) 1,160

(C) 20

(D) 1,032

(E) 50

25. Which one of the following compounds contains chlorine in a positive oxidation state?

(A) HCl
(B) KCl
(C) $HClO_3$
(D) $PCl_3$
(E) $NH_4Cl$

26. Which of the following is a superoxide?

(A) $KClO_3$
(B) $RbO_2$
(C) FeO
(D) $H_2O_2$
(E) $Na_2O_2$

27. What type of radiation is emitted when uranium decays by the nuclear reaction

$$^{238}_{92}U \rightarrow ^{234}_{90}Th + \text{? radiation}$$

(A) No radiation is emitted
(B) Beta
(C) Alpha
(D) Cosmic
(E) Gamma

28. How many times faster will hydrogen effuse from the same effusion apparatus than nitrogen at the same temperature?

(A) 14.0
(B) .3
(C) 5.3
(D) 3.8
(E) 2.7

29. What is the Normality of a solution that contains 23.2 g of $H_2SO_4$ and enough water to make 400 ml of solution?

(A) 2.4

(B) .60

(C) 1.2

(D) 6.0

(E) .50

30. If a certain number of moles of a gas occupied a volume of 200 ml at 25°C, what volume would it occupy if it was heated to 40°C at a constant pressure?

(A) 210 ml

(B) 320 ml

(C) 230 ml

(D) 500 ml

(E) 120 ml

31. Using molecular orbital theory predict the bond order of the $He_2^+$.

(A) 1/2

(B) 1

(C) 0

(D) 3

(E) 1/4

32. Which one of the following accounts for the disappearance of the purple color of $MnO_4^-$ ion when it reacts with $Fe^{2+}$ in acid solution.

(A) $MnO_4^-$ forms a complex with $Fe^{2+}$

(B) The $MnO_4^-$ is reduced to the colorless $Mn^{2+}$ ion

(C) The $MnO_4^-$ is oxidized

(D) $MnO_4^-$ is colorless in acid solution

(E) None of the above

33. Which is the correct electron configuration for a Cr atom?

(A) [Ar] $4s^2 3d^4$

(B) [Ar] $4s^1 3d^4$

(C) [Ar] $4s^2 3d^6$

(D) [Ar] $4s^1 3d^5$

(E) [Ar] $4s^2 3d^5$

34. For the reaction

$2A + 1B \rightarrow 2C$

The following rate data was obtained:

| Experiment | initial conc. of A mol/l | initial conc. of B mol/l | initial rate mol/l sec |
|---|---|---|---|
| 1 | .5 | .5 | 10 |
| 2 | .5 | 1.0 | 20 |
| 3 | .5 | 1.5 | 30 |
| 4 | 1.0 | .5 | 40 |

Which expression below states the rate law for the reaction above?

(A) Rate = $k[A]^2[B]$

(B) Rate = $k[A][B]$

(C) Rate = $k[A]^2[B]^2$

(D) Rate = $k[A]^3$

(E) Rate = $k[A]^2[B]^3$

35. The high boiling point of $H_2O$ relative to the boiling points of $H_2S$, $H_2Se$ and $H_2Te$ can be attributed to:

(A) The molecular weight of $H_2O$

(B) The covalent bonds between H and O

(C) The atomic number of oxygen

(D) The ability of water to absorb oxygen

(E) The ability of water to form hydrogen bonds

36. For the reaction

$$N_2\,(g) + O_2\,(g) \rightarrow 2\,NO\,(g)$$

$\Delta G°$ for the reaction is + 174 kj. This reaction would:

(A) be at equilibrium

(B) proceed spontaneously

(C) have a negative $\Delta H°$

(D) not proceed spontaneously

(E) exothermic

37. From the heats of reaction, $\Delta H°$, for

$C_{(graphite)} + O_2(g) \rightarrow CO_2(g) \quad \Delta H° = -394$ kj

$CO(g) + 1/2O_2 \rightarrow CO_2(g) \quad \Delta H_o = -284$ kj

The calculated $\Delta H°$ for the reaction:

$C_{(graphite)} + 1/2O_2(g) \rightarrow CO(g)$

is:

(A) 0

(B) +18kj

(C) –18kj

(D) –678kj

(E) –110kj

38. In the reaction

$HCN + H_2O \leftrightarrow H_3O^+ + CN^-$

which species are functioning as Brønsted bases?

(A) only $H_2O$

(B) HCN and $H_3O^+$

(C) $H_2O$ and $CN^-$

(D) $CN^-$ and HCN

(E) $H_3O^+$ and $CN^-$

39. How many grams of copper would be produced by the reduction of $Cu^{2+}$ if 3.0 amperes of current are passed through a copper (II) nitrate solution for 1 hour?

(A) 18.20

(B) 3.56

(C) 31.80

(D) 7.12

(E) 63.50

40. The half life of ${}^{224}_{88}Rn$ is 3.64 days. How many μg of ${}^{224}_{88}Rn$ would be left after 18.2 days if the starting amount weighed 2.0 μg?

(A) .06

(B) .40

(C) .19

(D) .69

(E) none

41. A saturated solution of $Ag_2CrO_4$ was found to have a $CrO_4$ concentration of $1.30 \times 10^{-4}$ M. The $K_{sp}$ for $Ag_2CrO_4$ is:

(A) $1.70 \times 10^{-8}$

(B) $4.4 \times 10^{-8}$

(C) $3.4 z 10^{-8}$

(D) $1.7 \times 10^{-12}$

(E) $8.8 \times 10^{-12}$

42. One atom of hydrogen weighs:

(A) 1.00 g

(B) 22.4 g

(C) $1.66 \times 10^{-24}$ g

(D) 2.00 g

(E) $6.02 \times 10^{23}$ g

43. Which one of the following is paramagnetic?

(A) He

(B) Be

(C) $Cl^-$

(D) $F^-$

(E) Li

44. What volume of a 1.3 M solution of NaCl contains 2.3 g of NaCl.

(A) 130 ml

(B) 30.2 ml

(C) 3.9 ml

(D) 177 ml

(E) none of the above

45. If oxygen is collected over water at 25°C and at a pressure of 760 tor, the pressure due to just the oxygen is: (the vapor pressure of $H_2O$ at 25°C is 19.0 torr)

(A) 779 torr

(B) 760 torr

(C) 19 torr

(D) 741 torr

(E) insufficient data to solve

46. Which of the following is not a colligative property of a solution?

(A) freezing point

(B) vapor pressure over the solution

(C) molecular weight of the solute

(D) boiling point

(E) none of the above

47. Which of the following solutions does not constitute a buffer pair?

(A) .2M $NH_4OH$ and .1M $NH_4Cl$

(B) .2M NaCl and .1M HCl

(C) .1M acetic acid and .05M sodium acetate

(D) .1M $NH_4OH$ and .1M $(NH_4)_2SO_4$

(E) .1M formic acid and .1M sodium formate

48. What is the hydrogen ion concentration of a buffer solution that is .05M in acetic acid and .1M in sodium acetate. (The $K_i$ for acetic acid is $1.8 \times 10^{-5}$.)

(A) $9.0 \times 10^{-6}$

(B) $1.8 \times 10^{-5}$

(C) $1.0 \times 10^{-14}$

(D) $1.8 \times 10^{-6}$

(E) $1.0 \times 10^{-7}$

49. What is the mole fraction of $CH_3OH$ in a solution that contains 53.0 g of water, 20.3 g of $CH_3OH$ and 15.0 g of $CH_3CH_2OH$?

(A) .33

(B) .48

(C) 1.22

(D) .96

(E) .16

50. For the reaction at 298K:

$$A(g) + B(g) \rightarrow C(g)$$

the value of $K_p$ is $1 \times 10^{10}$ atm$^{-1}$. The value for $\Delta G°$ expressed in kJ is:

(A) 0

(B) – 57.1

(C) – 27.3

(D) + 57.1

(E) + 83.7

51. For the following reaction at 500°K

$$C(s) + CO_2(g) \qquad 2CO(g)$$

the equilibrium mixture contained $CO_2$ and CO at partial pressures of 7.6 atm and 3.2 atm respectively. The value of the $K_p$ is:

(A) 2.4 atm

(B) 18.1 atm

(C) .6 atm

(D) 1.0 atm

(E) .4 atm

52. Which one of the following drawings represents the region of space an electron with a quantum number of $l = 1$ would be found?

(A) A

(B) B

(C) C

(D) D

(E) none of the above

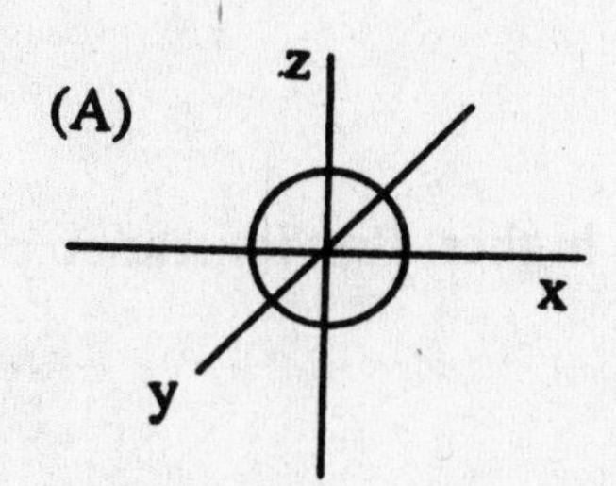


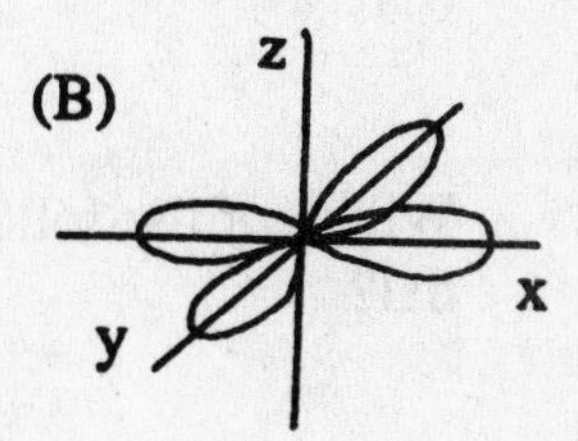


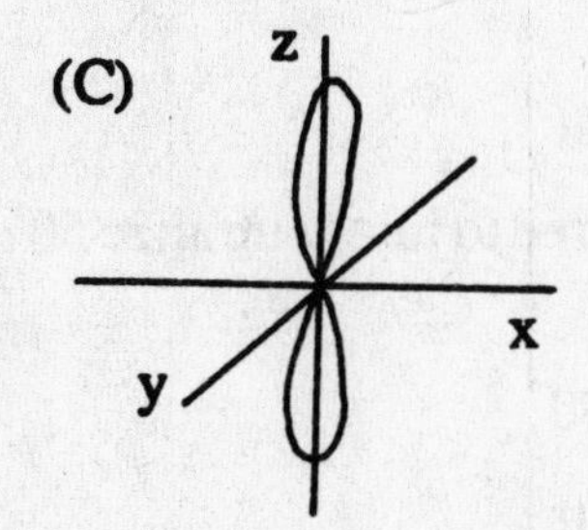


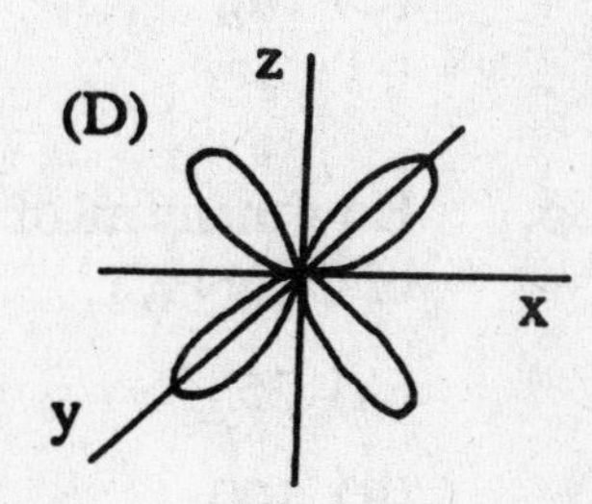

53. Which of the following is the weakest acid?

(A) HCl
(B) HBr
(C) HI
(D) HF
(E) none of the above

54. How many moles of $KClO_3$ must be decomposed to produce 3.20 g of $O_2$?

(A) .10
(B) .30
(C) .24
(D) .14
(E) .07

55. Which of the following has the highest first ionization potential?

(A) Na
(B) Ar
(C) Mg
(D) P
(E) Si

56. How many ml of 0.5 NaOH are required to just neutralize 50 ml of 0.2N HCl?

(A) 25
(B) 100
(C) 20
(D) 50
(E) 40

57. In the Van der Waal's equation:

$$\left(P + \frac{n^2a}{V^2}\right) \quad (V - nb) \quad = \quad nRT$$

The constant a is best described as a correction factor due to:

(A) temperature

(B) intermolecular attractions of real gases

(C) the molecular weights of gases

(D) the volume of the actual gas molecules

(E) the specific heat of the gas molecules

58. Which of the following statements is correct concerning the reaction:

$Fe^{2+} + 2H^{+} + NO_3^{-} \rightarrow Fe^{3+} + NO_2 + H_2O$

(A) $Fe^{3+}$ is oxidized and $H^{+}$ is reduced

(B) $Fe^{2+}$ is oxidized and nitrogen is reduced

(C) $Fe^{2+}$ and $H^{+}$ are oxidized

(D) oxygen is oxidized

(E) $H^{+}$ and oxygen are reduced

59. Thioacetamide is used as a reagent in qualitative analysis to form precipitates of $Fe^{3+}$, $Ni^{2+}$, $CO^{2+}$, $Mn^{2+}$, and $Zn^{2+}$. This reagent is used because it provides a convenient method of generating which of the following materials in aqueous solution?

(A) $H_2SO_4$

(B) $H_2S$

(C) $SO_3$

(D) $SO_2$

(E) both $SO_3$ and $SO_2$

60. The standard free energy, $\Delta G^{\circ}$ expressed in kj for the reaction below at 25°C is:

$Cd + Pb^{2+} \rightarrow Cd^{2+} + Pb \quad E^{\circ} = +.28V$

(A) – 27

(B) – 54

(C) + 27

(D) – 108

(E) 0

61. Which of the following salts would hydrolize to form a solution whose pH is below 7?

(A) $CaCl_2$

(B) $CH_3COONa$

(C) NaCl

(D) $NH_4Cl$

(E) both (D) and (B)

62. For the reaction:

$$\underset{(aq)}{FeCl_2} + \underset{(aq)}{KMnO_4} + \underset{(aq)}{HCl} \rightarrow \underset{(aq)}{FeCl_3} + \underset{(aq)}{MnCl_2} + \underset{(aq)}{KCl} + \underset{(l)}{H_2O}$$

the net ionic equation is:

(A) $Fe^{2+} \rightarrow Fe^{3+}$

(B) $Fe^{2+} + MnO_4^- + H^+ \rightarrow Fe^{3+} + Mn^{2+} + H_2O$

(C) $MnO_4^- + H^+ \rightarrow Mn^{2+} + H_2O$

(D) $FeCl_2 + MnO_4^- \rightarrow FeCl_3 + Mn^{2+}$

(E) none of the above

63. What do the following ions have in common?

$Mg^{2+}$ $O^{2-}$ $F^-$ $Na^+$

(A) they are isoelectronic

(B) the same number of protons

(C) they are metal ions

(D) they have the same atomic radius

(E) none of the above

64. VSEPR theory predicts that the geometry of the $ICl_4^-$ ion is:

(A) tetrahedral
(B) square planar
(C) octahedral
(D) linear
(E) trigonal pyramidal

65. Which of the following can best be described as a polar co-valent molecule?

(A) $Cl_2$
(B) HCl
(C) KCl
(D) $H_2$
(E) NaCl

66. The $K_{sp}$ for $PbI_2$ is $8.7 \times 10^{-9}$. What is the molar solubility of $PbI_2$?

(A) $1.3 \times 10^{-3}$
(B) $8.7 \times 10^{-3}$
(C) $9.3 \times 10^{-5}$
(D) $1 \times 10^{-9}$
(E) 0

67. Which of the following is an example of a hydride?

(A) HCl
(B) $H_2O$
(C) LiH
(D) HF
(E) $H_2S$

68. A catalyst will increase the rate of a chemical reaction by:

(A) shifting the equilibrium to the right
(B) lowering the activation energy
(C) shifting the equilibrium to the left
(D) increasing the activation energy
(E) none of the above

69. In the correctly balanced equation for the reaction:

$MnO_4^- + Cl^- + H^+ \leftrightarrow Mn^{2+} + Cl_2 + H_2O$

The coefficients for $Cl^-$ and $H^+$ are respectively:

(A) 10 and 8
(B) 10 and 16
(C) 5 and 8
(D) 5 and 5
(E) 5 and 10

70. Which one of the following hydroxides is amphoteric?

(A) $Ba(OH)_2$
(B) $Ca(OH)_2$
(C) $Mg(OH)_2$
(D) $Al(OH)_3$
(E) NaOH

71. Which of the following would have the lowest entropy at 25°C?

(A) NaCl(s)
(B) $H_2(g)$
(C) $H_2O(l)$
(D) He(g)
(E) both $H_2$ and He

72. From the data below, calculate the ΔH°, expressed in kj, for the reaction:

$2Na(s) + 2H_2O(l) \rightarrow 2NaOH(s) + H_2(g)$

| Substance | ΔH° kj/mol |
| --- | --- |
| $H_2O(l)$ | – 285.8 |
| NaOH(s) | – 426.7 |
| $H_2(g)$ | – 241.8 |

(A) – 281.8
(B) + 140.9
(C) – 712.5
(D) + 712.5
(E) zero

73. For which one of the following equilibrium equations will $K_p$ equal $K_c$?

(A) $PCl_5 \leftrightarrow PCl_3 + Cl_2$

(B) $COCl_2 \leftrightarrow CO + Cl_2$

(C) $H_2 + I_2 \leftrightarrow 2HI$

(D) $3H_2 + N_2 \leftrightarrow 2NH_3$

(E) $2SO_3 \leftrightarrow 2SO_2 + O_2$

74. The weak electrolyte is:

(A) $HNO_3$

(B) KI

(C) HCl

(D) NaCl

(E) $NH_4OH$

75. The following phase diagram was obtained for compound X:

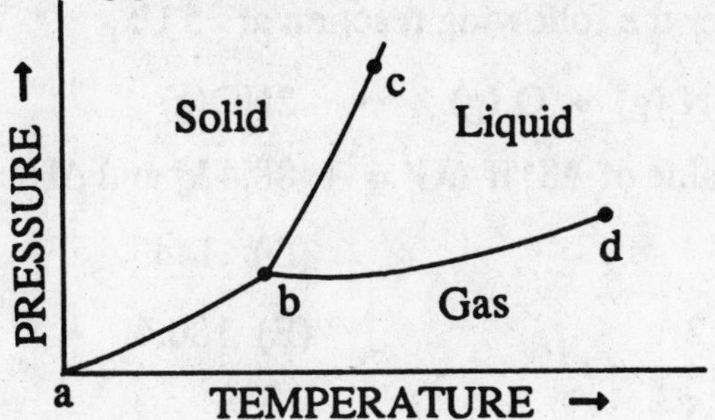


The melting point of X would:

(A) be independent of pressure

(B) increase as pressure increases

(C) decrease as pressure increases

(D) depends on the amount of X present

(E) both (C) and (D)

76. The standard electrode potential, E°, for the half reaction

$Sn^{4+} + 2\,e \rightarrow Sn^{2+}$

is + .150 V. What would be the electrode potential of this cell if the $Sn^{2+}$ concentration is five times the $Sn^{4+}$ concentration?

(A) + .129

(B) + .171

(C) – .171

(D) + .191

(E) + .091

77. Which of the following transition metal complexes or complex ions could exhibit geometric isomerism?

(A) octahedral $Co(NH_3)_5 I^{2+}$

(B) square planar $Pt(NH_3)_2Cl_2$

(C) tetrahedral $Zn(CN)_4 2^-$

(D) trigonal bipyramidal $Fe(CO)_6$

(E) both (A) and (B)

78 Consider the following reaction at 25°C:

$$N_2(g) + O_2(g) \rightarrow 2NO(g)$$

What is the value of $\Delta S°$ if $\Delta G° = + 88.4$ kj and $\Delta H° = 92.0$ kj?

(A) 0

(B) .012

(C) .065

(D) .144

(E) 180.4

79. The reddish color of bromine dissolved in $CCl_4$ rapidly disappears when a few drops of this reagent is added to the solution:

(A) $CH_4$

(B) $CH_2{=}CH{-}CH_2{-}CH_2{-}CH_3$

(C) $CH_3{-}CH_2{-}CH_2{-}CH_3$

(D) $CHCl_3$

(E)

```
      Cl     Cl
      |      |
  H - C ---- C - H
      |      |
      H      H
```

80. Which of the following compounds contains a chiral carbon?

(A)

```
      H
      |
  H - C - Cl
      |
      H
```

(B) $CH_3 - CH_2 - C(=O)OH$

```
                  =O
  CH3 - CH2 - C
                  \
                   OH
```

(C)

```
         OH
         |
  CH3 -  C - CH2 - CH3
         |
         H
```

(D) $Cl - CH_2 - CH_2 - Cl$

(E)

```
         O
         ||
  CH3 -  C - CH3
```

**QUESTIONS 81 – 82:**

(A) $Cl_2$

(B) Ne

(C) Na

(D) Pb

(E) Sn

81. Which element reacts with water, liberating hydrogen gas and forming a metal hydroxide?

82. Which element can react directly with hydrogen, $H_2$, to produce a gas which when dissolved in water results in an acid solution?

**QUESTIONS 83 – 84:**

(A) No change in the position of equilibrium

(B) Shifts position of equilibrium to the left

(C) Shifts position of equilibrium to the right

(D) Cannot predict from data given

(E) Will only shift position of equilibrium if a catalyst is added

Consider the following reaction:

$$N_2\,(g) + 3H_2\,(g) \leftrightarrow 2NH_3\,(g)$$

In which direction will the position of equilibrium shift if:

83. Additional $NH_3$ is added to the system?

84. The pressure on the system is increased from 1 atm to 2 atm?

85. The pH of a 1.0 M acetic acid solution, $C_2H_3O_2$ was found to be 3.5. The percent ionization of acetic acid is:

(A) 1.0

(B) $1.8 \times 10^{-5}$

(C) 3.5

(D) 0.031

(E) 35.0

# ADVANCED PLACEMENT CHEMISTRY EXAM VI

## SECTION II

**Directions:** The percentages given for each individual part indicates the scoring for this section. Spend approximately 35 minutes on Parts A and B together and 40 minutes on Part C.

The methods and steps used to solve a problem must be shown clearly. Partial credit will be given if work is shown. Pay attention to significant figures.

### Part A

(30 percent)

1. For the hypothetical reaction: at 25°C

$$A + 2B + C \rightarrow AB + BC$$

the following rate data was obtained:

| Run | Initial conc.[A] | Initial conc. [B] | Initial conc. [C] | Initial rate of formation of AB in sec. |
|---|---|---|---|---|
| 1 | 0.10 | 0.10 | 0.10 | $6.0 \times 10^{-8}$ |
| 2 | 0.20 | 0.15 | 0.10 | $2.4 \times 10^{-7}$ |
| 3 | 0.10 | 0.20 | 0.20 | $1.2 \times 10^{-7}$ |
| 4 | 0.10 | 0.15 | 0.10 | $6.0 \times 10^{-8}$ |

a. From the data above, derive the rate law for this reaction.

b. Calculate the rate constant, k.

c. At 35°C (308°K) the rate constant for this reaction doubles. Calculate the activation energy, Ea.

## Part B

(30 Percent)

**Solve ONE of the two problems presented. Only the first problem answered will be scored.**

2. For the weak triprotic acid $H_3PO_4$, calculate the hydrogen concentration that results from a 0.05 M aqueous solution of this acid:

   (The $K_1 = 7.0 \times 10^{-3}$, $K_2 = 6.0 \times 10^{-8}$ and $K_3 = 4.0 \times 10^{-13}$.)

3. What is the $H_3O^+$ ion concentration of a 0.1 M $NH_4OCl$ aqueous solution?

   Given: $NH_4^+ \rightarrow NH_2 + H^+$

   $OCl^- \rightarrow HOCl + OH^-$

## Part C

(40 Percent)

**Choose THREE of the following five topics. Only the first three answers will be scored. Your answers will be scored based on their accuracy, relevance of the details chosen, and appropriateness of the descriptive material used. Be as specific as possible and use illustrative examples and equations where helpful.**

4. The Born-Haber cycle is often used to predict the enthalpy of formation of ionic compounds.

   a. Construct the Born-Haber cycle for the formation of 1 mole of $NaCl_{(s)}$ from its elements:

   $$Na_{(s)} + 1/2\ Cl_{2(g)} \rightarrow NaCl_{(s)}$$

   b. Which terms in the Born-Haber cycle would you expect to be endothermic for the formation of $NaCl_{(s)}$.

c. Which term in the cycle is usually determined from theoretical calculations rather than by experimental methods.

5. Construct a flow chart that details the procedures for the separation and identification of the Group I metal ions: $Hg_2^{2+}$, $Ag^+$, $Pb^{+2}$. Your answer should include a discussion of why the separations are achieved as well as a specific listing of the reagents that are utilized in the final identification of the metal ions.

6. In the development of modern acid-base theory, various concepts of acids and bases have been postulated. These include the Arrhenius, Bronsted-Lowry and the Lewis concepts.

Describe the important principles of each of these theories and discuss any inherent limitations in regard to a generalized concept of acid-base reactions.

7. For each statement listed below, use either ionic or molecular formulas to describe both the reactants and the products of the chemical reactions that will be expected to occur. (Assume all reactions are occuring in aqueous solution except where other conditions are indicated).

a. A copper strip immersed in a solution of silver nitrate.

b. Carbon dioxide is bubbled through water.

c. Magnesium nitride is added to water.

d. Iron sulfide is reacted with hydrochloric acid.

e. Calcium carbide is added to water.

8. Give the scientific explanation for the following:

a. Real gases tend to behave non-ideally at low temperatures and high pressures.

b. Fish kills are more prevalent in the summer than in the winter.

c. Ethylene glycol-based anti-freezes offer protection both against freezing in the winter and boilover in the summer.

d. The endpoint determined by an indicator might or might not correspond to the equivalent point in an acid base titration.

e. Fresh water lakes freeze from the top down, and do not usually freeze solid.

# ADVANCED PLACEMENT CHEMISTRY EXAM VI

## ANSWER KEY

| | | | | | |
|---|---|---|---|---|---|
| 1. | B | 29. | C | 57. | B |
| 2. | E | 30. | A | 58. | B |
| 3. | A | 31. | A | 59. | B |
| 4. | B | 32. | B | 60. | B |
| 5. | D | 33. | D | 61. | D |
| 6. | C | 34. | A | 62. | B |
| 7. | D | 35. | E | 63. | A |
| 8. | B | 36. | D | 64. | B |
| 9. | E | 37. | E | 65. | B |
| 10. | C | 38. | C | 66. | A |
| 11. | A | 39. | B | 67. | C |
| 12. | A | 40. | A | 68. | B |
| 13. | D | 41. | E | 69. | B |
| 14. | D | 42. | C | 70. | D |
| 15. | C | 43. | E | 71. | A |
| 16. | C | 44. | B | 72. | A |
| 17. | A | 45. | D | 73. | C |
| 18. | A | 46. | C | 74. | E |
| 19. | E | 47. | B | 75. | B |
| 20. | B | 48. | A | 76. | A |
| 21. | D | 49. | E | 77. | B |
| 22. | A | 50. | B | 78. | B |
| 23. | B | 51. | B | 79. | B |
| 24. | D | 52. | C | 80. | C |
| 25. | C | 53. | D | 81. | C |
| 26. | B | 54. | E | 82 | A |
| 27. | C | 55. | B | 83. | B |
| 28. | D | 56. | C | 84. | C |
| | | | | 85. | D |

# ADVANCED PLACEMENT CHEMISTRY EXAM VI

# DETAILED EXPLANATIONS OF ANSWERS

## SECTION I

1. (B)

According to the equation:

$$2C_6H_6 + 15O_2 \rightarrow 12CO_2 + 6H_2O$$

for every 2 moles of $C_6H_6$, 15 moles of oxygen are required for the complete combustion of the $C_6H_6$. Therefore it is necessary to find the number of moles of $C_6H_6$ in 39.0 g. Since the number of moles is always equal to:

$$\text{moles} = \frac{g}{MW} = \frac{39.0}{78.0} = 0.5 \text{ moles of } C_6H_6$$

According to the balanced equation for every 2 moles of $C_6H_6$, 15 moles of oxygen are required. Therefore

$$\frac{2 \text{ moles } C_6H_6}{.5 \text{ moles } C_6H_6} \qquad \frac{15 \text{ moles } O_2}{x \text{ moles } O_2}$$

$x$ = 3.75 moles of oxygen are needed for .5 moles of $C_6H_6$. However the question asked for the answer expressed in grams, and so we must now convert 3.75 moles of $O_2$ to grams of $O_2$. Since moles = g/MW the number of grams can be calculated by

g = moles x MW

g = 3.75 moles x 32 g/mole

g = 120 g of oxygen.

2. (E)

The number of molecules in one mole is given by Avogadro's number which is: 1 mole of molecules contains $6.02 \times 10^{23}$ molecules. Since the amount of $CO_2$ given is 22 g it is first necessary to find the number of moles of $CO_2$ in 22 g. This is calculated from:

$$\text{moles} = \frac{g}{MW} = \frac{22\text{ g}}{44\text{ g/mol}} = 0.5 \text{ moles of } CO_2$$

It is now necessary to calculate the number of molecules in .5 moles of $CO_2$. This is obtained using Avogadro's number.

$$\frac{1 \text{ mole}}{0.5 \text{ moles}} \quad \frac{6.02 \times 10^{23} \text{ molecules}}{x \text{ molecules}}$$

$$x = 3.01 \times 10^{23} \text{ molecules}$$

3. (A)

The percent of an element in a compound is equal to the weight of that element divided by the total weight of the compound times 100. Therefore the percent carbon in sucrose is equal to

$$\%C = \frac{12 \times \text{atomic weight of C}}{\text{molecular weight of sucrose}} \times 100$$

$$= \frac{12 \times 12.0 \text{ g}}{(12 \times 12)_C + (22 \times 1)_H + (16 \times 11)_O} \times 100$$

$$= \frac{144}{342} \times 100 = 42.1\% \text{ carbon}$$

4. (B)

Atoms that have the same number of protons but differ in the number of neutrons they contain are referred to as isotopes. For the element hydrogen there are three forms, Hydrogen or Protium, Deuterium, and Tritium which occur naturally. In each of the three forms the number of protons and electrons is identical but they contain a different number of neutrons. The only difference between isotopes is the number of neutrons. Thus isotopes of hydrogen will all contain 1 proton and 1 electron but will have additional neutrons. Of the choices listed the only one that could be an isotope of hydrogen would be (B), since it has the same atomic number as hydrogen, but has two additional neutrons.

Isotopes of Hydrogen

| | Atomic number | protons | neutrons | electrons |
|---|---|---|---|---|
| Hydrogen | 1 | 1 | 0 | 1 |
| Deuterium | 1 | 1 | 1 | 1 |
| Tritium | 1 | 1 | 2 | 1 |

In this case the isotope of hydrogen would be tritium.

5. (D)

The empirical formula of a compound tells us the relative number of moles or atoms of each element it contains. From the data given above it is possible to calculate the mole ratio of each element. In a 100 g sample there would be 40 g of carbon, 6.7 g of hydrogen and 53.3 g of oxygen. Therefore the number of moles of each element would be:

$$\text{moles of C} = \frac{\text{g of C}}{\text{at. wt. C}} = \frac{40.0\text{g}}{12.0\text{g}} = 3.3$$

$$\text{moles of H} = \frac{\text{g of H}}{\text{at. wt. H}} = \frac{6.7\text{g}}{1.0\text{g}} = 6.7$$

$$\text{moles of O} = \frac{\text{g of O}}{\text{at. wt. O}} = \frac{53.3\text{g}}{16.0\text{g}} = 3.3$$

The molar ratio is then

$$C_{3.3}H_{6.7}O_{3.3}$$

The relative whole number ratio of each can be found by dividing each by the smallest number, which in this case is 3.3:

| C | H | O |
|---|---|---|
| $\frac{3.3}{3.3} = 1$ | $\frac{6.7}{3.3} = 2$ | $\frac{3.3}{3.3} = 1$ |

Therefore the empirical formula would be $CH_2O$.

6. (C)

A noble gas configuration is one in which all of the available energy levels contained in the atom are completely occupied by electrons. For example He has a configuration $1s^2$ and Ne has a configuration $1s^22s^22p^6$. In both cases these elements contain the maximum number of electrons that their available energy levels can contain. The configurations for the species given are:

| | | |
|---|---|---|
| $Be^+$ | 3 electrons | $1s^22s^1$ |
| $N^-$ | 8 electrons | $1s^22s^22p^4$ |
| $Mg^{2+}$ | 10 electrons | $1s^22s^22p^6$ |
| $H^+$ | 0 electrons | |
| $O^{2+}$ | 6 electrons | $1s^22s^22p^2$ |

The only one that has a noble gas configuration is $Mg^{2+}$ which has the same electron configuration as Ne which is also $1s^22s^22p^6$.

7. (D)

All oxides contain oxygen in an oxidation state of –2. Thus in all of the above compounds the oxidation state of the oxygen is –2. Since all compounds are neutral we may assign the oxidation state of the Iron as follows:

$FeO$ Since oxygen is –2 the iron must be +2 for the compound to be neutral.

$Fe_3O_4$ Each oxygen is –2 for a total charge of –2 x 4 = –8. Therefore the oxidation state of the iron must be

$$\frac{-8}{3} = -2\ 1/2$$

$Fe_2O_3$ Each oxygen is –2 for a total charge of –2 x 3 = –6, therefore the iron is +6/2 = 3.

$Fe_2O$ The oxygen is –2, therefore total charge due to the two iron atoms is +2 hence the oxidation state of the iron would be +1.

8. (B)

We can use the ideal gas law, PV = nRT to solve for the number of moles. By rearranging we have:

$$n = \frac{PV}{RT}$$

To solve this equation we must use the correct units. Using R = .082 $\frac{1\ atm}{mole \cdot K}$ we must express the pressure in atmospheres, the volume in liters and the temperature in °K.

$V = 500 \text{ ml} = .5 \text{ l}$
$P = .4 \text{ atm.}$
$T = 27°C + 273° = 300°K$

Substituting

$$n = \frac{.4 \text{ atm} \times .5 \text{ l}}{.082 \frac{\text{l atm}}{\text{mole°K}} \times 300°K} = .008 \text{ moles}$$

9. (E)

Using the ideal gas law $PV = nRT$ and rearranging:

$$n = \frac{PV}{RT}$$

and since $n = \frac{g}{MW}$

we now have $\frac{g}{MW} = \frac{PV}{RT}$

and solving for MW gives us $MW = \frac{g\,RT}{PV}$

Using the ideal gas law constant $R = .082 \frac{\text{l atm}}{\text{mole°K}}$
we make the substitutions
$g = 0.25$
pressure = 2.5 atm
volume (in liters) = 1
temperature = 25°C = 298°K

$$MW = \frac{0.25 \text{ g} \times .082 \frac{\text{l atm}}{\text{mole°K}} \times 298°K}{2.5 \text{ atm} \times .1 \text{ l}}$$

$MW = 24.4 \text{ g / mole}$

10. (C)

Covalent bonding occurs when an electron pair is shared between two atoms, whereas ionic bonding will occur when one atom has a tendency to lose one or more electrons and the other atom has a tendency to gain electrons. This process will result in the formation of a positive and a negative ion which will result in the formation of an ionic bond (i.e) $Na^+Cl^-$. In covalent bonding both atoms will have a tendency to gain electrons. In the simplest example of covalent bonding, the $H_2$ molecule is formed when one hydrogen atom H ($1s^1$) shares its electron with a second hydrogen atom,

$$H : H$$

In this process of sharing electron pairs, both hydrogen atoms now have effectively obtained their next nearest noble gas configuration which is a very stable configuration.

In the answers listed the only pair of atoms that can form a covalent bond is H and Cl.

Both the H. and $\cdot\ddot{\underset{\cdot\cdot}{Cl}}:$ would each like to gain 1 electron to achieve a pseudo noble gas configuration. This results in hydrogen achieving the pseudo He configuration and Cl achieving the pseudo argon configuration:

$$H:\ddot{\underset{\cdot\cdot}{Cl}}:$$

11. (A)

The hydronium ion concentration can be calculated from the ion – product constant for water, $K_w$,

$$K_w = [H_3O^+]\ [OH^-] = 1 \times 10^{-14}$$

Due to the auto ionization of water both the $OH^-$ and the $H_3O^+$ ions exist in acid or basic solutions. However the product of the $OH^-$ ion concentration and the $H_3O^+$ ion concentration will always be equal to $1 \times 10^{-14}$, therefore substituting the $OH^-$ concentration into the $K_w$ expression we can calculate the $H_3O^+$ ion concentration.

$$K_w = [H_3O^+]\ [OH^-] = 1 \times 10^{-14}$$

$$H_3O^+ = \frac{1 \times 10^{-14}}{1.4 \times 10^{-4}} = 7.2 \times 10^{-11}$$

12. (A)

The pH of a solution is defined as $pH = -\log [H^+]$

Therefore $pH = -\log (1.2 \times 10^{-4})$

$$= -\log 1.2 + \log 10^{-4}$$

The log of 1.2 is obtained either from a log table or from a calculator that has a log function key. The log of $10^{-4}$ is $-4$. Therefore:

$$\begin{aligned} pH &= -(.079 - 4.00) \\ &= -(-3.92) \\ &= 3.92 \end{aligned}$$

13. (D)

Since a $K_i$ is given, the acid is a weak acid and reaches a state of equilibrium. In this case $HCN \leftrightarrow H^+ + CN^-$ and the expression for the equilibrium constant is

$$K_i = \frac{[H^+][CN^-]}{HCN} = 4 \times 10^{-10}$$

The initial amount of HCN present is .01 Molar, however the amount at equilibrium is .01 minus the amount that has ionized ( $.01 - x$). For $x$ amount that ionizes there will be $x$ amount of $H^+$ and $CN^-$ ions:

$$\underset{HCN}{0.01 - x} \quad \leftrightarrow \quad \underset{H^+}{x} \quad \underset{CN^-}{x}$$

To solve for the $H^+$ concentration we substitute

$$\frac{(x)\ (x)}{0.01 - x} = 4 \times 10^{-10}$$

Since $x$ is very small compared to the total amount of acid present we may neglect it in the denominator without introducing a measurable error, thus

$$\frac{x^2}{0.01} = 4 \times 10^{-10}$$

$$x^2 = 4 \times 10^{-12}$$

$$x = [H^+] = 2 \times 10^{-6}$$

14. (D)

Since the velocity of light is expressed in meters per second and the wavelength is expressed in nm it is necessary to convert nm to meters before calculating the frequency form:

$$v = \underline{c}$$

Since $\frac{1 \text{ nm}}{600 \text{ nm}} = \frac{1 \times 10^{-9} \text{ m}}{x\text{m}}$

Solving for $x$ we find $x = 6.0 \times 10^{-7}$ m

Substituting this wavelength we have

$$v = \frac{3.0 \times 10^{8} \text{ m/s}}{6.0 \times 10^{-7} \text{ m}}$$

$$v = .5 \times 10^{15} \text{ s}^{-1}$$

or

$$5.0 \times 10^{14} \text{ s}^{-1}$$

15. (C)

The energy of electromagnetic radiation is equal to the frequency times Planck's constant

$$\begin{aligned} E &= hv \\ E &= (6.63 \times 10^{-34} \text{ j/s})(8.00 \times 10^{11} \text{ 1/s}) \\ &= 53.0 \times 10^{-23} \text{ j} \\ &= 5.3 \times 10^{-2} \text{ j} \end{aligned}$$

16. (C)

The graph in question 16 is a representation of Charles' Law which states that at constant pressure the volume of a given quantity of a gas varies directly with temperature. $V \propto T$

17. (B)

The graph in 17 is a representation of Boyle's Law which states that at constant temperature the volume of a fixed quantity of a gas is inversely proportional to the pressure:

$$V \propto \frac{1}{P}$$

It can be seen from the graph that as the pressure increases the volume decreases as stated in Boyle's Law.

18. (A)

The molarity of a solution is defined as the number of moles of solute in a liter of solution

$$\text{Molarity} = \frac{\text{moles of solute}}{\text{liters of solution}}$$

To solve this problem we must first determine the number of moles of KCl in 32.0 g of KCl, which is:

$$\text{moles} = \frac{\text{g of KCl}}{\text{MW of KCl}} = \frac{32.0 \text{ g}}{74.6 \text{ g/mole}} = .428 \text{ moles}$$

Substituting into this expression for molarity and expressing the volume in liters,

$$M = \frac{.428 \text{ moles}}{.425 \text{ liters}} = 1.0$$

19. (E)

The presence of a non-volatile solute such as $C_2H_6O_2$ will depress the freezing point of the solution according to Raoult's Law. The freezing point depression is given by the expression:

$$\Delta T = K_f \bullet m$$

Where $K_f$ is the freezing point depression constant, $\Delta T$ is the change in the freezing point, and m is the molality of the solution, or

$$m = \frac{\text{moles of solute}}{\text{kg of solvent}}$$

and since the number of moles $= \dfrac{\text{g of solute}}{\text{MW of solute}}$

we now have $m = \dfrac{\dfrac{\text{g of solute}}{\text{MW of solute}}}{\text{kg of solvent}}$

Substituting this expression for molality in the first equation we have,

$$\Delta T = K_f \; \frac{\dfrac{\text{g of solute}}{\text{MW of solute}}}{\text{kg solvent}}$$

$$\Delta T = 1.86 \left(\frac{\frac{8.0}{62.0}}{.1}\right)$$
$$\Delta T = 2.4$$

Since the normal freezing point of water is 0.0°C the freezing point of this solution will be 2.4°C lower than pure $H_2O$ or – 2.4°C.

20. (B)

Ozone, $O_3$ and oxygen, $O_2$ are allotropes. Allotropes are two forms of the same element that differ in their bonding or molecular structure. In the ozone form of oxygen the structure is

O
O‖ | O

Whereas in the diatomic form of oxygen the structure is O = O.

Other examples of allotropes are carbon in the graphite and diamond form, and sulfur in its $S_2$, $S_6$, and $S_8$ forms.

21. (D)

In a saturated solution of AgCl the following equilibrium is established

$$AgCl_{(solid)} \leftrightarrow Ag^+_{(aq)} + Cl^-_{(aq)}$$

and the expression for the $K_{sp}$ is:

$$K_{sp} = [Ag^+][Cl^-]$$

To calculate the $K_{sp}$ it is necessary to know both the $Ag^+$ and $Cl^-$ ion concentration. The $Ag^+$ concentration is given as $1.34 \times 10^{-5}$ M. Since for every mole of AgCl that ionizes there will be 1 mole of both $Ag^+$ and $Cl^-$. The $Ag^+$ concentration is $1.34 \times 10^{-5}$ M and the $Cl^-$ will also be $1.34 \times 10^{-5}$ M. Substituting the concentration of both $Ag^+$ and $Cl^-$

$$K_{sp} = [1.34 \times 10^{-5}][1.34 \times 10^{-5}]$$
$$K_{sp} = 1.80 \times 10^{-10}$$

22. (A)

The freezing point depression or boiling point elevation of a solution depends on the number of particles present in the solution. A solution that contains a strong electrolyte such as the salts listed above will depress the freezing point according to the number of particles that are produced when the salt dissolves and ionizes completely. When $Na_2SO_4$ ionizes it forms 2 $Na^+$ and 1 $SO_4^{2-}$ ions:

$$Na_2SO_4 \rightarrow 2\,Na^+ + 1\,SO_4^{2-}$$

So the effective molality of the solution will be three times the molality of the undissociated $Na_2SO_4$. The other salts will only produce an effective molality 2 times the undissociated form of the salt.

$$NaCl \rightarrow Na^+ + Cl^-$$

Therefore the freezing point depression would be the greatest for $Na_2SO_4$.

23. (B)

In molecular orbital diagrams all of the electrons contained in the molecule or ion are placed in the energy levels according to both the auf bau principle and Hund's rule. In the diagram above there are thirteen electrons. The number of electrons for the species listed as possible answers are:

| Species | Number of electrons |
|---|---|
| NO | N = 7, O = 8 total = 15 |
| $CO^+$ | C = 6, O = 8 However since it is a +1 ion, the species contains 1 less electron than the neutral molecule or a total of 6 + 8 – 1 = 13 |
| $NO^+$ | N = 7, O = 8 – 1 (since the species is a + 1 ion). 7 + 8 – 1 = 14 |
| CO | C = 6, O = 8 total = 14 |
| $F_2$ | each F has 9 for a total of 18 |

Since the molecular orbital diagram shows a total of 13 electrons, it represents the $CO^+$ species.

24. (D)

The specific heat of a substance is the amount of heat required to produce a change of 1°C in 1 g of the substance. For aluminum it requires .89 joules to raise 1 g of Al 1°C, therefore the heat required to raise 23.2 g from 30°C to 80°C, or a change of 50°, will be equal to

$$q = \text{mass x specific heat x temperature change}$$
$$q = 23.2 \text{ g x .89 j / °C g x } 50°$$
$$q = 1{,}032 \text{ joules}$$

25. (C)

Group VII A elements assume the – 1 oxidation state in all their binary compounds with metals and with the ammonium ion, $NH_4^+$. However, Cl, Br, and I can exist in positive oxidation states of +1, +3, +5, and +7 in covalently bonded species that contain more electronegative elements. An example of such species are:

| | Oxidation state of Chlorine |
|---|---|
| $ClO^-$ | +1 |
| $ClO_2^-$ | +3 |
| $ClO_3^-$ | +5 |
| $ClO_4^-$ | +7 |

The compound $HClO_3$ is the only compound listed that contains chlorine in a positive oxidation state.

26. (B)

Compounds that contain oxygen in the minus 2 oxidation state are called oxides. Oxygen can also react with active metals to form peroxides which contain the $O_2^{2-}$ ion, and the superoxides which contain the $O_2^-$ ion. $RbO_2$ is the only superoxide listed. Rb only forms the $Rb^+$ ion therefore the compound $RbO_2$ must contain the superoxide ion $O_2^-$.

27. (C)

Radioactive elements will spontaneously emit radiation of three principal types, alpha ($\alpha$), beta ($\beta$), and gamma ($\gamma$) rays. The properties of these three types are summarized below:

| | mass | charge | corresponds to |
|---|---|---|---|
| alpha | 4 | +2 | $He^{2+}$ |
| beta | | −1 | electrons |
| gamma | 0 | 0 | electromagnetic radiation |

In the decay of $^{238}_{92}U$ to $^{234}_{90}Th$ we recognize a change in both atomic number, 92 to 90, and mass number, 238 and 234. To account for this difference an alpha particle, $^{4}_{2}He$, which consists of 2 neutrons and two protons, must have been emitted.

28. (D)

Graham's law of effusion states that the rates of effusion of gases are inversely proportional to the square roots of their molecular weights or densities.

$$\text{rate of effusion} \propto \frac{1}{\sqrt{MW}}$$

or, $$\text{rate of effusion} \times \sqrt{MW} = t_0 \text{ a constant}$$

Hence when 2 gases effuse from the same apparatus at the same conditions:

$$\frac{\text{rate}_A}{\text{rate}_B} = \frac{\sqrt{MW_B}}{\sqrt{MW_A}}$$

Substituting the molecular weight for $N_2$ and $H_2$:

$$\frac{\text{rate } H_2}{\text{rate } N_2} = \frac{\sqrt{28}}{\sqrt{2}} = \frac{5.29}{1.41} = 3.8$$

29. (C)

Normality is defined as the number of equivalent weights of solute per liter of solution.

$$N = \frac{\text{Number of equivalent weights}}{1}$$

To solve this problem we must first determine the number of equivalent weights there are in 23.2 g of $H_2SO_4$. For acids the equivalent weight is defined as the mass of the acid that will furnish 1 mole of hydrogen ions. Since 1 mole of $H_2SO_4$ ( 98.0 g ) produces 2 moles of $H^+$ ions,

$$H_2SO_4 \rightarrow 2H^+ + SO_4^-$$

its equivalent weight is 1/2 the molecular weight or 49.0 g. The number of equivalent weights in 23.2 g of $H_2SO_4$ will be:

$$\text{The number of equivalent weights} = \frac{\text{g present}}{\text{equivalent weight of } H_2SO_4}$$

$$= \frac{23.2 \text{ g}}{49.0 \text{ g/eq wt}} = .47$$

and the Normality is then calculated by

$$N = \frac{\text{Number of eq wt}}{\text{liter}}$$

$$N = \frac{.47 \text{ eq wt}}{.40 \text{ l}} = 1.2 \text{ N}$$

30. (A)

Charles' Law states that the volume is directly proportional to temperature at constant pressure. Therefore

$$\frac{V}{T} = k$$

in this problem

$V_1 = 200\text{ml}$
$T_1 = 25°C + 273°C = 298°K$
$V_2 = \text{unknown}$
$T_2 = 40°C + 273°C = 313°K$

Thus

$$\frac{200\text{ml}}{298°K} = \frac{V_2}{313°K}$$

and

$$V_2 = \frac{200\text{ ml} \times 313°k}{298°k} = 210\text{ ml}$$

**Note:** In all problems involving the gas laws the temperature must be expressed in degrees Kelvin (°K).

31. (A)

The electron configuration of He is $1s^2$. In $He_2^+$ the only orbitals available for molecular orbital formation are the 1s. The 1s from each He will interact to form a 1s bonding orbital and a 1s antibonding orbital.

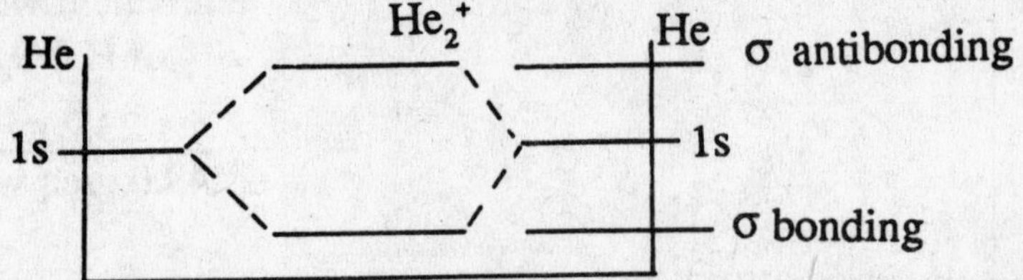


Placing the 3 electrons in the $He_2^+$ ion in the molecular orbital diagram we have

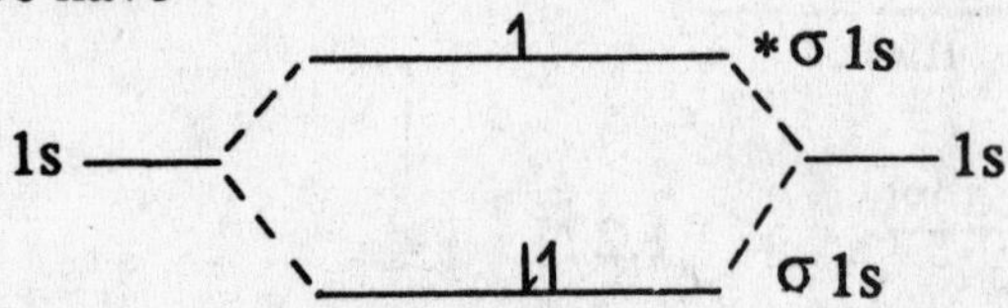


The bond order is then equal to

$$\text{Bond order} = \frac{\#\text{ of e in bonding MO} - \#\text{ of e in antibonding MO}}{2}$$

by substitution,

$$= \frac{2-1}{2} = 1/2$$

32. (B)

When $MnO_4^-$ reacts with $Fe^{2+}$ in acid solution, the following reaction occurs

$$MnO_4^- + Fe^{2+} + H^+ \rightarrow Fe^{3+} + Mn^{2+} + H_2O$$

The Mn in $MnO_4^-$ which is in the +7 oxidation state, is reduced to $Mn^{2+}$ which is almost colorless. The $Fe^{2+}$ is oxidized by the $MnO_4^-$ ion to $Fe^{3+}$ and in turn the $MnO_4^-$ is reduced to $Mn^{2+}$.

33. (D)

Electrons fill the energy levels according to the auf bau principle and Hund's rule. This means the electron will fill the lowest energy levels first and will tend to remain unpaired, unless there is an opportunity to achieve a 1/2 filled or completely filled d level by promotion of one s electron to the d level. The correct energy level diagram for the first transition series is

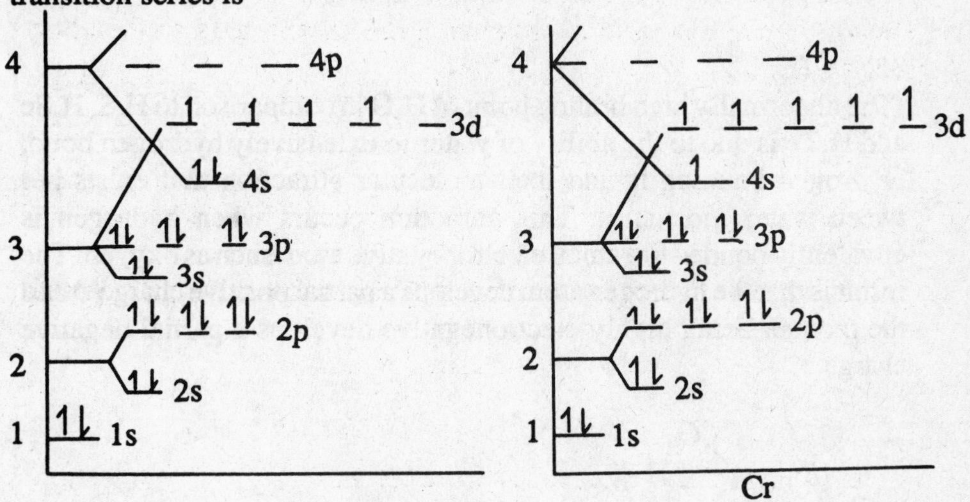


Since Cr has 24 electrons we would expect a configuration of $4s^2 3d^4$ as shown above, however in the case of Cr we could achieve a more stable 1/2 filled 3d level by promotion of one 4s electron to the 3d level. The more stable configuration of $4s^1 sd^5$ is achieved in this manner.

Note: Only 1 electron from an s level can be promoted to the d level to achieve the more stable 1/2 filled or completely filled d level.

34. (A)

The generalized rate for this reaction can be expressed as

$$\text{RATE} = k\,[A]^x\,[B]^y$$

The values of x and y are determined from the experimental data. In experiment 1 and 2 the concentration of A is constant but the concentration of B has been doubled. The effect of doubling B increased the rate by a factor of 2. Also in experiment 1 and 3 a tripling of B triples the rate. Hence the concentration of B is to the first power and the exponent y must be equal to 1. In experiment 1 and 4 the concentration of B is constant but the concentration of A is doubled. We see that by doubling the concentration of A the rate increases by a factor of 4. Therefore the value of the exponent x must be to the second power, x = 2. The overall rate is then

$$\text{RATE} = k\,[A]^2\,[B]$$

35. (E)

The abnormally high boiling point of $H_2O$ in comparison to $H_2S$, $H_2Se$ and $H_2Te$ is due to the ability of water to extensively hydrogen bond. Hydrogen bonding is and inter-molecular attraction that exists between water molecules. This attraction occurs when hydrogen is covalently bonded to a small electronegative atom such as oxygen. The result is that the hydrogen atom develops a partial positive charge $^+$ and the oxygen being highly electronegative develops a partial negative charge $^-$.

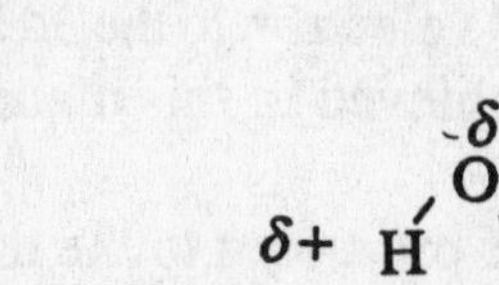

The water molecule will be attracted to other water molecules forming what is referred to as hydrogen bonds.

This dipole – dipole interaction is strong in water, and the molecules are highly attracted to one another. These forces are responsible for $H_2O$ being a liquid at room temperature rather than a gas.

36. (D)

$\Delta G°$ is the Gibb's free energy and may be calculated from the values of $\Delta H°$ (enthalpy) and $\Delta S°$ (entropy) by the relationship

$$\Delta G° = \Delta H° - T \Delta S°$$

The value of $\Delta G°$ will predict if the reaction is spontaneous, non-spontaneous or at equilibrium.

IF $\Delta G°$ is positive the reaction is non-spontaneous
$\Delta G° = 0$ system is at equilibrium
$\Delta G°$ is negative the reaction is spontaneous

Since the values of $\Delta G°$ for the reaction given is +174 kj the reaction will not proceed as written (non-spontaneous) but the reverse reaction,

$$2NO_{(g)} \rightarrow N_{2(g)} + O_{2(g)}$$

$$\Delta G° = -174kj$$

would be spontaneous.

37. (E)

According to Hess' Law the enthalpy change for a reaction is the same whether the reaction takes place in one step or a series of steps. We can utilize this law to determine the enthalpy change $\Delta H°$ for the reaction

$$C_{(graphite)} + 1/2O_2 \rightarrow CO_{(g)}$$

first by reversing the reaction

$$CO_{2(g)} + 1/2\ O_2 \rightarrow CO_{2(g)}\ \Delta H° = -284\ kj$$

we have

$$CO_{2(g)} \qquad CO_{(g)} + 1/2O_2 \qquad \Delta H^\circ = +284 \text{ kj}$$

and adding it to the reaction

$$C_{(graphite)} + 1/2O_{2(g)} \rightarrow CO_{2(g)} \qquad \Delta H^\circ = -394 \text{ kj}$$

$$CO_{2(g)} \rightarrow CO_{(g)} + 1/2\,O_2 \qquad \Delta H^\circ = +284 \text{ kj}$$

$$\text{We get } C_{(graphite)} + 1/2O_{2(g)} \rightarrow CO_{(g)} \qquad \Delta H^\circ = -110 \text{ kj}$$

Note: When reversing the reaction we must also change the sign of $\Delta H^\circ$.

38. (C)

The Danish chemist J.N. Brønsted defined an acid as a species that can donate a hydrogen ion, and a base as a hydrogen ion acceptor. In the given reaction in the forward direction

$$HCN + H_2O \rightarrow H_3O^+ + CN^-$$

the HCN is donating a hydrogen ion to the water molecule. Hence the HCN is a Brønsted acid and the water is functioning as a Brønsted base. In the reverse reaction

$$H_3O^+ + CN^- \rightarrow HCN + H_2O$$

the hydronium ion, $H_3O^+$ is donating a hydrogen to the $CN^-$ ion, hence it is functioning as a Brønsted acid and the $CN^-$ ion is functioning as a Brønsted base.

39. (B)

The equation for the reaction is

$$Cu^{2+} + 2e^- \rightarrow Cu$$

From Faraday's Law it is known that 96,500 coulombs will reduce 1 equivalent weight of a substance at the cathode. Since

1 coulomb = 1 ampere x seconds,

the number of coulombs passing through the solution is calculated by converting 1 hour to seconds and multiplying by the number of amperes.

$$\text{coulombs} = 3.0 \text{ amp} \times 3600 \text{ sec}$$
$$= 10{,}800 \text{ coulombs}$$

In the reaction copper is undergoing a 2 electron reduction so the equivalent weight of copper is

$$\text{eq wt of Cu} = \frac{\text{MW}}{\text{change in \# of } e^-}$$

$$= \frac{63.5}{2} = 31.8 \text{ g/eq wt}$$

Knowing 96,500 coulombs will deposit 1 eq wt. of copper or 31.8 g in this reduction we can calculate the weight of copper that results when 10,000 coulombs are passed through the solution:

$$\frac{96{,}500 \text{ coul}}{10{,}800 \text{ coul}} = \frac{31.8 \text{ g}}{x \text{ g}}$$

$$x = 3.56$$

40. (A)

For first order kinetics we can evaluate the rate constant from

$$T_{1/2} = \frac{.693}{k}$$

$$k = \frac{.693}{T_{1/2}} = \frac{.693}{3.64} = .190 \text{ day}^{-1}$$

The ratio of initial concentration of the Rn, $A_o$, to the concentration, A, after time T is given by

$$\log \frac{A_o}{A} = \frac{kT}{2.303}$$

Substituting the values of k and T

$$\log \frac{A_o}{A} = \frac{.190\ \text{days}^{-1}\ 18.2\ \text{days}}{2.303} = \frac{.345}{2.303}$$

$$\log \frac{A_o}{A} = 1.50$$

Taking the anti log of both sides

$$\frac{A_o}{A} = 31.7$$

and since the initial concentration, $A_o$ was 2.0 μg we can calculate the final concentration:

$$A = \frac{A_o}{31.7} = \frac{2.0\ \mu g}{31.7} = .06\ \mu g$$

41. (E)

In a saturated solution solid $Ag_2CrO_4$ is in equilibrium with its ions as shown by

$$Ag_2CrO_4 \leftrightarrow 2Ag^+ + CrO_4^{2-}$$

The $CrO_4^{-2}$ concentration is given as $1.30 \times 10^{-4}$. To calculate the $K_{sp}$ for $Ag_2CrO_4$ it is necessary to find the $Ag^+$ concentration. From the equation above it can be seen that in the dissolution of $Ag_2CrO_4$ will produce two moles of $Ag^+$ for each mole of $CrO_4^{2-}$. Hence if the $CrO_4^{2-}$ ion concentration is $1.3 \times 10^{-4}$ the $Ag^+$ must be twice that amount or $2.60 \times 10^{-4}$ M.

The expression for the $K_{sp}$ of $Ag_2CrO_4$ is:

$$K_{sp} = [Ag^+]^2 \quad [CrO_4^{2-}]$$

Substituting the concentration of both $Ag^+$ and $CrO_4^{2-}$

$$\begin{aligned} K_{sp} &= {2.60 \times 10^{-4}}^2 \quad 1.3 \times 10^{-4} \\ &= 6.76 \times 10^{-8} \quad 1.3 \times 10^{-4} \\ &= 8.8 \times 10^{-12} \end{aligned}$$

42. (C)

1 mole of atoms contains Avogadro's number of particles, $6.02 \times 10^{23}$. Since the atomic weight (the weight of one mole of hydrogen atoms) is 1.00 g, the weight of one atom of hydrogen can be calculated by:

$$\frac{1 \text{ atom of H}}{6.02 \times 10^{23} \text{ atoms/mol}} = \frac{X \text{ g}}{1.00 \text{ g/mol}}$$

Solving for X,

$$X = \frac{1 \text{ atom} \times 1 \text{ g/mol}}{6.02 \times 10^{23} \text{ atoms/mol}}$$

$$X = 1.66 \times 10^{-24} \text{ g}$$

43. (E)

Atoms or molecules that contain one or more unpaired electrons are said to be paramagnetic. Paramagnetic material will be weakly attracted into a magnetic field. Atoms in which all the electrons are paired are referred to as diamagnetic and will be weakly repelled from a magnetic field.

To solve this problem it is necessary to examine the electron configuration of each species and determine if it contains unpaired electrons. The configurations are:

| | |
|---|---|
| he | $1s^2$ |
| Be | $1s^2 2s^2$ |
| $Cl^-$ | $1s^2 2s^2 2p^6 3s^2 3p^6$ |
| $F^-$ | $1s^2 2s^2 2p^6$ |
| Li | $1s^2 2s^1$ |

From the above configurations it can be seen that only Li has an unpaired electron and is therefore paramagnetic.

The others all contain outer levels that are completely filled, and all electrons are therefore paired.

44. (B)

The molarity, M, is defined as the number of moles of solute per liter of solution.

$$M = \frac{\text{moles}}{V_l}$$

Rearranging and solving for volume, we have $V = \frac{\text{moles}}{M}$

Since the problem is asking for the number of grams it is necessary to substitute g/MW for the number of moles.

$$V = \frac{\frac{G}{MW}}{M} = \frac{\frac{2.3\text{ g}}{58.5\text{ g/mol}}}{1.3\text{ mol/l}}$$

$$V = .302\text{ l or }30.2\text{ ml}$$

45. (D)

According to Dalton's Law of partial pressures the total pressure is equal to the sum of the partial pressures ...

$$P_T = Pa + Pb + Pc + ...$$

In this case there are two gases present in the container, the oxygen and the water vapor. Since the vapor pressure of water is 19 torr at 25°C, the partial pressure of the water vapor is 19 torr and the total pressure in the container is 760 torr. From Dalton's Law we can calculate the pressure due to just the oxygen.

$$P_{total} = Po_2 + P_{H_2O\text{ vapor}}$$

$$760\text{ torr} = Po_2 + 19\text{ torr}$$

$$Po_2 = 760\text{ torr} - 19\text{ torr}$$

$$= 741\text{ torr}$$

46. (C)

Physical properties of solutions that only depend on the **number** of particles that are present, and not the **kind** of particles that are present are referred to as colligative properties.

It has been shown that a non-volatile solute in a solution will retard the rate of excape of solvent molecules in the solution. The total number of solvent molecules per unit area on the surface is reduced because of the presence of the solute. This results in a lowering of the vapor pressure of the solution. The lowering of the vapor pressure depresses the freezing point of the solution from that of only the pure solvent. The vapor pressure lowering is Raoult's Law, which states that the vapor pressure of a solvent in a solution decreases as the mole fraction of the solution decreases. The molecular weight of a solute does not effect the total number of particles in a solution. For example, 1 mole of NaCl and 1 mole of KCl will both produce the same number of particles when dissolved in a solvent, but NaCl and KCl have different molecular weights.

47. (B)

Buffer solutions contain conjugate acid – base pairs. They are usually prepared by mixing either a weak acid and a salt of the weak acid or a weak base and a salt of the weak base. In the answers given, only (B), .2M NaCl and .1M HCl, does not meet the requirement of a buffer. HCl is a strong acid. The others given are either a weak acid and a salt of the weak acid, or a weak base and a salt of the weak base.

48. (A)

Acetic acid ionizes,

$$HC_2H_3O_2 \leftrightarrow H^+ + C_2H_3O_2^-$$

and sodium acetate completely dissociates,

$$NaC_2H_3O_2 \rightarrow Na^+ + C_2H_3O_2^-$$

The solution is a buffer solution since it is composed of a weak acid and a salt of the weak acid. The hydrogen ion concentration of such a buffer can be calculated from the expression:

$$H^+ = \frac{[\text{acid}]}{[\text{salt}]} \times K_i$$

$$[H^+] = \frac{.05}{.1} \times 1.8 \times 10^{-5}$$

$$= .90 \times 10^{-5} \text{ or } 9.0 \times 10^{-6}$$

49. (E)

The mole fraction of a component in a solution is the number of moles of that component divided by the total number of moles,

$$X_{CH_3OH} = \frac{\text{moles of } CH_3OH}{\text{total number of moles present in solution}}$$

The number of moles of each component:

$$\text{moles } H_2O \ \frac{53.0}{18.0} = 2.90 \quad \text{moles } CH_3OH = \frac{20.3}{32.0} = .63$$

$$\text{moles } CH_3CH_2OH = \frac{15.0}{46.0} = .32$$

Therefore the mole fraction of $CH_3OH$ is:

$$X_{CH_3OH} = \frac{.63}{2.90 + .63 + .32}$$

$$= \frac{.63}{3.85} = .16$$

50. (B)

The relationship between $K_p$ and Gibb's free energy, $\Delta G°$, is given by the expression

$$\Delta G = -2.303 \ RT \log K_p$$

Where R is the ideal gas law constant, 8.314 expressed in joules/mole and T is the temperature expressed in degrees Kelvin.

$$\Delta G = -(2.303)(8.314)(298) \log 1 \times 10^{10}$$

$$\Delta G = -5706 [\log 1 + \log 10^{10}]$$

$$= -5706 [0 + 10]$$

$$= -57060 \text{ J} \approx 57.1 \text{ kJ}$$

51. (B)

$K_p$ is the notation for the equilibrium constant when the concentrations of both the reactants and products are expressed in terms of their partial pressures. The $K_p$ is defined for this reaction as

$$K_p = \frac{P^2 CO_2}{P_{CO}}$$

By convention concentrations and partial pressures for pure liquids and solids are omitted from the equilibrium constant expression. Therefore,

$$K_p = \frac{7.6^2}{3.2} = 18.1 \text{ atm}$$

52. (C)

The value of the azimuthal quantum number, l, may have values 0, 1, 2 up to a maximum of n – 1, where n is the principle quantum number. The l value tells the subshell or orbital that the electron is in.

| | |
|---|---|
| When l = 0 | the electron is in an s orbital |
| l = 1 | the electron is in a p orbital |
| l = 2 | the electron is in a d orbital |

The only p orbital that is shown is in diagram (C), which represents the $p_z$ orbital since the lobes are going through the z axis.

| | |
|---|---|
| Answer A | is an s orbital |
| B | is the $d_{x2-y2}$ orbital |
| D | i the $d_{xy}$ orbital |

The correct answer is (C).

53. (D)

All of the hydrogen halides are colorless gases that dissolve in water to give acid solutions. HCl, HBr, and HI ionize extensively and are strong acids. However HF only ionizes slightly in water. In order for ionization to occur the hydrogen – halogen bond must be broken:

$$HX \leftrightarrow H^+ + X^-$$

The H – F bond is considerably stronger than either the bonds in H – Cl, H – Br or H – I, and ionization is more difficult. In addition, when HF ionizes,

$$HF \leftrightarrow H^+ + F^-$$

a second ionization reaction also takes place due to hydrogen bonding. This results in the formation of the hydrogen difluoride ion:

$$F^- + HF \leftrightarrow HF_2^-$$

This hydrogen bonding effectively ties up some of the HF molecules and inhibits their dissociation.

54. (E)

$KClO_3$ readily decomposes to liberate $O_2$ by the reaction

$$KClO_3 \xrightarrow{\Delta} KCl + O_2$$

Before this problem can be solved, the equation must be balanced:

$$2KClO_3 \rightarrow 2KCl + 3O_2$$

Form the balanced equation it can be seen that for every 3 moles of oxygen that are produced, 2 moles of $KClO_3$ are required. It is necessary to determine the number of moles of $O_2$ in 3.20g:

$$\text{moles} = \frac{3.20\text{g}}{32.0\text{ g/mol}} = .10\text{ mol}$$

If 2 moles of $KClO_3$ will produce 3 moles of oxygen, then the amount of $KClO_3$ required can be determined:

$$\frac{2KClO_3}{3O_2} = \frac{X\ KClO_3}{.10\ O_2}$$

$$X = .07 \text{ mol of } KClO_3$$

55. (B)

Ionzation energies, I, of an atom is the amount of energy required to remove an electron from the gaseous atom. Electrons that are loosely held by the nucleus will have low ionization energies, consequently the ionization energies decrease as the size of the atoms increase. The size of atoms decrease moving from right to left on the periodic chart due to the increase in the effective nuclear charge. In the series given, Ar would have the highest first ionization potential. It has the stable noble gas configuration and is the smallest element in the series.

56. (C)

At the point of neutralization the number of equivalent weights of the acid is exactly equal to the number of equivalent weights of the base present.

Since, $\text{Normality} = \frac{\#\text{ of eq wt}}{V_L}$

then, $\#\text{ of eq wt} = N \times V_L$

and at the neutralization point,

$$\#\text{ eq wt}_{Acid} = \#\text{ eq wt}_{Base}$$

Therefore

$$V_B \times N_B = V_A \times N_A$$

$$V_B = \frac{V_A \times N_A}{N_B} = \frac{50ml \times 2N}{0.5}$$

$$V_B = 20ml$$

57. (B)

The Ideal gas law equation, $PV = nRT$ derived from kinetic and molecular theory neglects two important factors in regard to real gases. First it neglects the actual volume of gas molecules, and secondly, it does not take into account the intermolecular forces that real gases exhibit.

The factor b in the Van der Waal equation is a correction for the actual volume of the gas molecules and the factor a is a correction of the pressure due to the intermolecular attractions that occur in real gases.

58. (B)

To determine what is being oxidized and reduced the oxidation states of each species must be determined.

$Fe^{2+}$ is being oxidized to $Fe^{3+}$.

There is no change in either the oxygen or hydrogen, being –2 and +1 respectively on both sides of the equation.

Examining the nitrogen, we see it changes from +5 in $NO_3^-$ to + 4 in $NO_2$.

Therefore the $Fe^{2+}$ is being oxidized and nitrogen in $NO_3^-$ is being reduced.

59. (B)

Thioacetamide $CH_3CSNH_2$ will hydrolize in aqueous solutions according to the following equation:

$$H_3CSNH_2 + H_2O \rightarrow CH_3COO^- + NH_4^+ + H_2S$$

to produce hydrogen sulfide, $H_2S$, which is used as a reagent to form characteristic metal sulfides that are used in identifying the metal ion listed.

60. (B)

The standard free energy for an electrochemical cell is related to the standard electrode potential E° by:

$\Delta G^\circ$ = $-nFE^\circ$

– where n is the number of electrons involved in the oxidation - reduction reaction.
– $E^\circ$ is the standard electrode potential
– F is Faraday's constant, 96.487 kj/V

The value of $\Delta G^\circ$ is then equal to

$$\Delta G^\circ = -2 \times 96.487 kj/V \times .28V$$
$$\Delta G^\circ = -54 kj$$

61. (D)

Salts of weak acids and strong bases will hydrolize to produce basic solutions whereas salts of weak bases and strong acids will hydrolize to form acid solutions. Both $CaCl_2$ and NaCl are salts formed from the reaction of strong acids with strong bases and will not hydrolize, i.e.,

$$HCl + NaOH \rightarrow NaCl + H_2O$$

Sodium acetate $CH_3COONa$ is a salt of a weak acid and a strong base:

$$CH_3COOH + NaOH \rightarrow CH_3COO^-NA^+ + H_2O$$

The acetate ion, $CH_3COO^-$ will hydrolize;

$$CH_3COO^- + HOH \leftrightarrow CH_3COOH + OH^-$$

forming $OH^-$ ions which are basic. The salt of a weak base and a strong acid, will hydrolize to form an acid solution. $NH_4Cl$ is such a salt and the $NH_4^+$ ion will hydrolize:

$$NH_4^+ + HOH \quad H_3O^+ + NH_3$$

forming the Hydronium ion resulting in an acid solution whose pH is below 7.

62. (B)

In the net ionic equation only ions in solution that appear unchanged on both sides of the equation are omitted. If an ion changes in any way during the reaction it must be shown in the net ionic equation. In the reaction the chloride ions are not oxidized or reduced, and do not undergo any changes and are therefore omitted. On the other hand, the Fe in $FeCl_2$ is oxidized to $Fe^{3+}$ in $FeCl_3$ and must be shown in the net ionic equation as $Fe^{2+}$ going to $Fe^{3+}$. Also, the Mn in $MnO_4^-$ changes from $Mn^{7+}$ to $Mn^{2+}$ and must also be shown in the net ionic equation. Since the reaction is taking place in aqueous acid solution the $H^+$ ion in HCl is shown on the reactant side of the equation, and the water is shown on the product side. Both the reactants and products are soluble in water and they are written in their ionic form in the net ionic equation, e.g. $KMnO_4$ is written as $MnO_4^-$. In view of these changes in oxidation states, the only correct net ionic equation is choice (B).

63. (A)

The only commonality among the ions listed is they all contain 10 electrons and have the same election configuration

$$1s^2 2s^2 2p^6$$

Ions or atoms that have the same electron configurations are said to be isoelectronic.

64. (B)

Valence Shell Electron Pair Repulsion (VSEPR) is a convenient method of predicting molecular geometries based on the fact that electron pairs, either bonded pairs or lone pairs tend to orient themselves as far from one another as possible to reduce electron pair - electron pair repulsion. Iodine has seven valence electrons and since the charge on the ion, $ICl_4^-$ is negative, it is viewed as having eight electrons associated with it. Of the eight electrons 4 are involved in bonding (4 shared electron pairs) leaving 4 electrons or 2 pairs that are non-bonding):

total electron pairs = 2 (lone pairs) + 4 (bonding pairs) = 6

The possible structure for the arrangement of the six pairs of electrons (4 bonding pairs and 2 lone pairs) is then either (A) or (B).

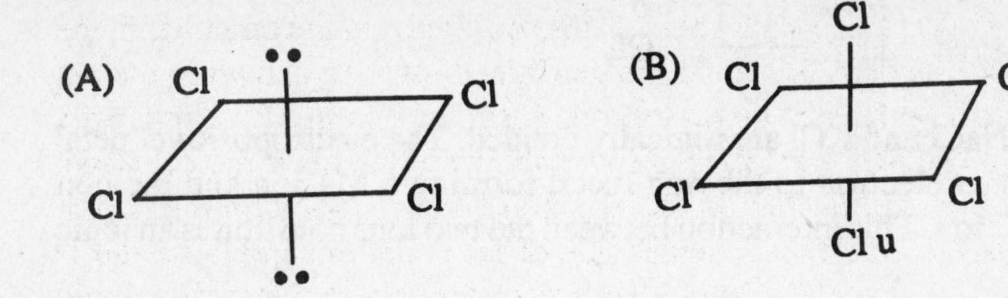


To chose the correct structure we evaluate the lone pair – lone pair interaction, and then the lone pair – bonding pair interaction.

| Structure (A) | Structure (B) |
|---|---|
| 0 LP – LP @ 90° | 1 LP – LP @ 90° |
| 8 LP – BP @ 90° | 6 LP – BP @ 90° |

Evaluating these interactions we see that structure (B) has a lone pair – lone pair interaction at 90°. This is a highly unfavorable situation and is not expected to occur. Hence structure (A) is favored. The atoms in $ICl_4^-$ are all in a plane and the geometry of the molecule is best described as square planar.

Note: Although structure (B) has two less lone pair – bonding pair interactions it is not sufficient to overcome the repulsion of the lone pair – lone pair interactions at 90°.

65. (B)

A polar covalent molecule must contain atoms of different electronegativities. Immediately it is recognized that $H_2$ and $Cl_2$ are not polar molecules. For a covalent bond to form there must be a sharing of electron pairs. Both hydrogen and chlorine will form covalent bonds by the sharing of an electron pair.

$$H:\ddot{\underset{\cdot\cdot}{Cl}}:$$

In doing so each element has achieved a stable inert gas configuration. Since chlorine is more electronegative than hydrogen it will develop a partial negative charge with respect to hydrogen. HCl is a polar covalent molecule.

$$\overset{\delta +}{H} \text{ —— } \overset{\delta -}{CL}$$

Both NaCl and KCl are ionically bonded. The electropositive metal donates an electron to the non metal forming a M+ ion and the non metal $X^-$ ion. This interaction between the two ions constitutes an ionic bond,

$K^+ \ Cl^-$ or $Na^+ \ Cl^-$

66. (A)

$PbI_2$ dissolves in water to form a saturated solution according to the equation:

$$PbI_{2(s)} \leftrightarrow Pb^{2+} + 2I^-$$

The expression for the $K_{sp}$ is

$$K_{sp} = [Pb^{2+}] + [I-]^2 = 8.7 \times 10^{-9}$$

The molar concentration of $Pb^{2+}$ will be equal to the molar concentration of $PbI_2$ since each mole of $PbI_2$ that dissolves produces the same number of moles of $Pb^{2+}$ ions. Hence if the concentration of $Pb^{2+}$ will also equal the molar solubility of $PbI_2$. If x equals the $Pb^{2+}$ concentration the $I^-$ concentration is then 2x. Substituting into the $K_{sp}$ expression:

$$[x]\ [2x]^2 = 8.7 \times 10^{-9}$$
$$4x^3 = 8.7 \times 10^{-9}$$
$$x = 1.3 \times 10^{-3}$$

67. (C)

Hydrides are compounds which contain hydrogen in the minus one oxidation state, $H^-$. They are formed when hydrogen reacts with group IA elements or the more active elements of Group IIA, Ca, Sr and Ba.

$$2\,Li + H_2 \rightarrow 2\,LiH$$

The other compounds listed all contain hydrogen in the plus one oxidation state and are not hydrides. In the case of LiH the hydrogen is in the minus one oxidation state since Li can only be in the plus one oxidation state. LiH is the only hydride of the possible choices, hence the correct answer is (C).

68. (B)

A catalyst increases the rate of a chemical reaction by lowering the activation energy. The activation energy of a reaction is the amount of energy the molecules or atoms must have, so that when they collide the collision will result in product formation. A lower activation energy means there will be a greater number of molecules possessing sufficient kinetic energy to form products. This will result in a greater reaction rate.

69. (B)

The first step in balancing this equation is to assign the oxidation numbers of each reactant and product:

$$\overset{+7}{MnO_4^-} + \overset{-1}{Cl^-}\ H+ \rightarrow \overset{+2}{Mn} + \overset{0}{Cl_2} + H_2O$$

Therefore: $Mn^{7+} \rightarrow Mn^{2+}$ 5 electron change
$2Cl^- \rightarrow Cl_2^{\circ}$ 2 electron change

To balance the number of electrons in the oxidation with the number in the reduction we multiply the manganese by 2 and the chloride by 5 and place these coefficients in the equation as shown below:

$$2MnO_4^- + 10Cl^- + ?H^+ \rightarrow 2Mn^{2+} + 5Cl_2 + ?H_2O$$

Next the charges on both sides of the equation are balanced by adding H+ ions to the left side. The charges are:

| Left | Right |
|---|---|
| (2x) (– 1) + (10x) (– 1) | (2x) (+ 2) |
| or | |
| –12 | +4 |

Therefore to balance the charges on each side we add 16 $H^+$ to the left.

$$2MnO_4^- + 10Cl^- + 16H^+ \rightarrow 2\,Mn^{2+} + 5Cl_2 + ?H_2O$$

Now 8 moles of $H_2O$ must be added to the right side, to balance the hydrogens on the left side. The oxygens should now also be balanced and the correct balanced equation is then:

$$2MnO_4^- + 10Cl^- + 16H^+ \rightarrow 2\,Mn^{2+} + 5Cl_2 + 8H_2O$$

70. (D)

Amphoteric compounds can function as either acids or bases. Hydroxides of Group I metals form solutions that are strongly basic when dissolved in water. Hydroxides of Group II metals although less soluble than Group I hydroxides also form basic solutions when dissolved in water. Hydroxides of metals that are of intermediate electronegativity and in relatively high oxidation states are usually amphoteric. $Al(OH)_3$ can function either as an acid or a base:

$$Al(OH)_3 + 3OH^- \rightarrow AlO_3^{3-} + 3\,H_2O \quad \text{(as an acid)}$$

or

$$Al(OH)_3 + 3H^+ \rightarrow Al^{3+} + 3H_2O \quad \text{(as a base)}$$

The metal hydroxides that exhibit amphoteric behavior are generally found along the diagonal of the periodic table that separates the metals from the non-metals.

71. (A)

The third law of thermodynamics states that the entrophy of a perfect cyrstal at 0°K is zero. Entropy is the randomness or disorder in a system. The greater the disorder, the higher is its entropy. In the choices above, both $H_2$ and He are gases. The molecules are highly disordered and the entropy would be high. In liquid water, the entropy would be lower than the gases but the molecules are essentially random. In the solid, NaCl, the molecules are held rigid in the cyrstal structure therefore it would be a more structured environment or less random, hence the entropy is low.

72. (A)

The standard heat of reaction, $\Delta H°$ is equal to the sum of the heats of the products minus the sum of the heats of the reactants:

$$\Delta H° = \Sigma_{\text{heats of products}} - \Sigma_{\text{heats of reactants}}$$

$$\Delta H° = (2(\Delta H°_{NAOH}) + \Delta H° H_2)) - (2(\Delta H° H_2O) + 2(\Delta H°_{Na}))$$

By convention the standard heats of formation of an element in its most stable form is zero. Thus both $H_{2(g)}$ and $Na_{(s)}$ equal zero, then:

$$\begin{aligned} \Delta H° &= 2(-426.7) - 2(-285.8)\ * \\ &= -853.4 - (-571.6) \\ &= -853.4 + 571.6 \\ &= -281.8 \text{ kj} \end{aligned}$$

***Note:** the enthalpies given in the table above are in units of kj/mol, therefore it is necessary to multiply by the number of moles of each substance as represented in the balanced equation to calculate the $\Delta H°$ for the overall reaction.

73. (C)

$K_c$ is the equilibrium constant when the concentrations are expressed in moles per liter. Kp is the equilibrium constant where the partial pressures of the gases is used in place of the molar concentrations. The relationship between $K_p$ and $K_c$ is:

$$K_p = K_c(RT)^{\Delta n}$$

where $\Delta n$ is the change in the number of moles of the gas upon going from reactants to products. The correct answer is (C), since there is no change in the number of moles in going from reactants to products. Thus $\Delta n = 0$ and $K_p$ will then be equal to $K_c$.

74. (E)

A weak electrolyte is one that when dissolved in water produces a low percentage of ions. Most of the material will remain in its undissociated form. NaCl, HCl, KI, and $HNO_3$ all dissociate completely when dissolved in water and are strong electrolytes. $NH_4OH$ on the other hand only ionizes to a very small extent according to the equation:

$$NH_4OH \leftrightarrow NH_4^+ + OH^-$$

where most of the $NH_4OH$ will remain in the undissociated form.

75. (B)

A phase diagram relates the pressure and the temperatures at which the gaseous, liquid and solid states of a material can exist. In the question above we are interested in the relationship between the temperature at which the material changes from a solid to a liquid, which is the melting point, and the pressure. This corresponds to the line indicated by b – c. It is apparent that this line slants to the right which shows that an increase in the pressure will cause an increase in the temperature at which this material melts.

**Note:** The point $T_c$ is called the triple point, and at that pressure and temperature the solid, liquid and gaseous states are in equilibrium with each other.

76. (A)

The Nernst equation,

$$E = E^\circ - \frac{.059}{n} \log Q$$

relates the cell potential E to the standard electrode potential, as a function of concentration of the species present, where

– E° is the standard cell potential at 25°C and 1 atm pressure and the molar concentrations of the ions is 1.

– n is the change in the number of electrons in the half reaction, $Sn^{4+} + 2e \rightarrow Sn^{2+}$, or in this case n = 2.

– Q is the reaction quotient which has the same form as the equilibrium constant K for the reaction. In this problem

$$Q = \frac{[Sn^{2+}]}{[Sn^{4+}]}$$

To calculate the electrode potential when the $Sn^{2+}$ concentration is five times the $Sn^{4+}$ concentration, Q is then equal to 5/1.

$$E = .150 - \frac{.059}{2} \log \frac{5}{1}$$

$$E = .150 - \frac{.059}{2} (.6989)$$

$$= .150 - .021 = .129 \text{ V}$$

77. (B)

When a compound exists in different geometric forms, each of which has the same molecular formula, the relationship between the forms are referred as geometric isomers. The only possible isomers of the compounds and geometries given are the cis and trans geometric isomers of $Pt(NH_3)_2Cl_2$. Writing the possible structural arrangements of the groups coordinated to the platinum we find there are only two distinct possibilities:

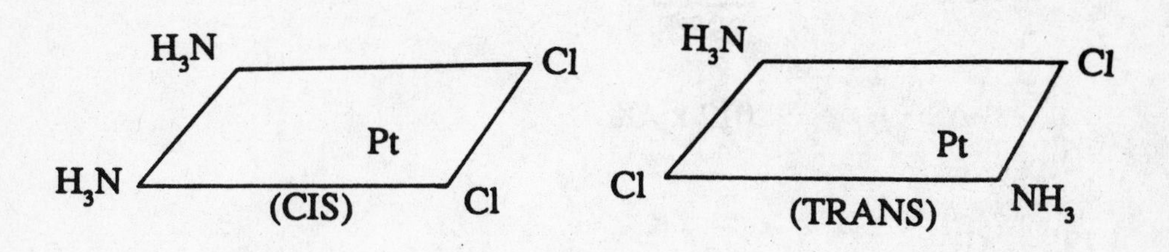


The one on the left is referred to as the cis isomer in which the two similar groups are adjacent to each other. Whereas in the structure on the right the similar groups are opposite each other, which is referred to as the trans isomer. For the other compounds listed, no geometric isomers exist since there exists only one possible structural arrangement of the groups around the metal.

78. (B)

$\Delta S°$ is the entropy change the reaction undergoes when the products and reactants are in their standard state. Entropy is a measure of the randomness or the amount of disorder in the system. A positive value of $\Delta S$ indicates an increase in the randomness of the system whereas a negative entropy indicates the system is becoming more ordered. To solve this problem we recognize that the entropy $\Delta S°$ is related to the $\Delta H°$ and $\Delta G°$ of the reaction by the expression

$$\Delta G° = H° - T\Delta S°$$

Substituting the values for $\Delta H°$ and $\Delta G°$ and solving for $\Delta S°$:

$$-\Delta S° = \frac{\Delta G° - \Delta H°}{T}$$

$$-\Delta S° = \frac{88.4 \text{ kj} - 92.0 \text{ kj}}{298°K}$$

$$-\Delta S° = \frac{-3.6 \text{ kj}}{298°K}$$

$$\Delta S° = .012 \text{ kj/°K}$$

79. (B)

When $Br_2$ reacts with unsaturated compounds it readily adds across the unsaturated bonds placing a bromine atom on each carbon.

$$\underset{\text{(red)}}{R - C{=}C - R + Br_2} \rightarrow \underset{\text{(colorless)}}{R - \underset{Br}{\underset{|}{C}} - \underset{Br}{\underset{|}{C}} - R}$$

This reaction is commonly employed as a test for the presence of double bonds in an unknown compound. As the reddish $Br_2$ solution reacts with the double bonds, the color of the $Br_2$ dissipates. The only compound listed above that is unsaturated is

$$CH_2 = CH - CH_2 - CH_2 - CH_3.$$

80. (C)

A chiral carbon, or a point of asymmetry, is a carbon which has four different groups attached to it. If a molecule contains a chiral carbon it can exist in two distinct forms that are non-superimposable mirror images. One form will rotate the plane polarized light to the right whereas the other form will rotate the plane polarized light to the left. The two distinct forms are said to be optically active and constitute what is referred to as a pair of enantiomers. The only compound listed in the answers that contains a chiral carbon is

$$CH_3 - \overset{\displaystyle OH}{\underset{\displaystyle H}{\overset{|}{\underset{|}{C}}}} - CH_2CH_3$$

(chiral carbon)

81. (C)

Metals of group IA and IIA are active metals. Active metals will react directly with liquid water or steam displacing hydrogen from the water and forming metal hydroxides. The only active metal listed is sodium, Na. When Na is added to water, the following reaction will take place:

| $2Na$ | + | $2H_2O$ | → | $H_2$ | + | $NaOH$ |
|---|---|---|---|---|---|---|
| active metal | | water | | hydrogen gas | | sodium hydroxide (base) |

Both Pb and Sn are relatively inactive metals and will not react with water to liberate hydrogen. They will however displace hydrogen from an acid liberating $H_2$:

$$Pb + HCl \rightarrow PbCl_2 + H_2$$

82. (A)

Fluorine, chlorine, and bromine react directly with hydrogen to produce the corresponding hydrogen halides, HF, HCL or HBr. When the hydrogen halides are dissolved in water they form acid solutions according to the following reaction:

$$HCL + H_2O \rightarrow H_3O^+ + Cl^-$$

The metals listed will not react with hydrogen to produce compounds that will form acid solutions when dissolved in water. In the case of sodium, it will react with $H_2$ to form a hydride

$$2Na + H_2 \rightarrow 2NaH$$

which will dissolve in $H_2O$ to produce a basic solution.

$$NaH + H_2O \rightarrow NaOH + H_2$$

Ne is a noble gas and does not react with hydrogen.

83. (B)

Le Chatelier's principle states that if stress is applied to a system at equilibrium the position of equilibrium will shift in a direction to reduce the effects of the stress. Stress can be a change in concentration, pressure or temperature. If $NH_3$ is added to the system the position of equilibrium will be shifted to the left to reduce the stress caused by the change in concentration of $NH_3$.

84. (C)

An increase in pressure will cause the position of equilibrium to be shifted to the side with the least number of moles. On the right side of this equation there are 2 moles of $NH_3$ whereas on the left side of the equation there is 1 mole of $N_2$ and 3 moles of $H_2$. Therefore an increase of pressure will shift the position of equilibrium to the right side.

**Note:** Le Chatelier's principle can be used to predict the factors that will enhance product formation in commercial processes. For example, in the reaction if the $NH_3$ is removed as it is being formed the equilibrium will be shifted to the right, and more $H_2$ and $N_2$ will react to attempt to re-establish equilibrium. The net result of this is the production of more $NH_3$.

85. (D)

The percent ionization of a weak acid is the molar $H_3O^+$ ion concentration divided by the molar concentration of the undissociated acid. In this case:

$$\% \text{ ionization} = \frac{H_3O^+}{HC_2H_3O_2} \times 100$$

To solve this problem the pH of 3.5 must be converted to $H_3O^+$ concentration.

Since, $pH = -\log [H_3O^+]$

$3.5 = -\log [H_3O^+]$

$H_3O^+ = \text{antilog of } -3.5$

and since, $-3.5$ is equal to $-4.0 + .5$

we write, $\log [H_3O^+] = +.5 - 4$

$H_3O^+ = \text{antilog of } .5 - 4$

$H_3O^+ = 3.16 \times 10^{-4}$

and to determine the percent,

$$\% \text{ ionization} = \frac{3.16 \times 10^{-4}}{1.0} \times 100 = 0.031\%$$

# SECTION II

## PART A

1. a. The generalized expression for the rate law is written as:

rate = $k[A]^x [B]^y [C]^z$

From the data given we can evaluate the exponents, x, y, and z.

Examining runs 1 and 4 it is observed that the concentration of A and C are held constant, hence any change in the rate is the result of a concentration change in B.

Comparing the rates of run 4 to run 1, we see that

$$\frac{Rate_4}{Rate_1} = \frac{k[B]_4^y}{k[B]_1^y} = \frac{k}{k}\left(\frac{[B]_4}{[B]_1}\right)^y$$

$$\frac{6 \times 10^{-8}}{6 \times 10^{-8}} = \left(\frac{0.15}{0.10}\right)^y$$

$$1 = 1.5^y$$

and y must equal 0, since any number to the zero power equals 1, hence the reaction is zero order in B.

Next, examining runs 1 and 3 and neglecting the concentration of B since the reaction is zero order in B, we observe that by holding A constant and doubling C, there is a two-fold increase in the rate.

$$\frac{Rate_3}{Rate_1} = \frac{k}{k}\left(\frac{[C]_3}{[C]_1}\right)^z$$

$$\frac{1.2 \times 10^{-7}}{6.0 \times 10^{-8}} = \left(\frac{.20}{.10}\right)^z$$

$$2 = 2^z$$

Therefore z must equal 1 and the reaction is first order in C.

To find the order for A we observe that in runs 1 and 2, C is held constant, and we already know the reaction rate is not dependent on B. We observe a doubling of A increases the rate by a factor of 4.

$$\frac{rate_2}{rate_1} = \frac{k}{k}\left(\frac{[A]_2}{[A]_1}\right)$$

$$\frac{2.4 \times 10^{-7}}{6.0 \times 10^{-8}} = \frac{k}{k}\left(\frac{.20}{.10}\right)$$

$$4 = 2^x$$
$$2^z = 2^x$$
$$x = 2$$

The reaction is thus second order in A and the rate law for this reaction is

$$rate = k[A]^2 [B]^0 [C]^1$$

or

$$rate = k[A]^2 [C]$$

b. To calculate the rate constant k we may substitute the rate for any particular experiment. For example in run 1

$$rate = k[A]^2 [C]$$

$$k = \frac{rate_1}{[A]^2 [C]}$$

$$k = \frac{6.0 \times 10^{-8} M/s}{[.1M]^2[.1M]}$$

$$\log \frac{{}^kT_2}{{}^kT_1} = \frac{Ea}{2.303\,R}\left(\frac{T_2 - T_1}{T_2 \times T_1}\right)$$

$$k = \quad 6.0 \times 10^{-5}\ M^{-2}\ sec^{-1}$$

To check this result you may calculate the rate constant k for each run and see that the value of k is actually a constant.

c. To calculate the activation energy at 35°C we use the Arrhenius equation:

$$\frac{Ea}{2.303\,R} \qquad \frac{T_2 - T_1}{T_2 \times T_1}$$

where $K_{T2}$ and $K_{T1}$ are the rate constants, at temperatures 2 and 1 (expressed in °K), and R is 8.314 j/mole °K.

The data given in problem C states that the rate constant doubles for an increase in temperature from 25°C to 35°C or 298°K to 308°K, hence the rate constant at 308°K is 2 ($6.0 \times 10^{-5} M^{-2} sec^{-1}$) or $1.2 \times 10^{-4} M^{-2} M^{-1}$.

Substituting into the Arrhenius equation we have equation

$$\log \frac{1.2 \times 10^{-4}}{6.0 \times 10^{-5}} = \frac{E_a\,(308 - 298)}{(2.303)\,(8.314)\,(308)\,(298)}$$

$$\log 2 = \frac{E_a\,(10)}{1.77 \times 10^{6}}$$

$$.3 = \frac{10\,E_a}{1.77 \times 10^{6}}$$

$$10\,E_a = 5.31 \times 10^{5}$$

$$E_a = 5.31 \times 10^{4}\ j/mole$$

## PART B

2. First we must write the three steps in the ionization process:

Step 1 $H_3PO_{4\,(aq)} \leftrightarrow H^+_{(aq)} + H_2PO_4^-{}_{(aq)}$ $K_1 = 7.0 \times 10^{-3}$

Step 2 $H_2PO_4^-{}_{(aq)} \leftrightarrow H^+_{2(aq)} + HPO_4^{-2}{}_{(aq)}$ $K_2 = 6.0 \times 10^{-8}$

Step 3 $HPO_4^{-2}{}_{(aq)} \leftrightarrow H^+_{3(aq)} + PO_4^{-3}{}_{(aq)}$ $K_3 = 4.0 \times 10^{-13}$

Examining the equilibrium expression, it is obvious that there are three ionization steps that produce $H^+$. The total $H^+$ concentration will be the sum of the $H^+$ in each step.

To calculate the $H^+$ in step 1, we write the expression for the equilibrium constant $K_1$.

$$K_1 = \frac{[H^+]\,[H_2PO_4^-]}{[H_3PO_4]} = 7.0 \times 10^{-3}$$

Substituting the original $H_3PO_4$ concentration .05M, hence the equilibrium concentration will be .05M minus x, the amount that reacts.

$$[H_3PO_4] = 0.05 - x$$

Therefore, both the $H^+$ and $H_2PO_4^-$ ion concentrations will also be x, the amount that dissociates

$$[H^+] = x$$

$$[H_2PO_4^-] = x$$

substituting into the expression

$$\frac{(x)\,(x)}{(0.05-x)} = 7.0 \times 10^{-3}$$

We could simplify this expression if we could drop the x in the denominator. As a rule of thumb if the value of $K_1$ is equal to or smaller than $1 \times 10^{-6}$, the x may be neglected since the total amount of $H^+$ is small compared to the molar concentration of the undissociated acid.

In this case we cannot neglect the x in the term .05–x since the value of the $K_i$ is 7 x $10^{-3}$. We must solve the equation by the quadratic equation

$$x = \frac{-b \pm \sqrt{b^2 - 4ac}}{2a}$$

Thus

$$\frac{x^2}{.05 - x} = 7.0 \text{ x } 10^3$$

$$x^2 = 3.5 \text{ x } 10^{-4} - 7.5 \text{ x } 10^{-3}$$

$$x^2 + 7.5 \text{ x } 10^{-3} - 3.5 \text{ x } 10^{-4} = 0$$

$$x = \frac{-(7.5 \text{ x } 10^{-3}) \pm \sqrt{(7.5 \text{ x } 10^{-3})^2 - 4(-3.5 \text{ x } 10^{-4})}}{2(1)}$$

Solving for x we get two answers from the quadratic solution. The negative answer has no physical significance and is rejected.

Thus, $x = 2.28 \text{ x } 10^{-2}$ or .023

where x is the hydrogen ion concentration that results from the first step in the ionization of $H_3PO_4$.

The next step is to calculate the additional $H^+$ ion that results from the second ionization step:

$$H_2PO_4^- \leftrightarrow H^+ + HPO_4^{-2}$$

From the calculation of the $H^+$ in the first step, we also recognize that the $H_2PO_4^-$ ion concentration is equal to the $H^+$, or .023M.

In the expression for $K_2$

$$K_2 = \frac{[H^+]\,[HPO_4^{-2}]}{[H_2PO_4^-]}$$

The hydrogen $H^+$ concentration will be .023 plus the $H^+$ formed in the second step or $(.023 + H^+_{i2})$ where $H^+_{i2}$ is the hydrogen ion concentration produced in the second step.

The $H_2PO_4^-$ concentration will be equal to .023 minus the amount that undergoes ionization in the second step or $.023 - H^+_{i2}$.

And since the $HPO_4^-$ concentrations will be identicalto that of the hydrogen ion formed in the second ionization we substitute to just the $H^+_{i2}$. We substitute

$$\frac{[.023 + H^+_{i2}]\ [H^+_{i2}]}{[.023 - H^+_{i2}]} = 6.0 \times 10^{-8}$$

Examining the $K_2$ we see that it is small. The amount of $H^+_{i2}$ will be insignificant in comparison to .023, and we may neglect the $H^+_{i2}$ concentration in the additive terms. Thus

$$\frac{(.023)\ (H^+_{i2})}{.023} = 6.0 \times 10^{-8}$$

$$H^+_{i2} = 6.0 \times 10^{-8}$$

The last step is to calculate the $H^+$ concentration that results for the third ionization step.

$$HPO_4^{-2} \leftrightarrow H^+ + PO_4^{-3}$$

$$K_3 = \frac{[H^+]\,[PO_4^{-3}]}{[HPO_4^{-2}]}$$

The $H^+$ concentration will essentially be equal to .023 plus the amount produced in step 3 or $(.023 + H^+_{i3})$.

The $PO_4^{-3}$ concentration will be identical to the $H^+$ produced in this step or $H^+_{i3} = PO_4^{-3}$.

The equilibrium amount of $HPO_4^{-2}$ will equal the $H^+$ formed in step 2 minus the amount of hydrogen that dissociates in the third step, or $6.0 \times 10^{-8} - H^+_{i3}$.

Substituting:

$$K_3 = \frac{[.023 + H^+_{i3}][H^+_{i3}]}{[6.0 \times 10^{-8} - H_{i3}^+]} = 4.0 \times 10^{-3}$$

Both the additive and subtractive terms may be neglected since their value will be insignificant since the value of the $K_3$ is $4 \times 10^{-13}$. Neglecting these terms we have

$$\frac{(.023)(H^+_{i3})}{(6.0 \times 10^{-8})} = 4.0 \times 10^{-13}$$

$$.023 \cdot H^+_{i3} = 24 \times 10^{-21}$$

$$H^+_{i3} = 1.04 \times 10^{-18}$$

The total $H^+$ concentration for the ionization of $H_3PO_4$ would then be equal to:

$$H^+ = H^+_{i1} + H^+_{i2} + H^+_{i3}$$

$$= .023 + 6.0 \times 10^{-8} + 1.04 \times 10^{-18}$$

Note: The hydrogen ion produced in the first step of the ionization of $H_3PO_4$ is for all practical purposes very close to the total hydrogen of an equilibrium solution of $H_3PO_4$. The amounts produced in Steps II and III are insignificant to the hydrogen produced in Step I. This is apparent by examination of the values of the three equilibrium constants for the ionization of $H_3PO_4$.

3. Salts of strong bases and weak acids, e.g. $Na^+C_2H_3O_2^-$ (sodium acetate) will hydrolyze to form **basic** solutions. For example:

$$C_3H_3O_2^- + HOH \leftrightarrow C_2H_3O_2H + OH^-$$

In this case the acetate ion reacts with water to form the undissociated acid, $C_2H_3O_2H$, plus the hydroxide ion, $OH^-$.

Salts of weak bases and strong acids eg. $NH_4Cl$ (ammonium chloride), will hydrolyze to for **acidic** solutions.

$$NH_4^+ + HOH \leftrightarrow NH_3 + H_3O^+$$

In this case the ammonium ion reacted with the water to form the ammonia molecule and a hydronium ion, $H_3O^+$.

In this problem we are asked to find the hydrogen ion concentration that results when a salt of a weak acid and a weak base hydrolyze.

In this case both species are hydrolyzed, one to produce an acidic solution, the other to produce a basic solution.

$$NH_4^+{}_{(aq)} + HOH \leftrightarrow NH_3 + H_3O^+{}_{(aq)}$$
$$OCl^- + H_2O \leftrightarrow HOCl + OH^-$$

The final $[H_3O^+]$ of this solution will depend on the degree of hydrolysis that each of these species are undergoing.

We must first calculate the $H_3O^+$ ion concentration that results with $NH_4^+$ hydrolysis.

The equilibrium expression for the hydrolysis of $NH_4^+$ ion:

$$K = \frac{[NH_3]\,[H_3O^+]}{[NH_4]\,[H_2O]}$$

and since the number of moles of water involved in the reaction is small compared to the total number of moles present. The $H_2O$ concentration essentially remains constant.

$$\frac{[NH_3][H_3O^+]}{[NH_4^+]} = K[H_2O] = K_h$$

where $K_h$ is the hydrolysis constant for the reaction. The value of the $K_h$ can be determined since

$$K_w = [OH^-][H_3O^+]$$

$$[H_3O^+] = \frac{k_w}{[OH^-]}$$

and substituting into the expression for the $K_h$,

$$\frac{[NH_3]\left[\frac{K_w}{OH^-}\right]}{NH_4^+} = K_h$$

and rearranging.

$$K_h = \frac{[NH_3] \cdot K_w}{[NH_4^+][OH^-]}$$

And because

$$\frac{[NH_3]}{[NH_4^+][OH^-]}$$

is the reciprocal of the ionization constant for the weak base $K_i$.

$$K_h = \frac{K_w}{K_i}$$

Thus, knowing the value of the $K_1$ for the weak base $NH_4OH$, the $K_h$ can be calculated. To solve for the amount of $H_3O^+$ produced when $NH_4^+$ hydrolyzes, we substitute into the expression:

$$K_h = \frac{[NH_3][H_3O^+]}{[NH_4^+]} = \frac{K_w}{K_i}$$

Since the original concentration of the $NH_4^+$ ion is given as 0.1 M, and we do not know the amount of either $NH_3$ or $H_3O^+$, we substitute accordingly.

$$\frac{(x)(x)}{(0.1 - x)} = \frac{1 \times 10^{-14}}{1.8 \times 10^{-5}}$$

And since x, the amount that hydrolyzes, is relatively small in comparison to the total amount of $NH_4^+$ in the original solution, it may be neglected. Thus

$$\frac{x^2}{.1} = 5.6 \times 10^{-10}$$

$$x^2 = 5.6 \times 10^{-11}$$

$$x = 7.5 \times 10^{-6} = H_3O^+$$

Next, it is necessary to compute the $OH^-$ ion concentration when the anion of the weak acid hydrolyzes, $OCl^-$.

$$OCl^- + H_2O \leftrightarrow HOCl + OH^-$$

$$K_h = \frac{[HOCl][OH^-]}{[OCl^-]} = \frac{K_{i(\text{for HClO})}}{K_w}$$

And since the original concentration of $OCl^-$ is 0.1 M and the amount that hydrolyzes to produce $HOCl + OH^-$ is unknown, we substitute

$$\frac{[x][x]}{[.1 - x]} = \frac{1 \times 10^{-14}}{3.0 \times 10^{-8}}$$

and neglecting the x in the .1 – x expression for the same reasons as before we have:

$$\frac{x^2}{.1} = 3.3 \times 10^{-7}$$

$$x^2 = 3.3 \times 10^{-8}$$

$$x = 1.8 \times 10^{-4} \text{ molar in } OH^-$$

Thus when a 0.1 M solution of $NH_4OH$ hydrolyzes there are produced:

.0001800 moles of $OH^-$
.0000075 moles of $H_3O^+$

Hence for all practical purposes, the $OH^-$ ion concentration is 1.8 x $10^{-4}$ and from the $K_w$, the hydrogen ion concentration can be calculated:

$$K_w = [OH^-][H_3O^+] = 1 \times 10^{-14}$$

$$[H_3O^+] = \frac{1 \times 10^{-14}}{1.8 \times 10^{-14}}$$

$$[H_3O^+] = 5.5 \times 10^{-11}$$

## PART C

4. a. To construct the Born-Haber cycle for the reaction

$Na_{(s)} + 1/2\ Cl_{2(g)} \rightarrow NaCl_{(s)}$ it must be realized that the following processes are involved in the formation of the ionic crystal $NaCl_{(s)}$ from its elements.

1. The sublimation of Na metal to form gaseous Na atoms

$$Na_{(s)} \rightarrow Na_{(g)} \quad \Delta H_{sublimation}$$

2. The ionization of 1 mole of Na atoms to form 1 mole of $Na^+_{(g)}$ ions

$$Na_{(g)} \rightarrow Na^+_{(g)} \quad \Delta H_{\text{first ionization energy for Na}}$$

3. The dissociation of 1/2 of a mole of $Cl_2$ to form gaseous chlorine atoms

$$1/2\ Cl_{2(g)} \rightarrow Cl_{(g)} \quad 1/2\ \Delta H_{\text{of dissociation for } Cl_2}$$

4. The addition of 1 mole of electrons (electron affinity) to the gaseous chlorine atoms.

$$Cl_{(g)} + e^- \rightarrow Cl^-_{(g)} \quad \Delta H_{\text{electron affinity of Cl}}$$

5. The crystal lattice energy for the formation of 1 mole of solid NaCl from its gaseous metal ions

$$Na^+_{(g)} + Cl^-_{(g)} \rightarrow NaCl_{(s)} \quad \Delta H_{\text{crystal lattice energy}}$$

Thus the $\Delta H_f$ (heat of formation) for 1 mole of NaCl from its elements is the sum of ΔH of these five reactions, since Hess law states that the enthalpy of a reaction is the same whether the reaction takes place in one step or in many steps, i.e.

$$\Delta H_f = 1/2\Delta H_{\text{diss of Cl2}} + \Delta H_{\text{electron affinity of Cl}} + \Delta H_{\text{subl. of Na}}$$

$$+ \Delta H_{\text{ionization energy of Na}} + \Delta H_{\text{crystal lattice energy}}$$

From the above energy terms, it is then possible to construct the Born-Haber cycle for this process, which is usually represented as:

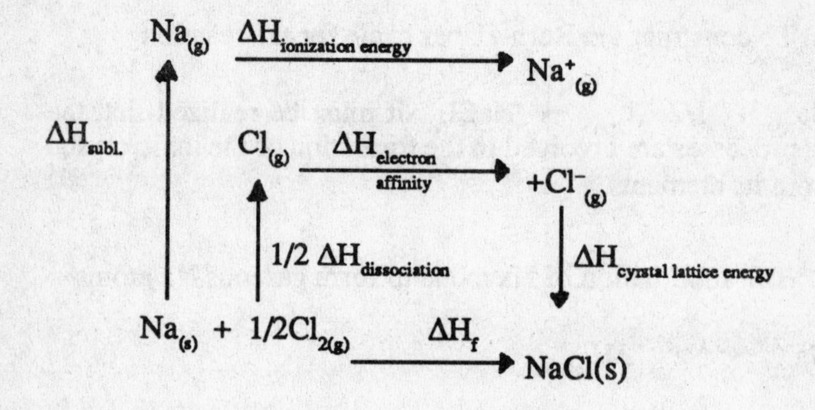


b. The endothermic terms are the ones that require energy. In the case of NaCl, the endothermic terms are as follows:

(a) $Na_{(s)} \rightarrow Na_{(g)}$ requires ~ + 110kJ of energy to convert 1 mole of $Na_{(s)}$ to 1 mole of $Na_{(g)}$ atoms

(b) The ionization of 1 mole of $Na_{(g)}$ to $Na^+_{(g)}$ requires + 496 kJ.

(c) To dissociate 1/2 of a mole of $Cl_{2(g)}$ molecules to gaseous chloride atoms Cl requires + 122kJ.

The two remaining heats, ΔH crystal lattice energy and ΔH(electron affinity for chlorine) are both exothermic (heat is given off) terms in the Born-Haber cycle.

c. Experimentally it is impossible to measure directly the crystal lattice energy. However, crystal lattice energies can be estimated from theoretical calculations. It is also possible to determine a fairly accurate value of the cryrstal lattice energy by utilizing the Born-Haber cycle, provided the other energy terms in the cycle are known, including the $\Delta H_f$, since

$$\Delta H_{(\text{crystal lattice energy})} = \Delta H_{(\text{formation})} - \Delta H_{(\text{subl. of Na})}$$

$$-\Delta H_{(\text{electron affinity of Cl})} - \Delta H_{(\text{dissociation of 1/2 Cl}_2)}$$

$$-\Delta H_{(\text{ionization energy of Na})}$$

5. In order to construct a flow chart detailing the separations of $Hg^{+2}_2$, $Ag^+$ and $Pb^{+2}$ it must be realized that each of these metal ions form insoluble chloride salts.

$$Ag^+ + Cl^- \rightarrow AgCl_{(s)}$$

$$Pb^{+2} + 2Cl^- \rightarrow PbCl_{2(s)}$$

$$Hg^{2+}_2 + 2Cl^- \rightarrow Hg_2Cl_{2(s)}$$

This is the basis of the separation of these group I metals from other metal ions that might be in solution. When the unknown solution is treated with 3M HCl the precipitate that forms will be be a mixture of $Hg_2Cl_2$, AgCl and $PbCl_2$.

In order to separate and identify these metal chlorides we must take advantage of the different chemical and physical properties of these precipitated metal chlorides.

Of the three metal chlorides, lead chloride is particularly soluble in boiling water. The $PbCl_2$ may then be separated by treating the precipitate with boiling water, and decanting off the water from the remaining solid. The decanted solution will contain only the $Pb^{+2}$ ions. The $Pb^{+2}$ ion can be confirmed by the presence of a yellow precipitate upon treatment with $CrO_4^{-2}$.

$$Pb^{+2} + CrO_4^{-2} \rightarrow PbCrO_4$$

yellow precipitate

Next the solid precipitate left behind after the treatment with boiling water will contain both the AgCl and $Hg_2Cl_2$, which must now be separated and identified. The separation is achieved by adding concentrated ammonium hydroxide, $NH_4OH$, to the precipitate containing both the $AgCl_{(s)}$ and the $Hg_2Cl_{2(solid)}$. If $Hg_2^{2+}$ ion is present a black precipitate, containing Hg + $HgNH_2Cl$, will result. The $AgCl_{(s)}$ will form the soluble salt $Ag(NH_3)_2^+Cl^-$ and will go back into solution. The filtrate is then separated from the Hg-$HgNH_2Cl$ precipitate. This soluble silver salt can then be reprecipitated from the filtrate with concentrated nitric acid, $HNO_3$, and a while precipitate of $AgCl_{(s)}$ will result, confirming the presence of the Ag+ ion.

The following flow chart summarized the procedure for the separation and identification of the group I metals:

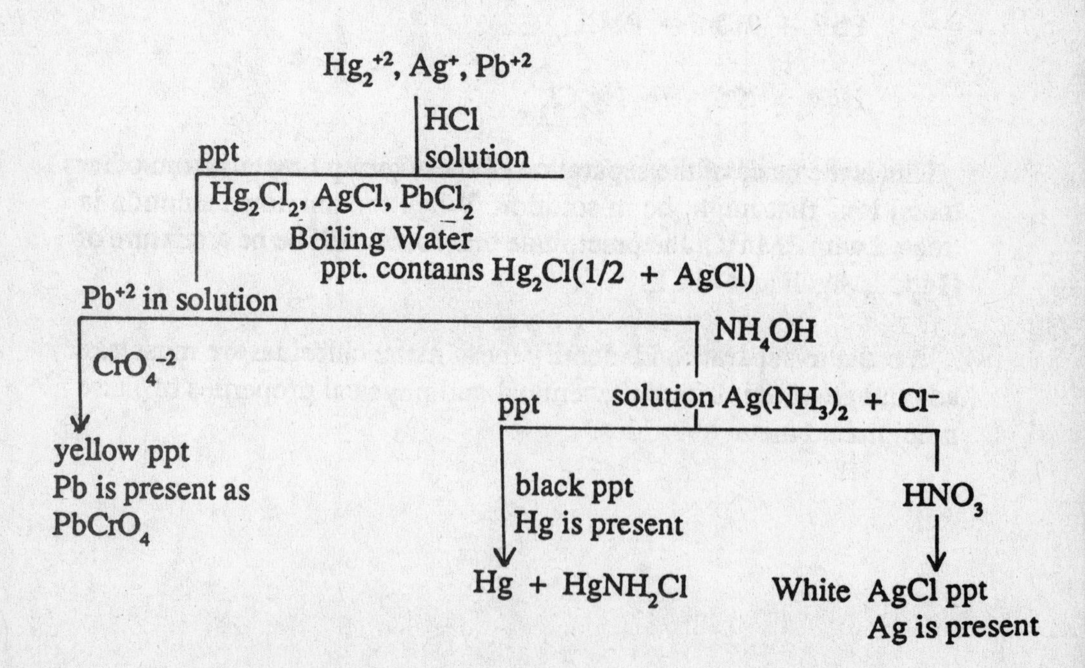

6. The first concept pertaining to acid base theory was presented by Arrhenius in 1884 when he defined an acid as any compound containing hydrogen that when dissolved in water produces hydrogen ions

$$\text{(An Acid)} \quad HA_{(aq)} \xrightarrow{H_2O_{(l)}} H^+_{(aq)} + A^-_{(aq)}$$

and a base as a compound that when dissolved in water produced a hydroxide ion:

$$\text{(A Base)} \quad NaOH_{(aq)} \xrightarrow{H_2O_{(l)}} Na^+_{(aq)} + OH^-_{(aq)}$$

Although the Arrhenius acid-base concept satisfactorily explained reactions of protonic acids (ie. HCl, $HNO_3$) and those of metal hydroxides (ie. NaOH, $Ca(OH)_2$) it was restrictive in its application because it limited acid-base reactions to reactions taking place in aqueous solutions.

Recognizing the limitation of the Arrhenius Concept, Thomas Lowry and Johannes Bronsted independently recognized that every reaction between an acid or a base involved the transfer of a proton from a molecule or an ion to another specie which accepted the proton. In 1923 they proposed that acids were proton donors and bases were proton acceptors.

For example, the reaction of HCl with $NH_3$, which can occur in solvents other than water, or in the gas phase.

$$HCl + NH_3 \rightarrow NH_4^+ + Cl^-$$

The HCl is donating a hydrogen ion (proton) to the ammonia molecule. According to the Bronsted-Lowry Concept, the HCl is functioning as an acid and the ammonia which is accepting the proton is the base.

The Bronsted-Lowry can also be applied to aqueous solutions. For example the ionization of HCl in water may be written as:

$$HCl_{(g)} + H_2O_{(l)} \rightarrow H_3O^+_{(aq)} + Cl^-_{(aq)}$$

In this case the HCl is donating a proton to the $H_2O$ molecule and is functioning as an acid and the water is now functioning as a base.

For the ionization of a weak acid, HX, which only occurs to a slight extent and reaches equilibrium, the reaction may be written as:

$$HX + H_2O \leftrightarrow H_3O^+ + X^-$$

$$(acid_1)\ (base_2) \qquad (acid_2)\ (base_1)$$

In this case, for the forward reaction, HX is functioning as an acid (proton donor) and the water is the base (proton acceptor) and in the reverse reaction, the hydronium ion $H_3O^+$ is the acid (proton donor) and the $X^-$ ion is the base (proton acceptor).

In the reaction above, the HX reacts to yield a base, $X^-$. These two species are related to each other by the loss or gain of a proton, and are said to be a conjugate acid-base pair.

Other examples of conjugate acid-base pairs are shown below.

| Conjugate acid | Conjugate base |
|---|---|
| HCl | $Cl^-$ |
| HF | $F^-$ |
| $HNO_3$ | $NO_3^-$ |
| $NH_4^+$ | $NH_3$ |

It is important to note in the Bronsted-Lowry Concept of an acid or base that a substance might act as an acid in one situation and as a base in another. For example, in the reactions below:

reaction 1: $HCl + \underset{base}{H_2O} \rightarrow H_3O^+ + Cl^-$

reaction 2: $NH_3 + \underset{acid}{H_2O} \leftrightarrow NH_4^+ + OH^-$

The $H_2O$ functions as a base in reaction 1 and as an acid in reaction 2. Whether it functions as an acid or a base depends on the species it is reacting with. The Bronsted-Lowry Concept is an important extension of the Arrhenius definition in that it explains acid-base theory in more general terms.

However, in the Bronsted-Lowry Concept of acid-base reactions, it limits acid-base reactions to ones that only involve proton transfer. There are many reactions that have all of the characteristics of acid-base reactions, but do not involve proton transfer.

In 1923, G.N. Lewis presented a more comprehensive definition of acid-base theory that expanded the Bronsted-Lowry Concept. He defined a base as a species that can donate a pair of electrons, and an acid as a substance that can accept a pair of electrons to form a bond. Thus in the Lewis definition, a base can donate an electron pair to species other than hydrogen ion, $H^+$. The Lewis definition covers all acid-base reactions that the previous theories included, plus many reactions that could not be explained in terms of acid-base behavior. For example when $NH_3$ reacts with $BF_3$:

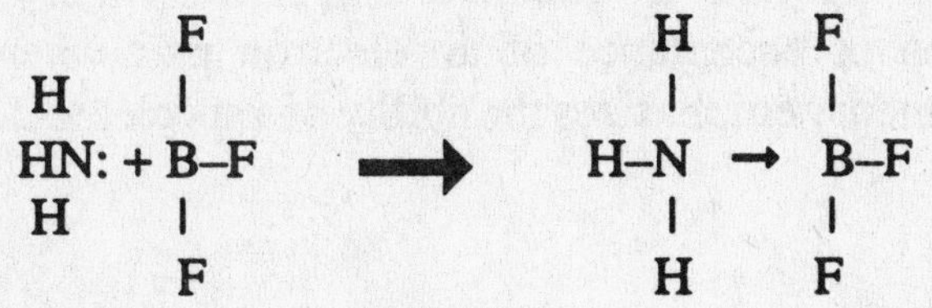

```
        F              H    F
 H      |              |    |
 HN: + B–F   ──►    H–N → B–F
 H      |              |    |
        F              H    F
```

The $NH_3$ is functioning as a base and the $BF_3$ is functioning as an acid in that it is accepting a pair of electrons.

Thus in the Lewis definition, many reactions that do not involve proton transfer can be classified as acid-base reactions as well as species that could be explained by the Bronsted-Lowry Concept. Other examples of Lewis acid-base reactions would include the following:

$$\underset{\text{acid}}{AlCl_3} + \underset{\text{base}}{Cl^-} \rightarrow AlCl_4^-$$

$$\underset{\text{acid}}{SnCl_4} + \underset{\text{base}}{2Cl^-} \rightarrow SnCl_6^{2-}$$

$Ag^+ + 2NH^3 \rightarrow Ag(NH_3)_2^+$
acid base

$H^+ + OH^- \rightarrow H_2O$
acid base

In the examples above, only the last reaction could be explained according to the more restrictive Bronsted-Lowry Concept.

The greatest disadvantage of the Lewis theory is that it is difficult to get quantitative data regarding the acid or base strengths of specific species, since it depends on the environment of the reacting species.

For example, steric effects can affect the strengths of acids or bases. For a base reacting either with the acid $(CH_3)_3B$ or the sterically hindered acid $(t\text{–}C_4H_9)_3B$, the base might appear to be a considerably weaker base when reacting with the acid that contains the bulky alkyl groups, or if the base is relatively small the base might not be affected by the size of the alkyl groups and could donate its electron equally strong to these Lewis acids.

In summary it is important to remember that Lewis Theory emphasizes the donation or acceptance of an electron pair whereas the Bronsted-Lowry theory emphasizes the ability of a species to donate or accept a proton.

7. a. Copper metal will reduce the $Ag^+$ ion:

$Cu^{+2}\ Ag^+ \rightarrow Cu^{+2} + 2Ag$ and metallic silver will be deposited on the copper metal. The net cell potential for this reaction is positive and the reaction will proceed as written below:

| | |
|---|---|
| $Cu \rightarrow Cu^{+2} + 2e^-$ | $E° = -.34V$ |
| $2Ag^+ + 2e^- \rightarrow 2Ag$ | $E° = +.80V$ |
| $Cu + 2Ag^+ \rightarrow Cu^{+2} + 2Ag$ | $E° = +.46V$ |

In general, the reduced form of an element, in this case, copper (Cu), will reduce the oxidized form of another element, in this case ($Ag^+$), that is below it in the electromotive series. Since the standard electrode reduction potential for copper is (+.34) and is above the standard electrode reduction potential of Ag (+.80), copper metal will displace silver ion from solution, depositing silver metal.

b. When $CO_2$ is bubbled through water some of it will dissolve in the water

$$CO_{2(g)} + H_2O_{(1)} \rightarrow H_2CO_{3(aq)}$$

to give carbonic acid. Since it is a relatively weak acid and a diprotic acid the following ionization will occur

$$H_2CO_{3(aq)} \rightarrow H^+_{(aq)} + HCO^-_{3(aq)} \qquad K_1 = 4.2 \times 10^{-7}$$
$$HCO_3^-{}_{(aq)} \rightarrow H^+_{(aq)} + CO^{-2}_{3(aq)} \qquad K_2 = 7 \times 10^{-11}$$

and the resulting solution will be acid.

c. Magnesium nitride, $Mg_3N_2$, which contains nitrogen in the –3 oxidation state, will react with water accordingly

$$Mg_3N_2 + 6H_2O \rightarrow 3Mg(OH)_2 + 2NH_3$$

to give a metal hydroxide and ammonia gas. All the nitrides of group I and II react similarly.

d. Iron sulfide FeS will react with hydrochloric acid HCl liberating hydrogen sulfide gas $H_2S$ and forming ferrous chloride $FeCl_2$ according to the following reaction:

$$FeS_{solid} + 2HCl_{(aq)} \rightarrow H_2S_{(g)} + FeCl_{2(aq)}$$

e. Calcium carbide $CaC_2$ is an ionic compound $[Ca]^{2+}$ $[C{=}C]^{2-}$ which readily hydrolyzes in water, producing calcium hydroxide $Ca(OH)_2$ and acetylene gas H–C=C–H:

$$CaC_{2(s)} + 2H_2O \rightarrow Ca(OH)_2 + C_2H_{2(g)}$$

8. a. An ideal gas is one that would theoretically describe the volume a gas occupies as the volume of the container the gas is contained in. In actuality, this is incorrect since it neglects the volume that the molecules themselves occupy in the space available.

Secondly, an ideal gas assumes that there are no intermolecular attractions between molecules. In real gases there are intermolecular attractions between molecules which in effect reduce the frequency and force of collisions on the container wall, thus causing the measured pressure to be lower than expected. Both of these effects are more pronounced when the molecules are close together, at high pressures and at low temperatures where the molecular motions of the gas molecules are diminished.

b. The molar solubility of a gas in a liquid is an exothermic process, thus the molar solubility of a gas in a liquid decreases as the temperature increases. Fish rely on dissolved oxygen in the water to survive. In the warmer summer months the oxygen content of the water is lower than in the colder months. When the oxygen level is reduced significantly, massive fish kills occur.

c. The addition of a non-volatile solute (ethylene glycol) to a solvent will reduce the vapor pressure above the solution, since the solute molecules will retard the rate of escape of the solvent molecules. The net effect is that the boiling point of the solution will be higher than that of the pure solvent. This will also cause a depression in freezing point. The freezing point depression or boiling point elevation is dependent on the concentration of the solute in the solution and can be expressed by the equations

$$\Delta T_{f.p.} = K_f m$$

$$\Delta T_{B.p.} = K_B m$$

where $\Delta T$ is the change in the boiling point from that of the pure solvent to that of the solution, $K_f$ and $K_B$ are the freezing point depression or boiling point elevation constants for the specific solvent and m is the molality of the solution. For example, the $K_f$ for water is 1.86°C/m, and the boiling point elevation constant for water is

$$K_b = .512°C/m.$$

For a 1.0 molal solution of ethylene glycol the freezing point depression and the boiling point elevation will be respectively:

$$\Delta T_{f.p.} = 1.86°C/m \times 1.0m$$

$$\therefore \Delta T_{f.p.} = 1.86°C,$$

and $$\Delta T_{B.P.} = .512°C/m \times 1.0m$$

$$\therefore \Delta T_{B.P.} = .512°C$$

Thus the 1 molal solution will boil at 100.000°C + .512 = 100.512°C and freeze at 0.00°C + –1.86°C = –1.86°C.

d. The choice of indicators is important when attempting to determine the equivalence point in an acid base titration. For example, when a 1 molar HCl solution is titrated with a 1 molar NaOH solution the pH at the equivalence point will be 7.0

$$HCl + NaOH \rightarrow NaCl + H_2O$$

since the salt NaCl does not hydrolyze.

In selecting an indicator for this titration it is necessary to choose one that has a color change as close to 7.0 as possible, which corresponds to the pH at the equivalence point. Examining the table below:

| indicator | color in acid range | color in base range | pH range |
|---|---|---|---|
| thymol blue | pink | yellow | 1.2 – 2.8 |
| phenolphthalein | colorless | red | 8.3 – 10.0 |
| alizarian yellow | colorless | yellow | 10.2 – 12.1 |

It is obvious that the best choice for the titration described above would be phenolphthalein, since a pronounced color change would occur at a pH of 8.3, which is close to the equivalence point of the titration. If either of the other two indicators listed were used the equivalence point would not correspond to the color change range of the indicators.

In summary it is always necessary to choose an indicator that has a color change as close to the pH at the equivalence point as possible.

e. When most liquids freeze the solid has a higher density and will not float on the liquid phase. However, water is unique in that its solid phase is less dense than its liquid phase. As liquid water beings to cool, its density does increase until 3.98°C. This is the temperature at which liquid water has its maximum density. Below this temperature the density decreases, thus the density of ice is less than that of liquid water and will float on the liquid phase. Hence when a fresh water lake freezes the ice will form on the surface. The surface ice will then tend to insulate the water below the surface, preventing further heat loss. This insulating effect accounts for the observation that freshwater lakes generally do not freeze solid in the winter.